D1868410

COSMOS MAGICUS

ETHICA HUMANA
OPUS 82

Werteschutz-Edition
Waardenbehoud-Editie
Edition for the preservation of values
Edizione per la tutela dei valori
Edition pour la protection des valeurs
Издание «Охрана ценностей»

Deutsch

Nederlands

English

Italiano

Français

Русский

COSMOS MAGICUS

ETHICA HUMANA
OPUS 82

Conception and photographic illustrations
by Anselm Spring
and text contributions
by Friedrich Abel
and Giuseppe Brunamontini

Roland Wolf
Publisher

Vorwort

Dieses Werk zählt zu einer Reihe von voraussichtlich 150 Bänden, die in den Jahren 1973 bis 2033 als „Ethica Humana" erscheinen. In der Darstellung von mehrheitlich anerkannten Werten unserer Kultur, in Kontext gestellt mit aktuellen Ereignissen oder Themen, sehen wir eine Möglichkeit, außerhalb philosophischer, religiöser oder ideologischer Darstellungsart, Werte zu sammeln, zu verbreiten, beziehungsweise zu bewahren, und damit wieder stärker in unser aller Bewußtsein zu rücken.
Durch die große Verbreitung dieser Bände in mehreren europäischen Ländern, bisher wurden mehr als zwei Millionen „Werte-Editionen" gekauft, und durch die Verbindung einer konkreten Wertvorstellung, in diesem Band zum Beispiel „Zuverlässigkeit", mit einem aktuell interessierenden Thema, wie „Cosmos", der, beschränkt auf die menschliche Dimension, geradezu ein Synonym für Zuverlässigkeit ist, sehen wir einen idealen Weg, auf populärer Basis Werten wieder eine größere Chance zur Entfaltung in unserer Gesellschaft zu geben.
Wir danken jedem Gastautor und allen Subskribenten für Ihre Unterstützung.

Roland Wolf

Prefazione

Questa Opera fa parte del Progetto Tutela dei Valori "Ethica Humana", una collana di circa 150 volumi pubblicati negli anni 1973 - 2033.
Nella presentazione dei valori maggiormente riconosciuti della nostra cultura che poniamo in un contesto di eventi o temi attuali, individuiamo la possibilità di raccogliere, diffondere e salvaguardare questi valori al di fuori di una collocazione filosofica, religiosa o ideologica per renderli più vivi nella nostra coscienza.
La grande diffusione di questi volumi in numerosi paesi europei, finora sono stati venduti oltre due milioni di "Edizioni sui Valori", nonché l'unione di un valore concreto in questo volume per esempio "l'affidabilità", con un tema di interesse attuale come il "cosmo" che, limitato alla dimensione umana, è proprio un sinonimo di affidabilità, ci offrono lo strumento ideale per divulgare con una maggiore chance i valori fondamentali nella nostra società.
Ringraziamo gli autori e tutti i sottoscrittori per il loro sostegno.

Roland Wolf

Woord vooraf

Dit werk behoort tot een reeks van waarschijnlijk 150 delen die in de jaren 1973 tot en met 2033 als "Ethica Humana" verschijnen. Met het uitbeelden van algemeen erkende waarden van onze cultuur, in context gebracht met actuele gebeurtenissen of onderwerpen, zien wij een mogelijkheid, en wel buiten een filosofische, religieuze of ideologische voorstellingswijze, waarden te verzamelen, te verspreiden ofwel te bewaren, en deze daarmee weer sterker in het bewustzijn van ons allen te brengen.
Door de grote verspreiding van deze delen in verscheidene Europese landen - tot nu toe werden meer dan twee miljoen "Waarden-edities" gekocht - en door het verbinden van een concrete waardevoorstelling, in dit boek bijvoorbeeld "betrouwbaarheid", met een actueel in de belangstelling staand onderwerp, zoals "Kosmos", die beperkt tot de menselijke dimensie eenvoudigweg een synoniem voor betrouwbaarheid is, zien wij een ideale weg om op een populaire basis waarden weer een grotere kans te geven zich in onze samenleving te ontplooien.
Wij danken iedere gastauteur en alle intekenaars voor hun steun.

Roland Wolf

Preface

This work is one of a prospective series of 150 volumes appearing in the years 1973 to 2033 as "Ethica Humana". In the depiction of generally recognized values of our culture and their links with the context of current events or topics, we see a possibility, indepen-dent of philosophical, religious or ideological manners of representation, of gathering values, and helping to propagate and preserve them and thus to bring them back again more strongly into everyone's consciousness.
As a result of the widespread distribution of these volumes in several European countries - to date more than two million of these "Value Editions" have been sold - and through the link with a concrete concept of value, in this volume for example "reliability" in connection with such a currently interesting topic as the Cosmos, which, limited to the human dimension, is virtually a synonym for reliability, we see an ideal means of giving such values a greater chance of evolving on a popular basis in our society.
We thank each guest author and all our subscribers for their support.

Roland Wolf

Préface

Cet ouvrage fait partie d'une série prévoyant 150 volumes qui paraîtront de 1973 à 2033 dans la collection "Ethica Humana". La présentation de valeurs reconnues par la majorité de nos contemporains et placées dans le contexte d'événements ou de thèmes actuels, offrent à notre avis la possibilité, en marge des exposés philosophiques, religieux ou idéologiques, de rassembler ces valeurs, de les diffuser et de les préserver, renforçant ainsi leur présence dans la conscience de tous les humains.
La grande diffusion de ces ouvrages dans plusieurs pays européens (plus de deux millions des "Werte-Editionen" ont été achetés jusqu'à présent) ainsi que la réunion de la présentation concrète d'une valeur (dans ce tome par exemple "la fiabilité") et d'un événement d'intérêt actuel comme "le Cosmos" qui, limité à la dimension humaine, est véritablement synonyme de fiabilité, nous paraissent paver une voie idéale pour donner aux valeurs, sur une base populaire, une nouvelle chance plus importante de se développer dans notre société.
Nous remercions les auteurs et tous les souscripteurs de leur appui.

Roland Wolf

Предисловие

Эта книга входит в серию «Этика гумана», состоящую из ок. 150 томов, которые выйдут в свет с 1973 по 2033 гг. Мы представляем читателю — без всякой философской, религиозной или политической оценки — признанные большинством людей ценности нашей культуры в контексте событий и тем современности и пользуемся этой возможностью, чтобы собрать, распространить и сохранить эти ценности и сделать их более осознанными для всех нас.
В широком распростанении этих книг в европейских странах — уже продано более двух миллионов экземпляров серии «Ценности» — и в соединении какого-либо понятия — в данном случае «надежность» — с такой современной и интересной темой, как космос, который, по человеческим масштабам, являетмя прямо-таки синонимом надежности, мы видим наилучший путь для того, чтобы на широкой основе способствовать новому раскрытию ценностей в нашем обществе.Мы благодарим каждого автора и всех подписчиков за их поддержку.

Роланд Вольф

IFK - Privat Institut für Sport, Cultur und Natur Verlagsgesellschaft für Werteschutz-Editionen Aktiengesellschaft München

SPORT E CULTURA SpA Segrate - Milano

Vorstand
Roland Wolf

Kaufmännische Leitung
Wolfgang Taubitz

Verkaufsdirektion
Renate Achtelik, Elisabeth Gutmann, Heidelinde Haider, Vera Hammerl, Gisela Hershoff, Christiane Mattow, Waltraud Nirschl, Ursula Seyfarth

Vertriebsleitung
Monika Assats, Elke Bandilla, Eleonore Biechely, Barbara Bischof-Eckstein, Marion Bodewell, Ingrid Hausigk, Christel Hoppe, Petra Orth, Marion Ortmann, Marie-Luise Richter, Ursula Riesenberg, Birgitt Schmidt, Viola Schneider-Miska, Jutta Seubert, Manuela Seyfarth, Sylvia Tepper, Dagmar Tipp-Stenzel, Karin Ullrich-Brox, Christa Wessien

Direktionsassistentin national
Angelika Corte

Direktionsassistentinnen international
Christel Reiter, Kornelia Rettinger

Mitarbeiterinnen
Renate Augustynek, Renate Aurin, Irene Böttcher, Andrea Babbe, Heidemarie Baesen, Doris Bierbaum, Dörte Blümel, Erika Boroske, Irene Bohne, Margit Bohnenstengel, Ulrike Brüsewitz, Anita Buch, Julia Burgeff, Hilde Bühler, Renate Clair-Giebls, Sibylle Cousin, Evelyn Dahl, Gisela Denzin, Christel Eckstein, Tina Eckstein, Rosa Maria Eichner, Martina Espeter, Eve Fischer, Monika Fleischmann, Claudia Frisch, Brigitte Hahn, Yvonne Hauck, Heide Haupt, Ina Hess, Dr. Evelin Hodek, Steffi Jubelt, Resi König, Bärbel Köppen, Hildegard Kokesch, Karin Krebs, Maria Kreul, Inge Kues, Heike Lohmann, Heike Mauß, Manuela Menk-Resch, Hiltraud Mentz, Hannelore Merz, Sieglinde Michalski, Barbara Müller, Cathrin Müller, Marion Niesen, Christa Obermeier, Ernestine Oertel, Marie-Louise Offergeld, Karin Preiser-Klein, Harald Probst, Heike Pyrzewski, Brigitte Rademacher, Sylvia Recknagel, Helga Rettweik, Gisela Reuter, Gisela Rietscher, Sylvia Roesler, Karin Ruloff, Rosemarie Säubert, Heidrun Schmeisser, Marion Schmieder, Liane Schulte, Renate Schäfer, Karin Schubring, Gisela Sehrndt, Marie-Luise Sprave, Leonore Stöckel, Sonja Strait, Ulrike Straub, Helga Strauß, Ilona Theiß, Beate Tryonadt, Monika Türk, Elke Unglaub, Dagmar Voigt, Bernadette Vorreiter, Edith Walbrodt, Iris Washington, Helga Weber, Ursula Weigmann, Angelika Wieneke, Marlene Wildner, Michael Zielonka

Amministratore Delegato
Antonio Ricciardelli

Direttore Vendite Nazionale
Giorgio Cristofori

Direttore Vendite
Alfio Sudano

Direttore Amministrativo
Lucio dell'Arciprete

Relazioni Esterne/Servizi Commerciali
Cinzia Trombetta

Zone Directors
Nora Calderone, Luisa Donsanto, Iva Faitelli, Anna Filippi, Flora Maritano, Gianna Natta, Paola Picotti, Wanda Saveriano

Area Directors
Maria Bruna Acquati, Nora Amadori, Rosita Baccelli, Antonietta Baso, Monica Bondi, Rosanna Cesaritti, Vincenza Eterno, Susanna Garofalo, Juccia Marchiori, Paola Romano, Olga Ventura, Marika Vezzosi

Managers
Rita Berti, Clara Bianchini, Giuliana Cammarata, Annabella Carpineto, Maria Grazia Cavo, Nicoletta Chiarappa, Rosa Conti, Carmela Genovese, Maria Gerardi, Eliana Ghiribelli, Gabriella Grilletti, Giuseppina Nardin, Maria Santa Prati, Giuseppina Quaini, Maria Assunta Raia, Maria Rita Romagnoli, Marisa Rossi, Alida Scaletta, Maria Assunta Speranza, Maria Rita Tarello, Lidia Tarlazzi, Rosalia Tomaselli, Rosarita Tremolada

Collaboratori
Gloria Bandini, Giuliana Barbieri, Gloria Bartole, Anna Maria Bartoli, Giuseppina Baruffi, Maria Pia Bassi, Donatella Battistini, Giancarla Belloli, Patrizia Beretta, Carla Bettega, Germana Bianco, Daniela Biava, Giulia Borali, Barbara Borghi, Florenza Bruni, Luciana Bruzzone, Carmela Calabrò, Vanna Caldieron, Rosanna Camolese, Simona Cappelletto, Nadia Carminati, Elena Casadei, Gloria Casini, Milene Casturi, Anna Cecchi, Loredana Celant, Anna Maria Chianese, Franca Cinel, Maria Conti, Gianna L. Cuccu, Fulvia D'Urso, Michela De Petris, Anna Del Chicca, Meri Franca Di Biagio, Emilia Dacò, Rossana Derocchi, Emanuela Diolaiti, Leonella Dolliana, Aurelia Ducoli, Liliana Fabietti, Loredana Fanin, Amelia Farese, Maurizio Ferri, Rina Filippi, Annamaria Folchini, A. Luigia Fossati, Emilia Fossati, Gloria Fusella, Isa Gai, Lucia Galletti, M. Luigia Gambino, Marilena Gatti, Loredana Gazzola, M. Antonietta Genestrone, Roberta Giannini, Gabriella Giusto, Alessandra Griggio, M. Italia Guadagnuolo, Antonella Ippolito, Brunella Lolli, Daniela Lonza, Marta Maestrelli, Edda Maggioni, Marilena Magrini, Patrizia Marveggio, M. Angelica Medolago, Antonia Meduri, Julia Mellina, Rita Merlo, Flavia Micheletto, Paola Michelini, Giuliana Micomonaco, Carla Miserocchi, Giuliana Missori, Paola Mondello, Marcella Mondin, Graziella Montano, Donella Morandini, Meris Morello, Elena L. Munno, Clara M. Mura, Alfonsina Muscetta, Enrica Muttoni, Alessandra Neri, Lucia Oggionni, Elena Oliva, Lucia Opalio, M. Grazia Palmieri, Giulia Pasquali, Elena Passi, Daniela Patricolo, Patrizia Plazzi, Dario Perli, Giovanna Porqueddu, Dominique Pronguè, Isabella Pyrzewicz, Giuliana Quaresima, Bianca Rebucci, Patrizia Renzi, Veronika Rogina, Luisella Ronchi, Anna Maria Rossignolo, Paolo Rubbioni, Rosanna Salvaderi, Anna Sangiovanni, M. Ines Soliani, Lucia Taglioni, Paola Talluto, Giovanna Tremosini, Luciana Visentin, Waltraud Wansch, Emanuela Zaninotto.

Sport, Cultuur en Natuur bv Amstelveen

Directie / Direction
Wolfgang Taubitz

Management
Mieke ten Berg, Marga Mus, Ria Nolten, Femmy Schinkel

Medewerksters
Trudy Deleij, Susan Geus, Annemarie Kannemans, Cornelia van Ketel, Nel van der Linde, Annerie Luykx, Aagje Meirik, Ellen van Nieuwstadt, Riky Nugteren, Anke van de Rijdt, Miep van Velzen, Claudia Vonk, Lindy van Wezel, Ineke de Wildt, Hinke Wissink

Gesellschaft für Sport- und Kultureditionen Verlagsges. m.b.H. Wien

Geschäftsführer
Roland Wolf

Vertriebsleitung
Melitta Kührer

Generalbevollmächtigte
Irmgard Baumgartner

Mitarbeiterinnen
Marlene Bäumchen, Ingrid Caiblinger, Tamara Epp, Isolde Felsberger, Edith Hofer, Xenia Peschel-Staf, Elisabeth Sousek

IFK - Institut für Kultur- und Sportförderung AG Kilchberg-Zürich

Geschäftsführer / Gérants
Antonio Ricciardelli

Vertriebsleitung
Brigitte Engler, Elsy Graeflein, Rosmarie Schweighauser, Sieglinde Strauss

Mitarbeiterinnen / Collaboratrices
Herta Bormann, Doriana Botta, Cornelia Caviezel, Elsbeth Eichholzer, Annarös Fink, Barbara Honegger, Astrid Kälin, Claudia Müller-Strauss, Hanni Roffler, Christine Schwander, Monique Vallon

Published by
Europoli & Eurolex
Herrsching

Publishers
Roland und Elfie E. Wolf

Photographic Editor and Conception
Anselm Spring

Editors-in-Chief
Friedrich Abel
Giuseppe Brunamontini

Editor-in-Chief Ethica Humana
Hans Dollinger

Art Director
Erwin Fieger

Production Management
Reinhard Hedrich

Managing Editor
Petra Daxauer

Translations
Lynn Beren
Dr. Véronique Brandner
Francesca Lopez
Gina Schloffer-Prins
Alexej Tchijov

Reproductions
FOTOLITO STAR,
Bergamo

Printing
MOHNDRUCK,
Gütersloh

© 1994 by Europoli & Eurolex
Elfriede Wolf GmbH & Co. KG, Herrsching
SPORT E CULTURA SpA
Segrate-Milano
Printed in Germany
ISBN 3-921412-34-X

CONTENTS

Inhalt
Inhoud
Indice
Table des matières
Содержание

MACRO

COSMOS

Magre Glut
der
Revolten
Rosenbrände
meiner
Dome
Feuerfester
Engel
dieser Erde

Aschenrabe
Friß
die Reste
des Vergessens
Vater
aller
wilden Flammen
Segne
deinen Feuersohn.

Magere gloed
van de
revoltes
rozenbranden
van mijn
gewelven
Vuurvaste
engel
dezer aarde

Asbeltraaf
vreet
de resten
van 't vergetene
Vader
aller
wilde vlammen
zegen
uw vuurzoon.

Meagre fervour
of
revolts
Rose fires
of my
domes
Fire-proof
angel
of this earth

Ash raven
Consume
the remnants
of forgetting
Father
of all
wild flames
Bless
your fire son.

Yvan Goll

Magre braci
delle rivolte
rosai
ardenti
dei miei duomi
angeli
tenaci al fuoco
di questa terra
corvo di cenere
divora
gli avanzi
dell'oblio
padre
di ogni
fiamma selvaggia
benedici
il figlio tuo
di fuoco.

Maigres braises
des révoltes
Roses ardentes
de mes
cathédrales
Ange ignifuge
de cette Terre
Corbeau
de cendres
Mange
les restes
de l'oubli
Père
de toutes
les flammes
farouches
Bénis
ton fils de feu.

Скудный жар
мятежей
Пожары роз
моих
соборов
Негорючий
ангел
этой
земли
Ворон
ест остатки
забвения
Отец дикого
пламени
благослови
своего
огненного
сына.

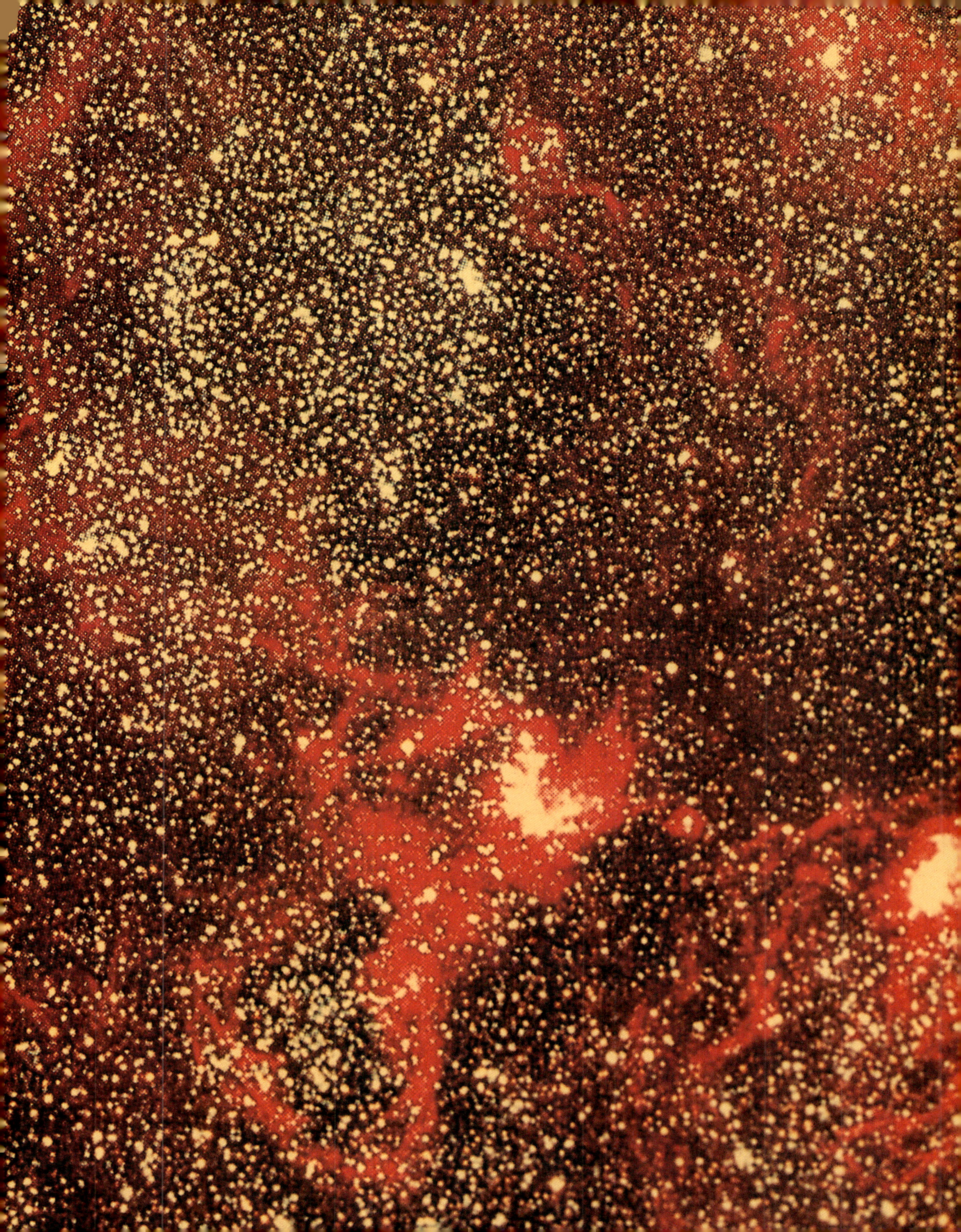

Die Ringe des Saturn machen eine Wandlung durch oder Saturn wird durch einen Mond verdunkelt. Er verschwindet, gesehen über einen Sonden-Arm, er taucht wieder auf, sein Untertauchen war nur von kurzer Dauer wie ein Boot, das verschwindet und wieder auftaucht.

De ringen van Saturnus gaan door hun schijngestalte of Saturnus wordt verduisterd door een maan. Gezien over de arm van een sonde, verdwijnt hij en duikt hij weer op, zijn onderduiken was slechts van korte duur, gelijk een boot die weer opduikt.Titan, zijn grootste maan.

The rings of Saturn go through their seeming change or Saturn is occulted briefly by a moon. It vanishes seen across a limb of a probe, it reappears, its immersion brief, like a boat that is eclipsed by the rough horizon, a boat that reappears. Titan with it moon its own.

Seymor Faust

Gli anelli di Saturno si trasformano apparentemente o Saturno viene oscurato da una luna. Si eclissa, lo vede solo il braccio di una sonda, riappare immediatamente, la sua scomparsa era solo di breve durata come una barca che riemerge dai flutti.

Les anneaux de Saturne semblent se transformer ou Saturne est obscurci par une lune. La planète s'éclipse, vue par le bras d'une sonde, puis réapparaît; son immersion ne fut que de courte durée, comme un bateau qui refait surface.

Кольца Сатурна видоизменяются или Сатурн затемняется луной. Он исчезает, увиденный антенной космической станции, опять появляется — он исчезал лишь на короткое время, будто лодка, которая вновь всплывает.

Seht Freunde, dieses Universum an, mit seinen spiralförmigen Sternenhaufen, die in alle Richtungen durchs All fliegen wie Sprungfedern, die plötzlich weggeschossen sind, und in dieser Galaxie die Sonne, die sich selbst verzehrt, umgeben von Planeten, herausragend in ihrer Würde und ungezähmt.

Kijk, vrienden, naar dit universum, met zijn spiraalvormige sterrenhopen die in alle richtingen door het heelal vliegen als springveren die plotseling zijn losgesprongen en in deze melkweg de zon die zichzelf verteert, omgeven door planeten, opvallend in hun waardigheid. In het midden de planeet met één enkele maan.

Look, friend, at this universe with its spiral clusters of stars flying out all over space like bedsprings suddenly busting free, and in this galaxy, the sun, fissioning itself away, surrounded by planets, prominent in their dignity, and bits and pieces running wild, and this middling planet with a lone moon circling round it.

Edward Field

Mirate o amici questo universo ed i suoi ammassi di stelle a spirale che volano in ogni direzione dello spazio come molle lasciate improvvisamente libere di scattare, ed in questa galassia il sole, circondato da pianeti, consuma se stesso, eminente nella sua dignità, indomabile nella sua missione vitale.

Regardez, mes amis, cet Univers, avec ses amas stellaires qui traversent l'espace dans toutes les directions comme des ressorts qui s'envolent soudain, et dans cette galaxie, le soleil, se consummant lui-même, entouré de planètes, et des fragments et cette planète au milieu, avec une seule lune qui tourne autour.

Посмотрите, друзья, на эту Вселенную со спиральными скоплениями звезд, летящими в ней по всем направлениям, как отпущенные пружины, и в этой галактике Солнце, пожирающее само себя, окруженное планетами, выдающееся в своем достоинстве.

Kazuaki Iwasaki 1037

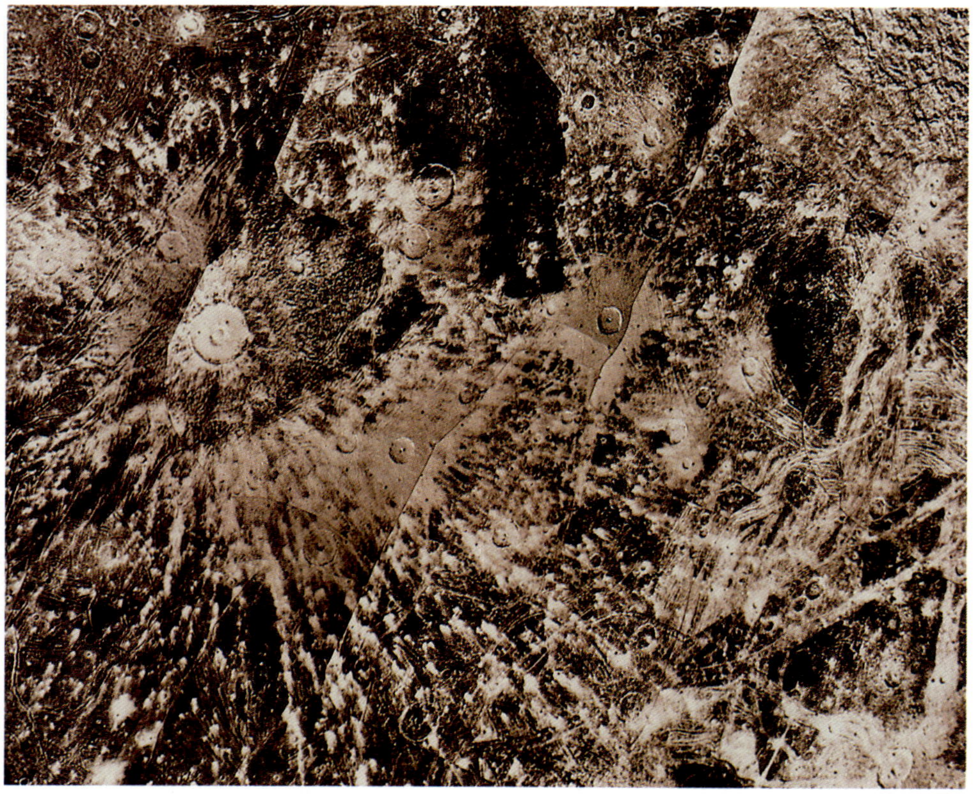

Hochhin führt eine Straße, sie trägt nach der Milch ihren Namen, sichtbar bei heiterem Himmel, am leuchtenden Glanze zu kennen. Vorn haben die starken Himmelsbewohner, die glänzenden, ihre Behausung errichtet. Dies ist der „Palatin des erhabenen Himmels".

Hoog naar boven leidt een weg, naar de melk draagt hij zijn naam, zichtbaar bij heldere hemel, te herkennen aan een schitterende glans. Voor aan hebben de sterke hemelbewoners hun behuizing. Dit is de plaats waarvoor de naam: "Paladijn van de grootse hemel" past.

In alto conduce una strada, del latte porta essa il nome, visibile quando il cielo è sereno, si conosce dal bagliore lucente. Le loro dimore davanti vi han costruito i forti abitanti del cielo, gli splendendi. Questo è il luogo degno del nome "Palatino del cielo sublime."

Loin dans le ciel s'élève une voie que l'on appelle lactée, visible quand le ciel est clair, reconnaissable à son éclat resplendissant. Devant, les habitants du ciel ont établi leur demeure forts et brillants. C'est l'endroit qui porte bien le nom de "Palatin de l'auguste Ciel".

A way leads up, named after milk, visible when the sky is clear, recognizable by her shining lustre. The strong inhabitants of the heavens have set up their dwellings in the forefront. This is the place for which the term "Palatine of the sublime heaven" suits.

Вверх ведет путь, он назван по молоку, видимый на ясном небе, узнаваемый по светлому блеску. Там сильные жители неба сияющие построили свое жилище. Это место, которому подходит название «Палатин возвышенного неба».

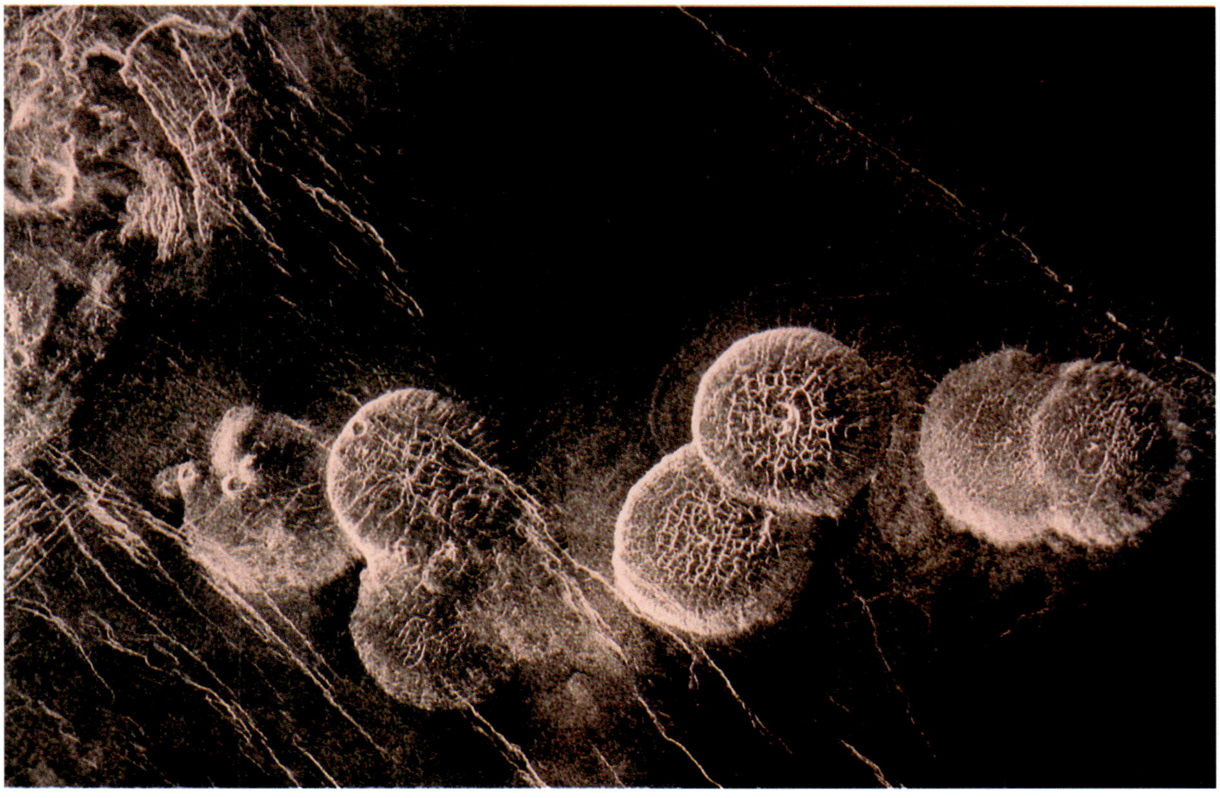

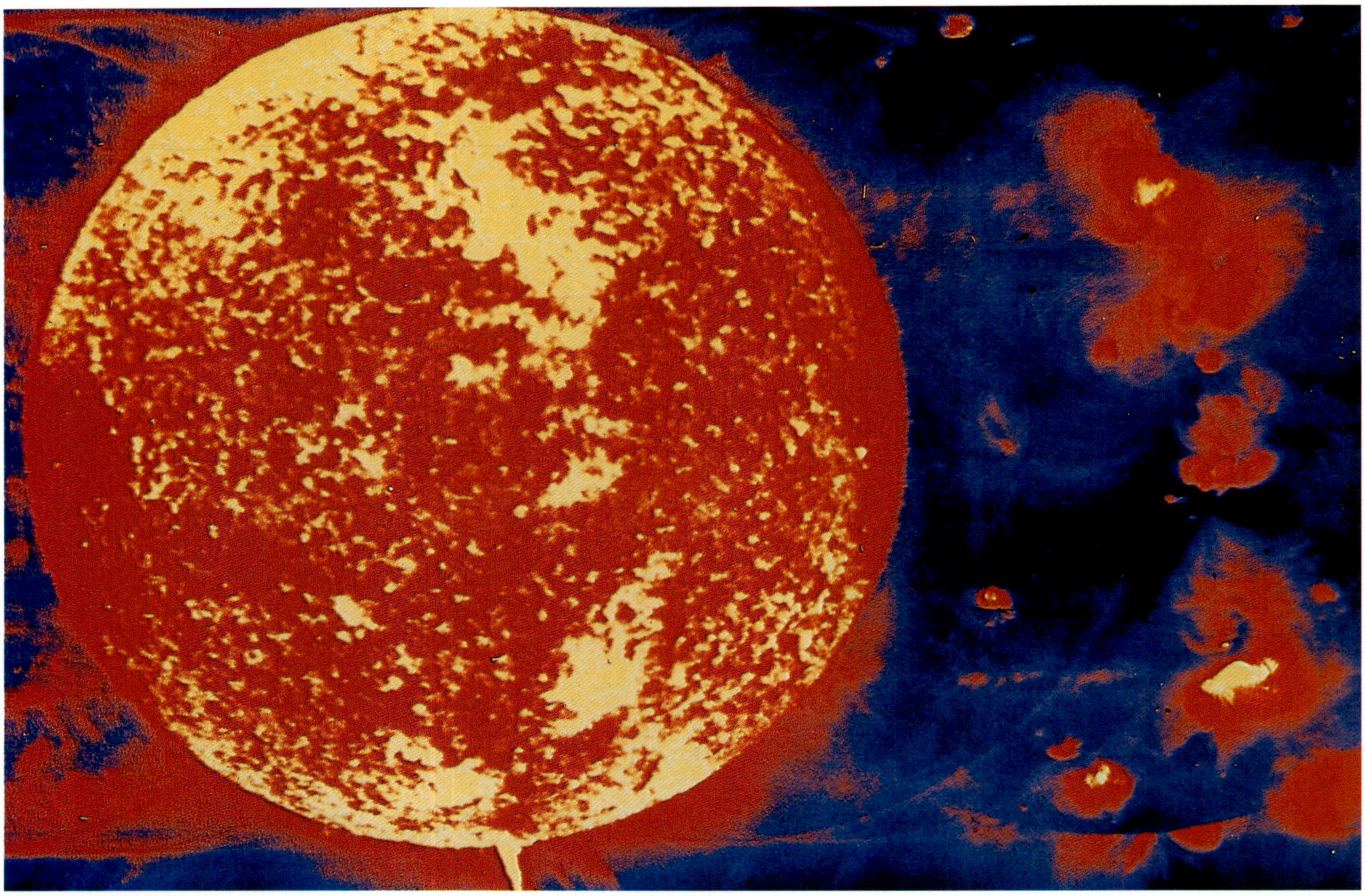

Meine Ohren schwimmen in Wellen von kodierten technischen Daten, Listen von Programmier-Schritten, um die Mondumlaufbahn zu erwischen, und ich sehe sie alle in der Anti-Atmosphäre schwimmen, wie sie Experimente vorbereiten wie Picknick-Tische... Ich fange an zu zweifeln, ob ich wirklich existiere außerhalb der Relais zum Mond und zurück.

Mijn oren dobberen in golven van gecodeerde technische data, lijsten van programmastappen om de juiste omloopbaan van de maan te bepalen, en ik zie ze allemaal in de anti-atmosfeer drijven hoe zij experimenten opzetten net als picknicktafels... Ik begin eraan te twijfelen of ik buiten de relaisverbindingen van en naar de maan besta.

All day commentators have been talking of voyages. My ears float in waves of coded engineering data, lists of steps to program lunar orbital insertion, and I see them swim in the anti-atmosphere, carefully laying out experiments like picnic tables... I begin to seriously doubt whether I really exist outside relay circuits to the moon and back.

Robert Vas Dias

Le mie orecchie nuotano tra le onde di dati tecnici codificati, errano tra liste di programmazione per riuscire a centrare l'orbita giusta intorno alla luna ed io li vedo nuotare tutti nell'anti-atmosfera mentre sono intenti a preparare degli esperimenti e inizio quasi a dubitare che io esisto veramente al di là della rotta di andata e ritorno dalla luna.

Mes oreilles flottent dans des vagues de données techniques codées, de listes d'étapes programmant l'insertion exacte dans l'orbite lunaire et je les voie nager dans l'antiatmosphère, préparant des expériences comme des tables de pique-nique... Je commence à douter d'exister vraiment en-dehors des circuits de relais entre la terre et la lune.

Мой слух плывет в волнах кодированных технических данных, перечней шагов программы для расчета правильной окололунной орбиты, я вижу их плывущими в антиатмосфере, как они готовят эксперименты, будто стол к пикнику. Я начинаю сомневаться, существую ли я помимо сигналов к Луне и обратно.

Sein und Nichtsein ist ungetrennt durcheinander, ehe Himmel und Erde entstehen. So still! So leer! Allein steht es und kennt keinen Wechsel. Es wandelt im Kreise und kennt keine Unsicherheit. Man kann es fassen als die Mutter der Welt. Ich bezeichne es als „Sinn". Mich mühend seine Arbeit zu künden, nenne ich es: „Groß". Groß, damit meine ich: immer im Flusse. Immer im Flusse, damit meine ich: in allen Fernen. In allen Fernen, damit meine ich: in sich zurückkehrend.

Zijn en niet-zijn ligt ongescheiden door elkaar voordat Hemel en Aarde ontstaan. Zo still! Zo leeg! Het staat alleen en kent geen wisseling. Het zwerft in een kring en kent geen onzekerheid. Men kan het vatten als de moeder der wereld. Ik ken zijn naam niet. Ik duid het aan als "Zin". Trachtend zijn aard te ontdekken noem ik het: "Groot". Groot, daarmee bedoel ik: altijd in beweging altijd in beweging, daarmee bedoel ik: in alle verten in alle verten, daarmee bedoel ik: in zich terugkerend.

Existence and non-existence is all one, all mixed up before Heaven and Earth come into being. So still, so empty. Alone it is and knows no change. It wanders in circles and knows no uncertainty. Man can view it as the mother of the world. I do not know its name, I term it "sense". Endeavouring to find out more about its workings, I call it "great". By great I mean: always flowing, always flowing. By it I mean: always moving far away, far away. By it I mean: returning to itself.

Essere e non essere sono uniti e confusi prima che si formino cielo e terra. Così tranquillo! Così vuoto! Il tutto non conosce mutamento alcuno. Si muove in cerchio e non conosce insicurezza. Lo si può intendere come madre del mondo. Lo chiamo "senso". Sforzandomi di conoscere il suo modo di manifestare lo chiamo: "grande", per grande intendo che scorre sempre, scorre sempre, con ciò intendo: in grandi lontananze, in grandi lontananze, e con ciò intendo: rientrando in sé.

Etre et ne pas être est un mélange inséparable avant la naissance du Ciel et de la Terre. Si calme! Si vide! Seul et sans changement aucun. Eternelle rotation et absence d'incertitude. On peut le comprendre comme la Mère du Monde. Je ne connais pas son nom. Je l'appelle "Sens". M' efforçant de révéler sa nature, je le qualifie de "grand". Par "grand", j'entends un "flux constant". Par "flux constant" j'entends: "Dans tous les espaces lointains", ce qui pour moi signifie: "Retour à soi-même".

Бытие и небытие неразрывно вместе, пока не возникнут Небо и Земля. Так тихо! Так пусто! Оно одно и не знает перемен. Оно блуждает по кругу и не знает неуверенности. Его можно осязать как мать мира. Я не знаю его имени. Я называю его «Смысл». Затрудняясь описать его вид, назову его: «Великое». Великое — я имею в виду: всегда в течении, я имею в виду: во всех далях, то есть я имею в виду: возвращаясь в себя.

Er hängt einfach da oben, dauernder als irgendetwas hier auf Erden. Einen Fuß darauf zu setzen, und wenn es nur einmal ist, heißt, ihn absolut zu verderben: wie wenn ein lustiger Tramper mit Fernglas und Käsebrötchen durch das Lächeln der Mona Lisa marschiert.

Zij hangt gewoon daar boven, bestendiger dan wat ook hier op aarde. Een voet op haar te zetten, ook als het maar eenmaal is, dat is haar totaal bezoedelen: als een vrolijke wandelaar die met verrekijker en broodje kaas jodelend door de glimlach van de Mona Lisa marcheert.

It just hangs up there more permanent than anything on earth. To set foot on it even once is to corrupt it utterly: a jolly hiker with binoculars and cheese sandwiches bursting with a yodel through the Mona Lisa's smile the canvas ripped beyond repair. Hi there!

Jack Anderson
(Apollo astronaut)

Pende semplicemente lassù, più durevole di qualsiasi cosa terrena. Porvi piede, anche solo una volta, significa rovinarla per sempre: come se uno spensierato vagabondo armato di binocolo e panini camminasse sul sorriso della Gioconda, rovinando irreparabilmente la tela.

Elle est accrochée là-haut, plus permanente que toute chose sur Terre. Y poser le pied, ne serait-ce qu'une fois, signifie la corrompre totalement: comme si un joyeux auto-stoppeur avec jumelles et sandwiches marchait en ioulant sur le sourire de la Joconde, déchirant la toile.

Она просто висит там наверху, более постоянная, чем любая вещь на Земле. Наступить на нее — значит безнадежно испортить, все равно как беззаботный турист с биноклем прошел бы по улыбке Моны Лизы, калеча холст. Hi there!

Magischer Kosmos: Aufbruch in die zehnte Dimension

Noch vor zwei, drei Jahren waren sie mir ziemlich suspekt, all die Hohepriester der Kosmologie mit ihren phantastischen Geschichten über Urknall, galaktischen Kannibalismus und schwarze Löcher. Ich wollte mir das Wunder des Sternenhimmels nicht wegerklären lassen, wollte nichts mehr hören von „Supersymmetrie" und „nackter Singularität". Nicht Stephen Hawkings Vision eines „Quanten-Universums" sprach mir aus der Seele, sondern Friedrich von Schillers zweihundert Jahre alte Himmelsforscher-Schelte:
„Schwatzet mir nicht so viel von Nebelflecken und Sonnen! Ist die Natur nur groß, weil sie zu zählen euch gibt?"
Heute hingegen lasse ich mich gern beschwatzen von Astronomen, Astrophysikern und Kosmologen, kann ich wie viele meiner Zeitgenossen gar nicht genug bekommen von Meldungen, die aus Observatorien, Teilchenbeschleuniger-Labors und Gelehrtenzimmern dringen. Ein Umbruch des Denkens findet dort statt, ein neues Weltbild scheint heraufzudämmern - von führenden Theoretikern im Moment noch tastend oder auch provozierend formuliert in Buchtiteln wie „Der Traum von der Einheit des Universums", „Die Physik der Unsterblichkeit" oder „Gleichungen der Ewigkeit". Angestachelt haben die Spekulationslust der Wissenschaftler zum einen ihre Erfolge bei der Suche nach den Urpartikeln der Materie - zuletzt das „Top Quark", ein Elementarteilchen aus der Geburtssekunde des Universums, zum anderen kommt derzeit aber auch eine Fülle verblüffender Daten von neuartigen „Superteleskopen", die vor allem im pazifischen Raum installiert worden sind oder als Satelliten die Erde umkreisen. Mit ihren schärferen Augen haben sie in kurzer Zeit bereits indirekte Belege für Phänomene registriert, deren Existenz in den theoretischen Konzepten eines Albert Einstein, Steven Weinberg oder Stephen Hawking vorausgesagt worden war. Jüngstes Beispiel: Die Sichtung eines Malstroms in der 50 Millionen Lichtjahre entfernten „M87"-Galaxie. Zwei Milliarden Sternen entspricht die Masse der dort mit einer Geschwindigkeit von 2 Millionen Stundenkilometern rotierenden Gaswolke. Was sie derart schnell kreisen läßt, ist offensichtlich die Saugkraft eines gigantischen „Schwarzen Lochs", das die Gasmassen in sich hineinzieht. Nicht weniger bedeutend sind etwa auch die Meßdaten des „Nachglühens" aus der Anfangszeit unseres Universums: ein Nachweis, daß der den Kosmos schaffende „Urknall" vor rund 15 Milliarden Jahren tatsächlich stattgefunden hat. Und schließlich funkte der europäische ROSAT-Satellit erste Befunde zur Frage der geheimnisumwitterten „dunklen Materie" aus dem All. Danach sehen wir Erdenbürger nur ein Hundertstel der Materie im All, wenn wir aufschauen zum Nachthimmel, die restlichen 99 Prozent, so scheint es nun, umlagern die Galaxien in Form gewaltiger unsichtbarer „Auren". Freilich ist das noch lange nicht alles. Unser Planet und die gesamte Milchstraße sowie tausende anderer Galaxien „unserer Gegend" des Kosmos' sind mit einer Geschwindigkeit von etwa 580 Kilometer pro Sekunde zum „Great Attractor" hin unterwegs, einem unsichtbaren Giganten, dessen Gravitationskraft 20 000 Milliarden Sonnen entsprechen soll. Aber selbst der „Große Anziehende" hat seine Probleme. Jüngsten Daten zufolge treibt er, einem Floß auf einem kosmischen Fluß vergleichbar, ins All hinaus - angezogen von einem „Etwas" von noch viel, viel größerer Masse.
Je mehr die Himmelsforscher über das Universum herausfinden, desto deutlicher wird, daß seine Realität oft phantastischer ist als alle Science-Fiction. So haben zum Beispiel viele der führenden „reinen Denker" unter den Kosmologen - darunter Stephen Hawking, Steven Weinberg und Murray Gell-Mann - den uns geläufigen dreidimensionalen Raum als Operationsfeld ihrer Weltentwürfe längst verlassen. Die Forscher sind in einen zehndimensionalen „Hyperspace" umgestiegen, einen bildhaft nicht mehr vorstellbaren, aber mathematisch begehbaren „Überraum" voll überraschender Eigenschaften und nicht immer klaren Grenzen zwischen Physik und Mystik. Ein „Multiversum" zehndimensionaler Kosmen - mit jeweils neun räumlichen Ausdehnungen und einer zeitlichen Dimension - gibt es in dieser anderen Welt; wie Seifenblasen schweben sie in einem zeitlosen „Nichtraum" und können via „Wurmlöcher" miteinander verbunden sein. Alle Materie besteht dort aus winzigen Fäden unterschiedlicher Frequenz, die Naturgesetze sind in der lange erträumten Harmonie vereint: „Der Hyperspace scheint das natürliche Habitat aller Naturkräfte zu sein." Allerdings gibt es noch ein großes Problem: Auch die besten mathematischen Köpfe von Cambridge, Yale oder CERN vermögen die den Hyperraum beschreibenden Gleichungen noch nicht befriedigend zu lösen. Die potentielle „Weltformel" - modelliert nach „Modularfunktionen" eines begnadeten indischen Laienmathematikers - ist ein „Werkzeug des 21. Jahrhunderts, das uns zufällig in die Hände gefallen ist"; wie der amerikanische Physiker Michio Kaku schreibt: „Wir müssen erst lernen, richtig damit umzugehen."

Magische kosmos: Op weg naar de tiende dimensie

Twee, drie jaar geleden vond ik hen nog tamelijk suspect, al die hogepriesters der kosmologie met hun fantastische verhalen over oerknal, galactisch kannibalisme en zwarte gaten. Ik wilde mij het wonder van de sterrenhemel niet laten wegverklaren, wilde niets meer horen van "supersymmetrie" en "naakte singulariteit". Niet Stephen Hawkings visioen van een "quantumuniversum" sprak naar mijn hart, maar Friedrich von Schillers tweehonderd jaar oude verwijten naar de hemelonderzoekers:
"Praat mij niet zoveel over nevelvlekken en zonnen! Is de natuur alleen maar groot, omdat zij ook te tellen biedt?"
Tegenwoordig praat ik wel graag met astronomen, astrofysici en kosmologen, kan ik net als veel tijdgenoten van mij helemaal niet genoeg krijgen van berichten die uit observatoria, deeltjesversneller-laboratoria en studeerkamers doordringen. Een ingrijpende verandering in het denken vindt er plaats, een nieuw wereldbeeld schijnt aan te breken - momenteel door vooraanstaande theoretici nog voorzichtig of ook provocerend geformuleerd in boektitels als "De droom van de eenheid van het universum", "De fysica van de onsterfelijkheid" of "Vergelijkingen van de eeuwigheid". Enerzijds is de speculatielust van de wetenschappers aangemoedigd door hun successen bij het zoeken naar de oerdeeltjes van de materie - het laatst de "top quark", een elementair deeltje uit de seconde van de geboorte van het universum.
Anderzijds komen er thans echter ook een heleboel verbluffende gegevens uit nieuwe "supertelescopen" die vooral in het gebied van de Grote Oceaan zijn geplaatst of als satellieten om de aarde draaien. Met hun scherpe ogen hebben zij binnen korte tijd al indirecte bewijzen geregistreerd voor fenomenen waarvan het bestaan was voorspeld in de theoretische concepten van een Albert Einstein, Steven Weinberg of Stephen Hawking. Het jongste voorbeeld: de waarneming van een maalstroom in het 50 miljoen lichtjaren van ons verwijderde galactische stelsel "M87". De massa van de daar met een snelheid van 2 miljoen kilometer per uur roterende gaswolk is zo groot als twee miljard sterren. Wat deze gaswolk zo snel laat draaien, is kennelijk de zuigkracht van een gigantisch "zwart gat" dat de gasmassa's naar zich toe naar binnen trekt. Niet minder belangrijk zijn bijvoorbeeld ook de meetgegevens van het "nagloeien" uit de begintijd van ons universum: een bewijs dat de "oerknal" waardoor de kosmos is ontstaan, rond 15 miljard jaren geleden inderdaad heeft plaatsgevonden. En tenslotte seinde de Europese satelliet ROSAT uit het

heelal eerste resultaten over de kwestie van de mysterieuze "donkere materie". Zo zien wij aardbewoners, als wij omhoogkijken naar de nachtelijke hemel, slechts één honderdste van de materie in het heelal. De resterende 99 procent, zo lijkt het nu, omringen de galactische stelsels in de vorm van geweldige, onzichtbare "aura's". Dat is uiteraard nog lang niet alles. Onze planeet en de gehele melkweg, en ook duizenden andere galactische stelsels van "onze buurt" van de kosmos zijn met een snelheid van rond 580 kilometer per seconde onderweg naar de "Great Attractor", een onzichtbare gigant, waarvan de aantrekkingskracht zo groot moet zijn als die van 20000 biljoen zonnen. Maar zelfs de "Grote Aantrekker" heeft zijn problemen. Volgens recente gegevens drijft hij als een vlot op een kosmische rivier het heelal in - aangetrokken door een "iets" met een nog veel, veel grotere massa.
Hoe meer de hemelonderzoekers over het universum te weten komen, des te meer blijkt dat zijn realiteit vaak fantastischer is dan alle science fiction. Zo hebben bijvoorbeeld veel van de vooraanstaande "reine denkers" onder de kosmologen - daaronder Stephen Hawking, Steven Weinberg en Murray Gell-Mann - de voor ons vertrouwde driedimensionale ruimte als operatieveld voor hun wereldconcepten al lang verlaten. De onderzoekers zijn overgestapt naar een tiendimensionale "hyperspace", een niet meer voor te stellen, maar mathematisch begaanbare "superruimte" vol verrassende eigenschappen en met niet altijd duidelijke grenzen tussen fysica en mystiek.
In deze andere wereld is er een "multiversum" van tiendimensionale kosmossen - telkens met negen dimensies van ruimte en één dimensie van tijd. Als zeepbellen zweven zij naar een tijdloze "niet-ruimte" en kunnen via "wormgaten" met elkaar verbonden zijn. Alle materie bestaat er uit heel kleine draadjes van verschillende frequentie; de natuurwetten zijn verenigd in de harmonie waarvan men al zolang gedroomd heeft: "De hyperspace schijnt de natuurlijke habitat van alle natuurkrachten te zijn." Weliswaar is er nog een groot probleem: ook de knapste mathematische koppen van Cambridge, Yale of CERN kunnen de vergelijkingen die de hyperruimte beschrijven, nog niet bevredigend oplossen.
De mogelijke "wereldformule" - gemodelleerd volgens de "modulaire functies" van een begenadigde Indische lekewiskundige - is een "werktuig uit de 21ste eeuw dat ons toevallig in handen is gevallen"; zoals de Amerikaanse natuurkundige Michio Kaku schrijft: "Wij moeten eerst leren er goed mee om te gaan."

Magic cosmos: Setting off to the tenth dimension

Two or three years ago I had my doubts about them, about the high priests of cosmology with their fantastic stories about the Big Bang, galactic cannibalism and Black Holes. I did not want to lose all the sense of wonder that I had when I looked at the starry sky, I did not want to hear anything more about "supersymmetry" and "naked singularity". It was not Stephen Hawking's vision of a "quantum universe" that expressed my innermost thoughts but Friedrich von Schillers two-hundred-year-old criticism of the astronomers:
"Don't go on so much about nebulae and suns! Is nature only great because she gives you enough to count?"
Today, on the other hand, I like to hear astronomers, astro-physicists and cosmologists go on about their fields of interest; like many of my contemporaries, I cannot get enough of the reports that come from observatories, particle accelerator laboratories and scholars' rooms. A revolution in thought is taking place there, a new picture of the world is emerging - at present still gropingly formulated by the leading theoreticians or provocatively formulated in book titles such as "The Dream of the Unity of the Universe", "The Physics of Immortality" or "Equations of Eternity". The scientists' desire to speculate has been spurred on by their successes in the search for the original particles of matter - the last find being the "top quark", an elementary particle dating back to that second in which the universe came into being - and by the wealth of startling data which is at present coming from the new "super telescopes", which have been set up in the Pacific area in particular, or are orbiting the Earth as satellites. With their even sharper eyes, they have within a short space of time already registered indirect proof of the existence of phenomena which fits in with the predictions made by the likes of Albert Einstein, Steven Weinberg or Stephen Hawking in their theoretical concepts. The latest example is the sighting of a maelstrom in the "M87" galaxy 50 million light years away. The mass of the cloud of gas rotating there at a speed of two million kilometres a second is equivalent to two thousand million stars. What makes it circulate at such a speed is obviously the suction force of a gigantic Black Hole which is sucking in the masses of gas. No less significant are also the data on the "after-glow" from the period when our universe came into being: proof that the Big Bang which created our cosmos did indeed take place 15 thousand million years ago. And, finally, the European ROSAT satellite has beamed back its first findings in connection with "dark matter" in space, a subject that was always shrouded in mystery. According to the information received, we Earthlings see only a hundredth of the material in space when we look up into the night sky, the remaining 99 percent, so it would seem, surround the galaxies in the form of gigantic, invisible "auras". But this is by no means all: our planet and the entire Milky Way and thousands of other galaxies from "our area" of the cosmos are moving at a speed of about 580 kilometers a second towards the "Great Attractor", an invisible giant whose power of gravitation is equivalent to 20,000 billion suns. But even the "Great Attractor" has its problems. According to the latest data, it is drifting, like a raft on a cosmic river, out into space - attracted by "something" of much, much greater mass.
The more the researchers find out about the universe, the clearer it becomes that its reality is more fantastic than all science fiction. Thus many of the "pure thinkers" among the cosmologists - among them Stephen Hawking, Steven Weinberg and Murray Gell-Mann - have for example long since left behind the three-dimensional space with which we are acquainted as the field of operations for their concepts of the universe. Scientists have switched to a ten-dimensional "hyperspace", a mathematically accessible world which cannot be depicted in visual terms, full of surprising qualities and not always with clear boundaries between physics and mysticism.
A "multiversum" of ten-dimensional cosmoses - with nine space dimensions and one time dimension - is to be found in this other world, like soap bubbles they float in a timeless "non-space" and are sometimes connected with each other via "worm holes". All matter here consists of tiny strings of differing frequency, the laws of nature are united in the long sought-after harmony: "Hyperspace seems to be the natural habitat of all natural forces." There is however still one big problem: even the best mathematical brains of Cambridge, Yale or CERN have not yet been able to satisfactorily balance the equations used to describe this hyperspace. The potential "world formula" - modelled according to the "modular functions" of a gifted Indian lay mathematician - is a "tool of the 21st century which has only by chance fallen into our hands" as the American physicist Michio Kaku writes. "We must," he continues, "still learn how to deal with it."

Mi chiamavano "Luna", come fabbricare una nebulosa spirale, il finito illimitato di Einstein, il cosmo sacro

„Mi canzonavano chiamandomi "Luna" / quand'ero un ragazzo di Spoon River / assetato di conoscere le stelle. / Mi schermivano quando parlavo delle montagne lunari, e del caldo e del freddo pungente, / e delle vallate d'ebano accanto alle vette d'argento, / e della Spica quadrilioni di miglia lontano, / e della piccolezza dell'uomo", così inizia una delle più belle poesie di Edgar Lee Masters compresa nella famosa "Antologia di Spoon River" scoperta da Cesare Pavese negli anni Trenta. La quale così prosegue: "Ma adesso, amici, che la mia tomba è onorata, / fate che non lo sia perché insegnai / la scienza delle stelle nel Collegio di Knox, / ma piuttosto per questo: che attraverso le stelle / io predicai la grandezza dell'uomo, / che è una parte del disegno del cosmo / nonostante la distanza della Spica o delle nebulose spirali ...".
Va ricordato che la Spica è la stella dominante della costellazione della Vergine; mentre le nebulose spirali sono agglomerati di stelle disposti a spirale raccolti probabilmente in altrettante galassie. Fabbricarne una è semplice: basterà disporre di circa diecimila milioni di stelle ed avere la possibilità di distribuirle nello spazio con l'avvertenza di distanziarle di tre o quattro anni luce l'una dall'altra. Si sparga una quantità analoga di sostanza gassosa e si schiacci il tutto in una sorta di disco da far rotare sul proprio piano. Ci si allontani di molti passi, proprio molti, e si potrà ammirare un oggetto somigliante ad una nebulosa spirale.
Più di ottanta sono quelle che si trovano intorno alla nostra Galassia e sono solo un aspetto minimo di un Universo infinito, quasi incomprensibile perfino dalla fantasia umana. Einstein, paradossalmente postulava: "Lo spazio è finito ma illimitato".
Due millenni e mezzo prima, il filosofo e scienziato greco Anassimandro esprimeva pressappoco le stesse convinzioni, ponendo come principio dell'universo l'illimitato collegato con i cicli cosmici. Questo lo porterà ad ammettere l'infinità dei mondi. Vale a dire il macrocosmo.
In una leggenda della Polinesia una dea di passaggio sulla terra, confida al guerriero Uenuku: "Mi piace questo mondo. Non è così freddo e vuoto come lo spazio lassù", esprimendo una concezione ben precisa del cosmo. Non a caso, nella notte lunare che è pari a 14 giorni terrestri, la temperatura scende a 156° C. sotto lo zero. Questo non muta la considerazione divina che esprime Santo Francesco quando dice: "Laudato si, mi Signore, per sora nostra luna e le stelle; / in celu l'ai formate clarite et preziose et belle." Luna, da tempo raggiunta dall'uomo e che ha aperto l'esplorazione planetaria con sonde su Marte e su Venere ed il previsto "gran viaggio" interplanetario di un satellite scientifico senza equipaggio che dovrebbe esplorare in un'unica missione Giove, Saturno, Urano e Nettuno, quando i quattro pianeti risulteranno allineati.
"Che vantaggio se ne trae?". Il geniale inventore e statista americano Benjamin Franklin, quando nel 1783 assisteva ad uno dei primi voli in pallone aerostatico, ebbe a dare questa risposta ad una richiesta probabilmente analoga: "Che vantaggi può dare un bambino appena nato?"
L'esplorazione, comunque, difficilmente chiarirà la genesi del cosmo rispetto alla sua creazione. Talune cosmogonie, infatti, si basano sul concetto che l'universo sia nato, anziché dal nulla dal caos. È una storia priva di testimoni e di essa si ha solo la realtà terminale, ossia il frutto della creazione, non la creazione in sé. L'indagine, quindi, è stata sempre indiziaria, avvolta nella sacralità dei miti. Quello greco, fra i più importanti, assicura che gli dei apparvero nell'universo quando questo già esisteva. Non ne provocarono i natali e neppure vi assistettero. Un fatto, d'altronde, plausibile in quanto gli dei dell'antica Grecia erano immortali, non eterni; ossia vivi sempre per il futuro, ma non vivi da sempre. "Prima di ogni cosa fu il Caos, poi la Terra dal vasto petto, / incrollabile sede per tutti per tutto il tempo futuro ..." scriveva Esiodo nella sua „Teogonia", almeno sette secoli prima di Cristo.
La Bibbia, al contrario, dichiara che l'universo è stato creato da Dio ed esiste per sua sola volontà. Ma dopo questa affermazione di principio accenna ripetutamente all'immane battaglia primordiale con la quale Dio annientò le forze del caos confinandole entro i limiti invalicabili e distruggendo i mostri terrificanti che le componevano. E qui la contraddizione è evidente in quanto si conviene che il caos esisteva prima dell'intervento supremo. Il quesito, perciò, non è in assoluto quello delle origini, bensì quello del principio organizzatore, che taluni vogliono identificare con il soffio vitale. In ogni modo, come scrisse Mircea Eliade, "Tutto ciò che assomiglia ad un Cosmo, è sacro."

Giuseppe Brunamontini

Cosmos magique: départ pour une dixième dimension

Il y a encore deux ou trois ans, à mes yeux, ils étaient assez suspects, tous ces grands prêtres de la cosmologie avec leurs histoires d'explosion originelle, de cannibalisme galactique et de trous noirs. Je n'étais pas disposé à me laisser ravir le miracle du ciel étoilé par des explications scientifiques; je ne voulais plus entendre parler de "supersymétrie" et de "singularité pure". Ce n'est pas la vision de Stephen Hawking d'un "univers quantique" qui rejoignait mon opinion mais plutôt le blâme qu'écrivit Schiller il y a deux cents ans à l'attention des astronomes: "Ne discourez pas tant sur les nébuleuses et les soleils! La nature est-elle seulement grande à vosyeux parce qu'elle vous fait compter?"
Par contre, aujourd'hui, je me laisse volontiers persuader par les astronomes, les astrophysiciens et les cosmologues et, comme bon nombre de mes contemporains, je ne me lasse pas des compte-rendus émanant des observatoires, des laboratoires d'accélérateurs de particules et des salles d'étude des hommes de sciences. C'est là que la pensée est révolutionnée; une nouvelle vision du monde semble poindre à l'horizon, formulée certes de manière encore tâtonnante par des théoriciens renommés, dans "Le rêve de l'unité de l'Univers", "La physique de l'immortalité" ou "Equations de l'éternité". Le plaisir de la spéculation qui anime les scientifiques a été d' une part stimulé par leurs succès dans la recherche des particules primordiales de la matière, en dernier lieu le "Top Quark", une particule élémentaire apparue à la seconde où l'Univers est né. D'autre part, de nouveaux "supertéléscopes" qui sont installés principalement dans la zone pacifique ou qui tournent autour de la terre comme des satellites, livrent actuellement une masse de données stupéfiantes. Grâce à leur acuité visuelle, ils ont enregistré des preuves indirectes de phénomènes dont l'existence avait été prédite dans les concepts théoriques d'Albert Einstein, Steven Weinberg ou Stephen Hawking.
L'exemple le plus récent: le repérage d'un malstrom dans la galaxie "M87" à 50 millions d'années-lumière de distance. La masse de la nébuleuse gazeuse qui y tourne à une vitesse de 2 millions de kilomètres-heure équivaut à deux milliards d'étoiles. La rapidité de sa rotation est apparemment provoquée par la force d'aspiration d'un gigantesque "trou noir" qui avale les masses gazeuses. L'importance des données de mesure de l'"incandescence résiduelle" au début de notre univers n'en est pas moindre: c'est une preuve que l'"explosion initiale" dont est né le cosmos il y a environ quinze milliards d'années a effectivement eu lieu. Pour finir,

Магический космос: прорыв в десятое измерение

citons que le satellite européen ROSAT a envoyé de premiers résultats pour élucider la question de la mystérieuse "matière noire". D'après ces informations, nous ne verrions qu'un centième de la matière dans l'Univers quand nous regardons le ciel nocturne; les 99 pourcent restants semblent donc envelopper les galaxies sous forme d'immenses "auras" invisibles. Bien entendu, il demeure encore beaucoup à découvrir. Notre planète et l'ensemble de la Voie lactée ainsi que des milliers d'autres galaxies "de notre région" du cosmos se dirigent à une vitesse d'environ 580 kilomètres-seconde vers le "Great Attractor", un géant invisible dont la force gravitationnelle est estimée à 20 000 billions de soleils. Cependant, même le "Grand Attireur" a ses problèmes. D'après les données récentes, tel un radeau sur un fleuve cosmique, il vogue à la dérive dans l'Univers, attiré par un "quelque chose" ayant une masse encore beaucoup plus grande.
Plus les astronomes en découvrent sur l'Univers, plus il est clair que sa réalité est souvent plus fantastique que toute la science-fiction. Ainsi, un grand nombre des grands "penseurs purs" parmi les cosmologues, dont Hawking, Weinberg et Gell-Mann, ont quitté depuis longtemps l'espace tridimensionnel qui nous est familier. Les chercheurs sont passés dans un "hyperespace" décadimensionnel, un "surespace" que l'on ne peut plus se représenter mais que l'on peut parcourir mathématiquement, un espace rempli d'étonnantes propriétés mais dont les limites entre la physique et la mystique ne sont pas toujours bien définies. Dans cet autre monde, il existe un "univers multiple" de cosmos décadimensionnels comprenant chacun neuf extensions dans l'espace et une dimension dans le temps; ils flottent comme des bulles de savon dans un "non-espace" intemporel et peuvent être reliés entre eux par l'intermédiaire de "trous de ver". Toute la matière se compose là-bas de fils minuscules de différente fréquence; les lois de la nature sont réunies dans l'harmonie dont on rêvait depuis longtemps: "L'hyperespace semble être l'habitat naturel de toutes les forces de la nature." Un grand problème demeure cependant: solutionner de manière satisfaisante les équations décrivant l'hyperespace. La "formule universelle" potentielle, modelée d'après les "fonctions modulaires" d'un mathématicien génial, un Hindou surdoué, est un "outil du XXIe siècle tombé par hasard entre nos mains"; comme l'écrit le physicien américain Michio Kaku: "Il nous faut d'abord apprendre à nous en servir correctement".

Еще два-три года назад я относился к ним с недоверием — ко всем этим апостолам космологии с их фантастическими рассказами о Большом взрыве, «галактическом каннибализме» и «черных дырах». Я не желал, чтобы меня лишали чуда звездного неба, не желал слушать про «суперсимметрию» и «голую сингулярность». Не «квантовая Вселенная» Стивена Хаукинга будила мои чувства, а возмущенная тирада шиллеровского Звездочета:
«Не болтайте мне тут о туманностях и солнцах! Разве величие природы в том, что она дает вам возможность что-то посчитать?»
А сегодня я с удовольствием слушаю астрономов, астрофизиков и космологов, зачитываюсь сообщениями из обсерваторий, ускорительных центров и кабинетов ученых. Там происходит переворот в мышлении и медленно вырисовывается новая картина мира, которую ведущие теоретики пока еще очень осторожно или, наоборот, провокационно формулируют в названиях книг: «Мечта о единстве Вселенной», «Физика бессмертия» или «Уравнения вечности». Фантазию ученых подстегнули, с одной стороны, их собственные успехи в поисках фундаментальных частиц, например, «т-кварка» — элементарной частицы, родившейся в первую секунду жизни Вселенной. С другой стороны, сейчас поступает огромное количество поразительных данных с новейших «супертелескопов», установленных в зоне Тихого океана и на спутниках. Своим острым глазом они в короткое время нашли косвенные доказательства феноменов, предсказанных в теоретических выкладках Энштейна, Вайнберга или Хаукинга. Новейший пример: наблюдение «космического водоворота» в галактике М87, отстоящей от нас на 50 млн. световых лет. Газовое облако, вращающееся там со скоростью 2 млн. км/час, имеет массу двух миллиардов звезд. Заставляет его вращаться с такой бешеной скоростью, скорее всего, гигантская «черная дыра», которая всасывает газовые массы. Не менее важны сведения о «догорании» вещества, образовавшегося на самой ранней стадии Вселенной — доказательство того, что Большой взрыв, положивший ей начало 15 млрд. лет назад, действительно был. А вот европейский спутник РОСАТ передал из космоса первые данные о таинственной «темной материи». Получается, что мы на Земле видим только сотую часть материи во Вселенной, а 99 % ее, как невидимый ореол, окружает галактики. Конечно, это еще далеко не все. Наша планета, как и весь Млечный путь, как и тысячи других галактик из наших «космических окрестностей», летит со скоростью ок. 580 км/сек к невидимому гиганту, сила притяжения которого соответствует 20000 млрд. солнц. Но и у этого «супермагнита» есть свои проблемы. Согласно новейшим сведениям, он, подобно плоту на космической реке, плывет в дальний космос под воздействием Чего-то, имеющего уж совершенно невероятную массу.
Чем больше ученые узнают о Вселенной, тем яснее становится, что ее реальность превосходит любую фантастику. Так, ведущие «чистые мыслители» из числа космологов — С. Хаукинг, С. Вайнберг и М. Геллмен давно покинули в своих моделях мира привычное нам трехмерное пространство. Исследователи перебрались в 10-мерное «гиперпространство», которое невозможно себе образно представить, но в котором можно путешестовать с помощью математических моделей; оно имеет немыслимые свойства, а границы между физикой и мистикой размыты. В этом мире существует «мультивселенная» 10-мерных космосов с девятью пространственными и одной временной компонентой; как мыльные пузыри, плавают они в безвременном «непространстве» и соединяются друг с другом тоннелями. Вся материя состоит там из крошечных нитей разной частоты, а законы природы объединены в желанную гармонию: «Гиперпространство, как кажется, есть естественная среда всех природных сил«. Правда, есть существенная проблема: самые лучшие головы в Кембридже, Йелле и Европейском Центре ядерных исследований пока не могут удовлетворительно решить уравнения, описывающие гиперпространство. Потенциальная «универсальная формула», смоделированная по «модулярным функциям» одного гениального индийского математика-любителя, это — «инструмент 21 века, который случайно попал нам в руки»; как пишет американский физик Мичио Каку, «мы сперва должны научиться правильно им пользоваться».

MICRO

COSMOS

.1/3

Das Innere eines Beschleunigers, wie die Teilchen ihn sehen: Metallröhren in einem Vakuumtank, die mit einem Hochfrequenzgenerator verbunden sind, ziehen Teilchen an und beschleunigen sie fast auf Lichtgeschwindigkeit.

L'interno di un acceleratore come lo vedono le particelle: tubi di metallo in un serbatoio sotto vuoto collegati con un generatore di alta frequenza, attirano particelle e le accelerano quasi fino alla velocità della luce.

Het binnenste van een versneller, zoals de deeltjes hem zien: glimmende metalen buizen in een vacuumtank, die met een hoogfrequentiegenerator verbonden zijn, trekken deeltjes aan en versnellen ze bijna tot lichtsnelheid.

L'intérieur d'un accélérateur tel que le voient les particules: dans un réservoir à vide, des tubes en métal poli reliés à un générateur haute fréquence attirent des particules et leur font presque atteindre la vitesse de la lumière.

The interior of an accelerator as the particles see it: polished metal tubes in a vacuum tank connected with a high frequency generator attract the particles and accelerate them close to the speed of light.

Вид ускорителя изнутри, как его видят частицы: полированные металлические трубы в вакууме, соединенные с ГВЧ, притягивают частицы и ускоряют их почти до скорости света.

Teilchenbeschleuniger und Detektoren: wozu? In Wirklichkeit sind es die leistungsfähigsten Mikroskope der Welt. Bei CERN in Genf, dem Europäischen Laboratorium für Teilchenphysik, werden u. a. Protonen, oder Elektronen und Positronen in kilometerlangen Tunnels beschleunigt. Über 600 Physiker aus aller Welt betreiben dort Teilchendetektoren und untersuchen die Rätsel des Mikrokosmos'.

Deeltjesversnellers en detectoren: waarvoor? In werkelijkheid zijn dit de sterkste microscopen ter wereld. Bij CERN in Genève, het Europese laboratorium voor deeltjesfysica, worden o.a. protonen, of elektronen en positronen, in kilometerlange tunnels versneld. Meer dan 600 fysici uit de gehele wereld bedienen er deeltjesdetectoren en onderzoeken de raadsels van de microkosmos.

Particle accelerator and detectors - for what purpose? In reality these are the most powerful microscopes in the world. At CERN in Geneva, the European Laboratory for Particle Physics, protons, electrons and positrons are accelerated in kilometer-long tunnels. More than 600 physicists from all over the world work with particle detectors and look into the mysteries of the microcosm.

Acceleratori di particelle e detettori: a che quale fine? In realtà questi sono i microscopi più potenti della terra. Al CERN di Ginevra, il Laboratorio Europeo di Fisica delle Particelle, in tunnel lunghi chilometri si accelerano per esempio protoni o elettroni e positroni. Qui oltre 600 fisici di tutto il mondo controllano detettori di particelle e cercano di risolvere gli enigmi del macrocosmo.

Accélérateurs de particules et détecteurs: pourquoi faire? En réalité, ce sont les microscopes les plus performants du monde. A Genève, au CERN, le laboratoire européen de physique des particules, on accélère protons, ou électrons et positrons dans des tunnels de plusieurs km. Plus de 600 physiciens y travaillent sur des détecteurs de particules et sondent les mystères du microcosme.

Ускорители и детекторы — для чего они? По сути, это самые сильные микроскопы в мире для изучения внутренней структуры материи. В ЦЕРНЕ (Женева) ускоряют в многокилометровых тоннелях, например, протоны, электроны и позитроны. Более 600 физиков со всего мира изучают там загадки микрокосмоса.

19A
20A
21A
22A
23A

ALEPH ist einer der vier Detektoren bei LEP im CERN, wo Elektronen mit deren Antiteilchen, den Positronen, bei höchsten Energien zum Zusammenstoß gebracht werden. Dabei entstehen neue Teilchen, die meist sofort wieder zerfallen.

ALEPH is een van de vier detectoren bij LEP in het CERN, waar elektronen met hun antideeltjes met zeer hoge energieën in botsing worden gebracht. Daarbij ontstaan nieuwe deeltjes die meestal onmiddellijk weer uiteenvallen.

ALEPH is one of four detectors at the LEP at CERN where electrons and their antiparticles the positrons are made to collide at very high energies. New particles come into being in the process which then usually decay again immediately.

ALEPH è uno dei quattro detettori presso il LEP nel CERN dove gli elettroni urtano contro i protoni, le loro antiparticelle, con la massima energia. Dallo scontro nascono nuove particelle che in genere si disintegrano subito.

ALEPH est l'un des 4 détecteurs au LEP au CERN où l'on provoque la collision d'électrons avec leurs antiparticules, les positons, qui possèdent des énergies extrêmes. En général, les nouvelles particules formées se désintègrent immédiatement.

АЛЕФ — один из четырех детекторов в ЦЕРНе; на нем с огромной энергией сталкивают электроны с их античастицами позитронами. Возникают новые частицы, почти сразу распадающиеся.

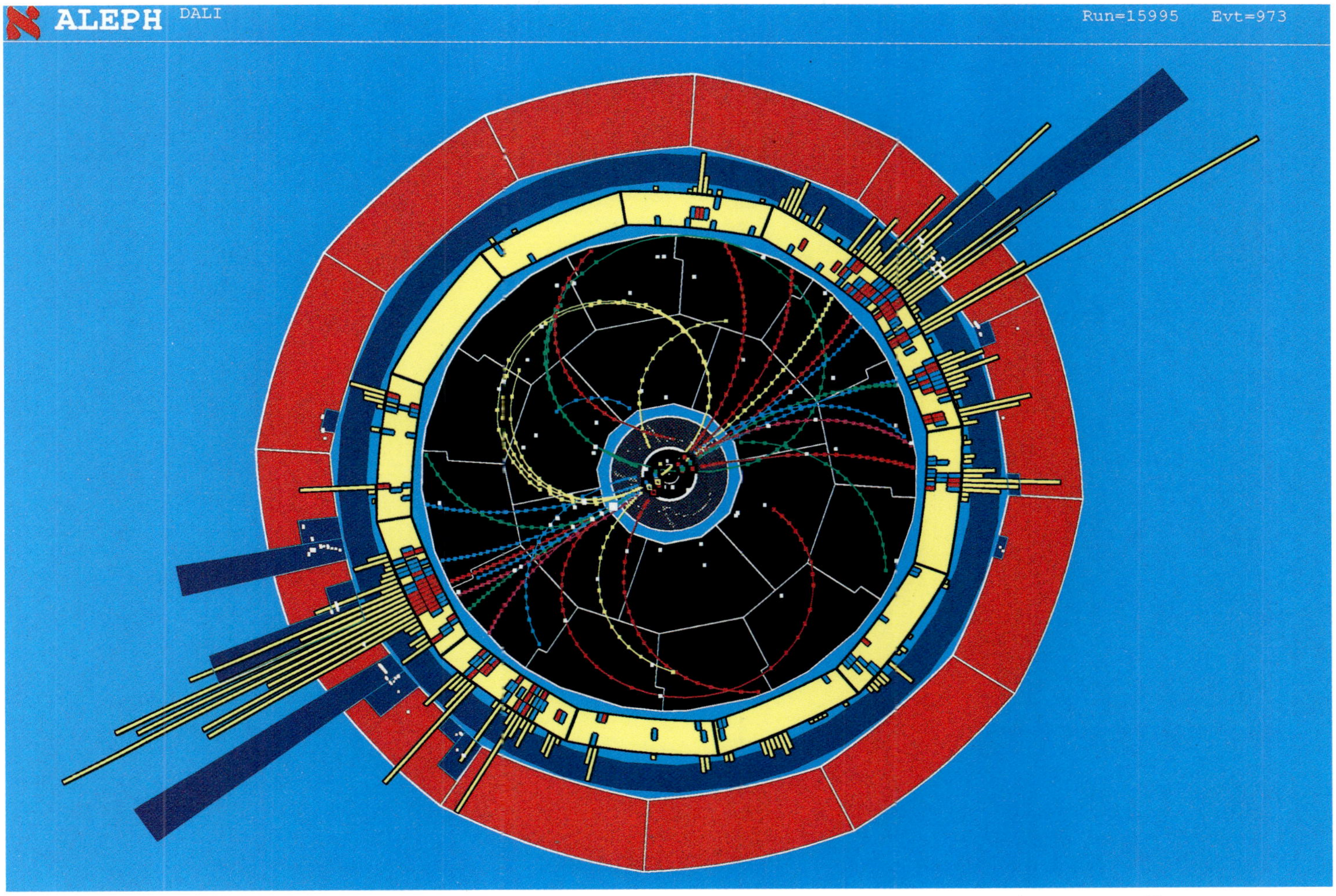

Die entstandenen Teilchen hinterlassen in ALEPH Abertausende von elektrischen Signalen. Computer rekonstruieren daraus die Teilchenbahnen und geben sie auf dem Bildschirm in Farbe wieder.

De ontstane deeltjes laten in ALEPH vele duizenden elektrische signalen achter. Computers reconstrueren de deeltjesbanen en geven ze op het beeldscherm weer. Kleuren helpen de fysici bij de interpretatie.

The particles that come into being leave behind thousands and thousands of electric signals in ALEPH. Computers reconstruct the particle tracks and reproduce them on a screen in colour.

Le particelle che si formano in ALEPH lasciano migliaia di segnali elettrici. I computer li usano per ricostruirne le orbite e rappresentarle sul schermo. I fisici le interpretano in base ai colori.

Les particules ainsi formées laissent dans ALEPH des milliers de signaux électriques. Des ordinateurs reconstruisent les trajectoires des particules qui apparaissent sur l'écran en couleur.

Новые частицы оставляют в АЛЕФе тысячи тысяч электросигналов. Компьютеры вычисляют по ним треки и выводят на экран. Цвет помогает определять их.

ALEPH
DALI
Run=10890
Evt=570

Mikrowelten: Quarks, Squarks und Quantenzauber

Der Professor für Physik war wieder einmal tief in Gedanken versunken - so tief, daß er das große Katzentier auf dem Urwaldpfad vor ihm gar nicht wahrnahm. Murray Gell-Mann, Nobelpreisträger und einer der brillantesten „Vordenker" seines Fachgebietes, grübelte über ein Problem der Quantenmechanik nach und wurde sich seiner Umgebung erst wieder bewußt, als er dem Jaguarundi schon ziemlich nahe gekommen war. „Ich blieb natürlich erschrocken stehen", erinnert sich der Forscher, „aber die Katze schien völlig entspannt zu sein. Wir standen bewegungslos und starrten einander an, es muß Minuten gedauert haben."
Wie die meisten seiner Kollegen, ist auch Gell-Mann ein leidenschaftlicher Grübler, der nicht aufhört, nach dem Woher, Wohin und Warum der Welt zu fragen - bewegt vom Glauben an eine urgesetzliche „Weltformel", ein fundamentales Prinzip des Kosmos', dem alle physikalischen Prozesse gehorchen. So kann es nicht verwundern, daß ihn selbst die Begegnung mit einer Großkatze im Regenwald von Belize zu einschlägigem Philosophieren anregte: „Ich mußte daran denken, daß jedes Objekt im Prinzip aus Quarks und Elektronen besteht. Auch ein Jaguarundi, so kraftvoll und wild, ist ein Bündel von Quarks und Elektronen, und er ist auch ein Ergebnis der fundamentalen quantenmechanischen Gesetze der Physik."
Seit unsere steinzeitlichen Vorfahren Sternkonstellationen in Mammutknochen ritzten, haben Menschen nach dem Warum gefragt. Die Frage führte sie von der Welt der Planeten, der Berge und Flüsse in die Welt der Atome und Moleküle. Sie hat sie in den Atomen Teilchen und in diesen wiederum Quarks entdecken, aber auch die Einheit allen Seins ahnen lassen. So wissen wir heute, daß im Himmel wie auf Erden und überall im vierzig Milliarden Lichtjahre weiten Kosmos die gleichen Gesetze gelten und die gleiche Materie existiert: Stoff, der sich für kurze Zeit zu einem Kristall, einem Regenbogen, einem Kind verdichtet, um sich dann wieder aufzulösen und neue Strukturen anzunehmen. Auf die Frage nach dem inneren Prinzip dieses Kreislaufs des Seins antworten Naturforscher bis ins 19. Jahrhundert mit Vorstellungen eines kosmischen „Uhrwerks". Erst Physiker wie Albert Einstein, Max Planck und Werner Heisenberg tasteten sich näher heran an die elementaren Wurzeln unserer Existenz - vor allem mit Hilfe radikal neuer Denkansätze, des „kreativen Prinzips Mathematik" (Einstein) und Ergebnissen experimenteller Forschung.
Materie war nun Energie, Kraft war Geometrie, Raum und Zeit ein Etwas, das sich unter dem Einfluß kosmischer Kräfte bog, krümmte und sogar zum Stillstand kam. Im Mikrokosmos der subatomaren Teilchen wiederum herrschten die seltsamen Regeln der „Quantenmechanik": Licht entstand aus winzigen Energiepaketen, Quanten und Materieteilchen konnten gleichzeitig hier und dort sein, sich im selben Moment schnell und langsam bewegen, solange niemand versuchte, ihre Position zu bestimmen. So verrückt, aber auch harmlos das alles klingen mochte - es gab die Wirklichkeit beängstigend genau wieder und ermöglichte später die Atombombe.
Seit den Tagen Einsteins ist natürlich viel neues Wissen hinzugekommen, etwa über die kleinsten Teilchen normaler Materie, die Quarks, die Physiker beim Zertrümmern von Materie in Beschleunigern nachgewiesen haben. Nach wie vor suchen die Forscher mit ihren Protonenschleudern in der Nähe von Genf und Chicago nach Teilchen in „höherem Energiezustand" - das sind Squarks, Gluinos oder Higgsbosonen genannte Partikel aus der Geburtssekunde des Kosmos'. Jüngster Fahndungserfolg ist das „Top Quark", das die Physiker in einer Art Mini-Urknall erzeugten. „Top" ist ein Elementarteilchen, das es seit 15 Milliarden Jahren gar nicht mehr gibt. Nur im Feuerball nach dem Urknall kam das Teilchen vor, dann verwandelte es sich in Strahlung.
Experimentelle Ergebnisse dieser Art beflügeln die Kosmologen bei ihrer Suche nach einem „vollkommenen" Naturgesetz, einer „Weltformel", die Relativitätstheorie und vereint. Stephen Hawking versucht dies zur Zeit mit einem „Quantenkosmologie" genannten Ansatz: Er denkt sich den gesamten Kosmos als ein Quantenteilchen mit seinen „unbestimmten" Verhaltensweisen. Er schafft damit eine Welt, die aus unzähligen zehndimensionalen Kosmen besteht. In einem davon, dem unsrigen, brechen die zehn Dimensionen in einer gewaltigen Explosion auseinander: Es entsteht das Universum, das wir kennen. Das andere, sechsdimensional, schrumpft hingegen bis zur Unsichtbarkeit, steht aber als „Rettungsinsel" bereit, sollte es eines Tages zum feurigen Untergang unserer Welt kommen. Albert Einstein benutzte gern die Metapher von der Natur, die „uns nur den Schwanz des Löwen zeigt, nie aber das ganze Tier".
Kosmische Theorien, so scheint es, haben einen ähnlichen Wesenszug: Mehr als der Schwanz ist bisher nicht sichtbar.

Microwerelden: quarks, squarks en quantumtoverij

De professor in de natuurkunde was weer eens diep in gedachten verzonken - zo diep dat hij het grote kattebeest vóór hem op het junglepad helemaal niet waarnam. Murray Gell-Mann, Nobelprijswinnaar en een van de briljantste "voordenkers" op zijn gebied, liep te piekeren over een probleem van de quantummechanica en werd zich pas weer van zijn omgeving bewust, toen hij al tamelijk dicht bij de jaguarundi gekomen was. "Ik bleef natuurlijk van schrik staan", herinnert de onderzoeker zich, "maar de kat scheen geheel ontspannen te zijn. Wij stonden roerloos stil en staarden elkaar aan, het moet wel enkele minuten geduurd hebben."
Ook Gell-Mann is net als de meeste collega's van hem een hartstochtelijke piekeraar die niet ophoudt te vragen naar het waarvandaan, waarheen en waarom van de wereld - bewogen door het geloof in een oorspronkelijke wetmatige "wereldformule", een fundamenteel principe van de kosmos, waardoor alle natuurkundige processen zich laten leiden. Zo is het niet te verwonderen dat hij zelfs door de ontmoeting met een grote kat in het regenwoud van Belize tot filosofische gedachten hierover werd aangezet: "Ik moest eraan denken dat ieder voorwerp in principe uit quarks en elektronen bestaat. Ook een krachtige, wilde jaguarundi is een bundel van quarks en elektronen, en ook hij is een resultaat van de fundamentele quantummechanische wetten van de natuurkunde."
Sinds het stenen tijdperk, toen onze voorouders constellaties van sterren in mammoetbeenderen sneden, hebben mensen naar het waarom gevraagd. Deze vraag heeft hen van de wereld der planeten, bergen en rivieren naar de wereld der atomen en moleculen geleid; deze heeft hen niet alleen in de atomen deeltjes en in deze deeltjes wederom quarks laten ontdekken, maar ook de eenheid van al het zijn laten vermoeden. Zo weten wij vandaag dat er in de hemel als op aarde en overal in de 40 miljard lichtjaren grote kosmos dezelfde wetten gelden en dezelfde materie bestaat: stof die voor korte tijd tot een kristal, een regenboog, een kind verdicht, en dan weer oplost en nieuwe structuren aanneemt. De vraag naar het inwendige principe van deze kringloop beantwoordden natuuronderzoekers tot in de 19de eeuw met voorstellingen van een kosmisch "uurwerk". Het waren pas fysici als Albert Einstein, Max Planck en Werner Heisenberg, die voorzichtig probeerden de elementaire wortels van onze existentie te ontraadselen - vooral met radicaal nieuwe gedachtengangen, het "creatieve principe mathematica" (Einstein), en met resultaten van experimenteel onderzoek.

Micro-worlds: Quarks, squarks and quantum magic

Materie was nu energie, kracht was geometrie, ruimte en tijd een "iets" dat zich onder invloed van kosmische krachten boog, kromde en zelfs tot stilstand kwam. En in de microkosmos van de subatomaire deeltjes heersten de merkwaardige regels van de "quantummechanica": licht ontstond uit heel kleine energiebundeltjes, quanta en materiedeeltjes konden tegelijkertijd hier en daar zijn, op hetzelfde moment zich snel en langzaam bewegen, zolang niemand probeerde hun positie te bepalen. Zo dwaas, maar ongevaarlijk dat alles ook mocht klinken - het was een beangstigend nauwkeurige weergave van de werkelijkheid en maakte later de atoombom mogelijk. Sinds de dagen van Einstein is er natuurlijk veel nieuwe kennis bij gekomen, bijvoorbeeld over de kleinste deeltjes van de normale materie, de quarks, die fysici bij het stukslaan van materie in versnellers hebben aangetoond. Nog altijd zijn de onderzoekers met hun protonenversnellers in de buurt van Genève en Chicago op zoek naar deeltjes in een "hogere energietoestand" - dat zijn deeltjes uit de seconde van de geboorte van de kosmos, squarks, gluïno's of Higgsbosonen genaamd. Het meest recente succes bij het speurwerk is de "top quark" die de fysici in een soort mini-oerknal lieten ontstaan. "Top" is een elementair deeltje dat sinds 15 miljard jaren helemaal niet meer bestaat. Het deeltje kwam alleen voor in de vuurbal na de oerknal, daarna veranderde het zich in straling. Dit soort experimentele resultaten bevleugelen de kosmologen bij het zoeken naar een "volmaakte" natuurwet, een "wereldformule" die relativiteitstheorie en quantummechanica verenigt. Stephen Hawking probeert dit op het ogenblik met een methode die "quantumkosmologie" heet: hij stelt de gehele kosmos voor als een quantumdeeltje met zijn eigen "onbepaalde" gedrag. Hij schept daarmee een wereld die uit talloze tiendimensionale kosmossen bestaat. In een van deze kosmossen, de onze, vallen de tien dimensies met een geweldige explosie uiteen: er ontstaat het universum dat wij kennen. Het andere, zesdimensionaal, krimpt daarentegen, wordt onzichtbaar, maar staat als "reddingsvlot" gereed, mocht het op een dag tot een vurige ondergang van onze wereld komen. Albert Einstein gebruikte graag de metafoor van de natuur die "ons alleen de staart van de leeuw toont, maar nooit het gehele dier".
Kosmische theorieën hebben naar het schijnt net zo'n karaktertrek: meer dan de staart is tot nu toe niet zichtbaar.

The professor of physics was once again absorbed in his thoughts, so absorbed that he did not even notice the big cat on the jungle pathway. Murray Gell-Mann, Nobel prize winner and one of the most brilliant advanced thinkers in his field, was pondering over a quantum mechanics problem and only became aware of his surroundings when the jaguarundi had already come quite close. "Of course I came to a halt full of fright," the research scientist recalls, "but the cat seemed to be completely at ease. We both stood motionless, staring at each other; this must have gone on for minutes."
Like most of his colleagues, Gell-Mann is a passionate muser, one who never ceases to ask where we came from and where we are going to, who never ceases to put the question "Why?" - inspired as he is by his belief in an underlying "world formula", a fundamental principle behind the cosmos, one which determines all physical processes. And so it is of no surprise that even this encounter with a big cat in the rain forest of Belize inspired him to philosophize on this subject; "I couldn't help thinking that, in principle, every object consists of quarks and electrons. The jaguarundi, too, wild and strong though he may be, is but a bundle of quarks and electrons, is also the outcome of the fundamental physical laws of quantum mechanics."
Since our Stone Age ancestors scratched designs of constellations of stars in mammoth fossils, people have been asking "Why?". The attempt to answer the question has led them from the world of the planets and the mountains and the rivers to the world of atoms and molecules. It has led to the discovery of particles within the atoms and quarks within the particles; it has given rise to presentiments of the unity of all creation. Today, we know that in Heaven as on Earth and everywhere else in the 40 thousand million light years that our cosmos extends, the same laws apply and the same material exists: material which condenses for a short time to become a crystal, a rainbow or a child and which then dissolves again to form new structures. In their attempt to determine the inner principle behind this cycle of being, the natural scientists up until the 19th century reacted with ideas of a cosmic "clockwork". It was only later that physicists like Albert Einstein, Max Planck and Werner Heisenberg got closer to the elementary roots of our existence - with the help above all of the radical new approaches of the "creative principle of mathematics" (Einstein) and the results of experimental research.
Matter was now energy, force was geometry, space and time were something that bent under the influence of cosmic forces, which became warped or stood still. In the micro-cosmos of the subatomic participles the strange rules of "quantum mechanics" applied: light was the product of tiny little packets of energy; quanta and participles of matter could be simultaneously here and there and everywhere, could move quickly and slowly at the same time, as long as no one attempted to ascertain their position. As crazy or as harmless as that may sound, it was a frighteningly true picture of reality and it made it possible later to construct the atomic bomb.
Since Einstein's time, a lot of new knowledge has been added to the pile, for example about the smallest particles of normal matter, the quarks, the existence of which has been proved by scientists as a result of splitting matter in accelerators. With their proton accelerators in the vicinity of Geneva and Chicago research scientists are still searching for particles in a "higher state of energy" - for squarks, gluinos or Higgbosons, particles dating back to the second in which our cosmos came into being. The latest success story in the quest is the "top quark" which physicists produced in a kind of mini-Big Bang. This "top quark" is an elementary particle which has not existed for the last 15 thousand million years. It only occurred in the fireball after the Big Bang and then it changed into radiation. Experimental results of this kind lend wings to cosmologists in their search for a "perfect" law of nature, a "world formula", one combining the theory of relativity and quantum mechanics. At present Stephen Hawking is attempting this with his "quantum cosmology" approach: he imagines the entire cosmos to be a quantum particle with its "indeterminate" mode of behaviour. In the process he postulates a world which consists of countless ten-dimensional cosmoses. In one of them - ours - the ten dimensions are breaking apart - in a gigantic explosion - the universe that we know is coming into being as a result. The other, six-dimensional, is, on the other hand, shrinking to the point of invisibility, but it is there as a "rescue station" should, one day, our world go up in flames. Albert Einstein liked to describe nature by using the metaphor that it "only shows us the tail of the lion but never the whole animal". Cosmic theories, so it seems, share this characteristic: to date no more than the tail is visible.

La mitologia astrale, perché non sostituire i nomi dello Zodiaco con quello dei 12 Apostoli? L'eclissi e Cristoforo Colombo

L'ardita concezione della pluralità degli universi, al di fuori dei confini del nostro sostenuta dagli astronomi, probabilmente non circoscrive il concetto del "piccolo mondo" che, secondo l'antica filosofia ed il Rinascimento, vedeva nell'uomo la rappresentazione dello universo. Il nostro, in ogni caso. Cioè l'organizzazione umana, il microcosmo, non differisce da quella universale, ossia il macrocosmo. Anzi, il termine originario è il primo e l'altro ne è la contrapposizione, seppur convergente.
Cominciò a pensarci Democrito, ma pure Anassimene ed Empedocle, per non parlare di Aristotele, di Platone e dello stoicismo. È una parentela, quella dell'uomo con il cosmo, sulla quale si è sempre insistito per evidente e permanente convenienza del genere umano. Così che si possono imputare alla simbiosi microcosmo e macrocosmo quelle speculazioni astrologiche che hanno sempre voluto collegare l'esistenza dell'uomo sulla terra al moto celeste.
Le costellazioni, dunque, furono fonti di mitologia astrale, tanto che l'uomo ne stabilì l'anagrafe nel tempo, secondo la sua realtà e le sue esigenze psichiche ed esistenziali. È il permanere di quel paganesimo greco-latino tanto fascinoso e suggestivo, pure nei giorni nostri, dal quale l'umanità non sembra intenzionata a discostarsi. Il Venerabile Meda, grande enciclopedista inglese, intuì il pericolo di tale astrologia e tentò di deviarne l'impatto, alleggerendo il peso della superstizione che ne derivava, rimpiazzando i nomi pagani delle costellazioni e dei segni con quelli della Bibbia e del Vangelo.
I dodici segni dello Zodiaco, tanto per incominciare, avrebbero dovuto assumere i nomi dei dodici apostoli: così l'Ariete diventava San Pietro, il Toro Sant'Andrea e via dicendo, mentre la costellazione della Lira si sarebbe dovuta chiamare Culla di Cristo, Andromeda il Sepolcro, Ercole i Magi. Un'industria fiorente passata e futura di oroscopi che si avviava al fallimento. Ma non solo questa ragionevolmente si oppose: in realtà il dissenso venne dai suoi stessi discepoli.
Eppure è innegabile la consistenza di taluni fenomeni presagistici di altri, in cui eclissi e comete agivano da protagoniste. Nelle "Note geo-astro-fisiche" stralciate dalle "Cronache Normanne" che vanno dal 1004 al 1254, si legge, datata 1011: "Il sole si eclissò e fu gran carestia". Del 1095, invece, è l'annotazione: "Parve che cadessero le stelle dal cielo. Il dì 4 di aprile di mercoledì fu gran rivoltura nella Gerusalemme dei Cristiani"; mentre nel 1098 avvenne che "Apparve una cometa. In questo anno la città di Antiochia fu presa dai Cristiani il dì 3 di giugno nella quinta feria".
Chi, invece, fosse stato consapevole di tali fenomeni, li avrebbe potuti usare a proprio vantaggio, come fece Cristoforo Colombo che superò la diffidenza degli indigeni ottenendone l'aiuto predicendo l'ora esatta di una eclissi di luna.
Così, dunque, il microcosmo occupa di simboli e di significati quel macrocosmo, quasi cercando di portarlo a sua misura. E questo per le antiche regole che assimilano i due concetti ad intero vantaggio umano. Del microcosmo, Leonardo Da Vinci, scriveva: "L'uomo è detto da li antiqui mondo minore, e certo la dizione d'esso nome è bene collocata imperò che, sì come l'omo è conposto di terra, acqua, aria e foco, questo corpo della terra è il simigliante ...". Una vera e propria garanzia.
E non fu il solo perché le testimonianze hanno avuto una presenza costante nel tempo. Fra le più recenti, ricordiamo quella di Thomas Browne che, nel suo „Religio Medici", scriveva: "V'è certo un tratto di divinità in noi, qualcosa che esisteva prima degli elementi e che non deve nulla al sole"; concetto amplificato da F. Quarles che in "Hieroglyphics" dichiara: "Non v'è nessuno che possa leggere bene Iddio, se prima non ha compitato l'uomo!".
Le dimensioni di questo microcosmo diventano così confortanti e documentate, da consentire passaggi ripetuti nella sfera celeste di poeti e narratori, immensamente prima degli astronauti, a partire da quel Dante Alighieri, forse unico testimone de "L'amor che muove il sole e l'altre stelle". Va bene che lo fece con una invocazione propiziatoria ad Apollo, stupenda certamente, e che si avventurò nel cielo seguendo le teorie di Tolomeo che immobilizzava la terra al centro dell'universo ed attorno ad essa, in orbite circolari e concentriche i cieli della Luna, di Mercurio, di Venere, del Sole, di Marte, di Giove, di Saturno, mentre l'ottava sfera è delle stelle fisse; però raggiunge l'immobile Empireo verso il quale, trasportato dall'energia stessa che muove i cieli, si eleva anche lui spinto dagli accesi occhi di Beatrice.

Giuseppe Brunamontini

Microcosmes: quarks, squarks et la magie quantique

Une fois de plus, le professeur de physique était plongé dans ses pensées, et ceci à un point tel qu'il ne perçut pas le grand félin qui venait vers lui sur le sentier de la jungle. Murray Gell-Mann, lauréat du Prix Nobel, réfléchissait à un problème de mécanique quantique et ne reprit conscience de son entourage que lorsque le jaguar se fut approché fort près. "Bien entendu, tout à ma frayeur, je ne fis plus un mouvement", se souvient le chercheur, "mais le fauve semblait très décontracté. Immobiles tous les deux, nous nous regardions fixement; cela a certainement duré plusieurs minutes."
Comme la plupart de ses collègues, Gell-Mann est un penseur passionné qui n'arrête pas de chercher le pourquoi du monde, essayant de savoir d'où il vient et où il va, mû par la foi en une "formule universelle" basée sur la loi originelle, un principe fondamental du cosmos auquel obéissent tous les processus physiques. Il n'est donc pas étonnant que même cette rencontre avec un fauve dans la forêt de Belize l'ait incité à philosopher sur ce sujet: "J'ai pensé que chaque objet était en principe composé de quarks et d'électrons. Un jaguar, aussi puissant et sauvage qu'il soit, est également un ensemble de quarks et d'électrons et un résultat des lois fondamentales quantummécaniques de la physique."
Depuis que nos ancêtres de l'âge de pierre ont gravé des constellations dans des os de mammouth, l'humanité a cherché le pourquoi. Cette question a fait passer les humains du monde des planètes, des montagnes et des fleuves dans le monde des atomes et des molécules. Dans les atomes, elle leur a fait découvrir des particules et dans ces dernières des quarks, mais elle leur a aussi fait pressentir l'unité de ce qui existe. Ainsi, nous savons aujourd'hui qu'au ciel comme sur terre et partout dans ce cosmos d'une ampleur de 40 milliards d'années-lumière, les mêmes lois sont valables et la matière existante est la même: une substance qui se condense peu de temps pour former un cristal, un arc-en-ciel ou un enfant avant de se désagréger et d'élaborer de nouvelles structures. Jusqu'au XIXe siècle, les experts en sciences naturelles solutionnaient la question du principe interne du cycle de l'existence par la représentation d'un "mécanisme d'horloge". On dut attendre des physiciens comme Einstein, Planck et Heisenberg pour s'approcher lentement des racines élémentaires de notre existence, grâce à des modes de pensée radicalement nouveaux, au "principe créatif que sont les mathématiques" (Einstein) et aux résultats de la recherche expérimentale.
La matière n'était qu'énergie, la force était

Микромиры: кварки, с-кварки и квантовые чудеса

géométrie, espace et temps, quelque chose qui, sous l'influence des forces cosmiques, se pliait, se tordait et arrivait même à s'arrêter. Les étranges règles de la "mécanique quantique" dominaient par contre dans le microcosme des particules subatomiques: la lumière naîssait de minuscules paquets énergétiques, les quanta et les particules de matière pouvaient être à la fois ici et là, se mouvoir au même moment vite ou lentement tant que personne n'essayait de définir leur position. Aussi fou et inoffensif que cela pouvait paraître, cette vision reproduisait la réalité avec une exactitude angoissante et permit plus tard de construire la bombe atomique.
Depuis Einstein, bien des découvertes ont été faites, sur les infimes particules de la matière normale par exemple, les quarks, dont les physiciens ont prouvé l'existence en fragmentant la matière dans des accélérateurs.
Aujourd'hui encore, les savants cherchent, avec leurs accélérateurs de protons près de Genève et de Chicago, des particules d'un "état énergétique supérieur": squarks, gluinos ou higgsbosons qui ont surgi à l'instant où le cosmos est né. Le succès le plus récent de la recherche est le "Top Quark" généré par les physiciens au cours d'une sorte d'explosion originelle en miniature. "Top" est une particule élémentaire qui n'existe plus depuis 15 milliards d'années. Elle apparut dans la boule de feu après l'explosion originelle, puis se transforma en rayonnement. De tels résultats expérimentaux donnent des ailes aux cosmologues à la recherche d'une loi "parfaite" de la nature, d'une "formule universelle" réunissant la théorie de la relativité et la mécanique quantique. Stephen Hawking essaie actuellement d'y parvenir avec une "cosmologie quantique"; il se représente l'ensemble du cosmos comme une particule quantique aux "comportements" indéterminés. Il créé ainsi un monde composé d'innombrables cosmos décadimensionnels. Dans l'un deux, le nôtre, les dix dimensions se disloquent dans une violente explosion d' où naît l'univers que nous connaissons. L'autre monde, hexadimensionnel, se rétrécit par contre jusqu'à devenir invisible, mais demeure en alerte comme un "îlot de sauvetage" pour le cas où notre monde périrait un jour par le feu. Albert Einstein aimait évoquer la métaphore de la nature qui "nous montre la queue du lion mais jamais l'animal en entier". Les théories cosmiques ont apparemment une caractéristique similaire: jusqu'à présent, on n'en voit pas plus que la queue.

Профессор физики опять погрузился в свои мысли, и на этот раз так глубоко, что не заметил на тропе среди сельвы дикую кошку. Маррей Геллмен, лауреат Нобелевской премии и один из лучших умов в своей области, задумался над одной проблемой квантовой механики и вернулся к действительности только тогда, когда увидел ягуарунди прямо перед собой. «Я, конечно, в страхе остановился, — вспоминает ученый, — но она казалась совершенно спокойной. Мы стояли неподвижно и смотрели друг на друга. Это длилось, наверное, несколько минут.»
Как и большинство своих коллег, Геллмен постоянно задается вопросами о мироздании, твердо веря в фундаментальную «универсальную формулу», этот основополагающий принцип космоса, которому подчиняются все физические процессы. Поэтому неудивительно, что даже встреча с дикой кошкой в тропическом лесу на берегу Белизе настроила его на философский лад: «Я постоянно должен думать о том, что, в принципе, любой объект состоит из кварков и электронов. Даже ягуарунди, такая сильная и дикая, — всего лишь пучок кварков и электронов, она тоже есть результат фундаментальных законов квантовой механики».
С тех пор, как наши предки в каменном веке впервые нацарапали на мамонтовой кости расположение звезд, люди задают вопрос «почему?». Этот вопрос привел их из мира планет, гор и рек в мир атомов и молекул, заставил открыть в атомах частицы, в частицах кварки и задуматься о единстве всего бытия. И вот сегодня мы знаем, что на небе, как и на земле, как и за сорок миллиардов световых лет отсюда, действуют те же законы и существует та же материя — материал, уплотняющийся на короткое время в кристалл, в радугу в ребенка, чтобы распасться и принять новую структуру. Внутренний принцип этого круговорота бытия натуралисты вплоть до 19 в. видели в некоемом «космическом часовом механизме». И только такие физики, как А. Эйнштейн, М. Планк и В. Гейзенберг приблизились к изначальным корням нашего существования, прежде всего благодаря совершенно новому образу мышления, «творческому принципу математики» (Эйнштейн) и результатам экспериментальных исследований. Тогда материя стала энергией, сила — геометрией, пространство и время — чем-то, что под воздействием космических сил может изгибаться, деформироваться и даже останавливаться. В микрокосмосе субатомных частиц царят странные законы квантовой механики: свет возник из крошечных сгустков энергии, кванты и материальные частицы могут одновременно находиться в разных местах, в один и тот же момент двигаться быстро и медленно, если не пытаться определить их положение. Как бы непонятно и в то же время невинно это ни выглядело — это была пугающе верная картина мира, которая и позволила в конце концов сделать атомную бомбу. Со времен Эйнштейна объем знаний, конечно, увеличился. Взять хотя бы кварки, эти мельчайшие частицы, которые физики обнаружили, разрушая материю в ускорителях. Ученые неутомимо ищут на своих протонных пушках близ Женевы и Чикаго частицы «более высоких энергий», которые называются с-кварки, глюино или Н-бозоны. Самый последний их успех — «т-кварк», который физики получили, устроив Большой взрыв в миниатюре. «Т-кварк» — это элементарная частица, которая не существует вот уже 15 млрд. лет. Она родилась в огне Большого взрыва, а потом превратилась в излучение. Подобные результаты окрыляют космологов в поисках совершенного закона природы, «универсальной формулы«, которая объединит теорию относительности и квантовую механику. С. Хаукинг пытается сделать это с помощью т. н. квантовой космологии: он представляет весь космос в виде квантовой частицы с «неясным» поведением. Тем самым он конструирует мир, состоящий из бесчисленных 10-мерных космосов. В одном из них, нашем, после мощнейшего взрыва десять измерений распались, и образовалась Вселенная, какой мы ее знаем. Другая, шестимерная, сжалась до невидимых размеров, но всегда «стоит наготове», подобно спасательному плоту, если в один прекрасный момент наш мир погибнет в огне. Альберт Эйнштейн любил повторять метафорическое определение природы, которая «всегда показывает нам только львиный хвост, но никогда — льва целиком». Космические теории, как кажется, в этом сходны: пока, кроме хвоста, ничего не видно.

CREATIVE

COSMOS

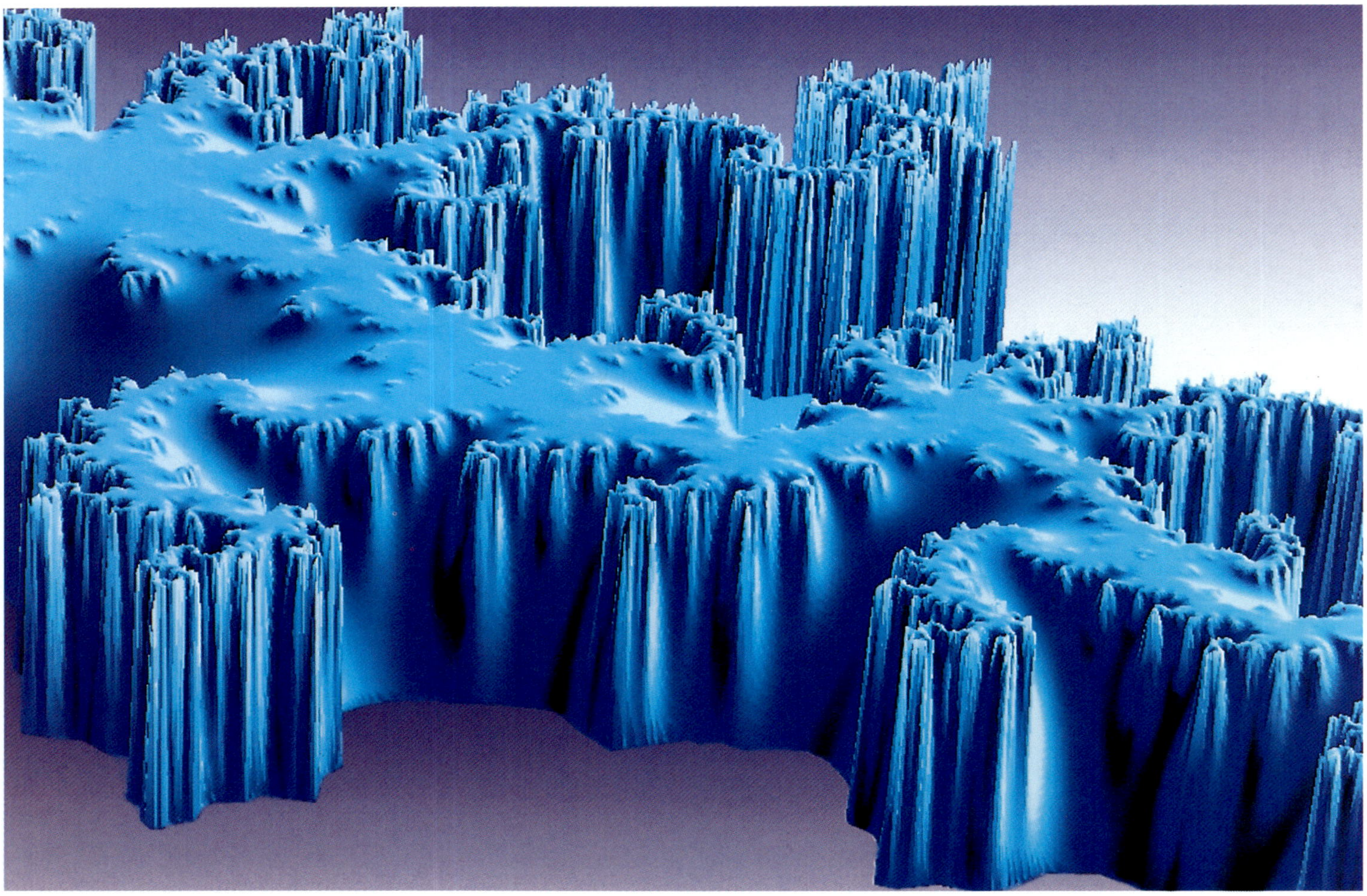

Mit seinen detaillierten Erosionsnarben wirkt der „Fractal Mountain" fast wie ein echter Gebirgszug, obwohl er auf dem Computerschirm entstanden ist. Mit der „fraktalen Geometrie" entdeckten Mathematiker Regeln in allen Formen der Natur und entschlüsselten damit sogar komplexe „Chaosprozesse", wie sie im Kosmos oder im menschlichen Körper ablaufen können. „Chaos" repräsentiert gewissermaßen das kreative Prinzip in der Natur, denn es führt dazu, daß sich die Welt immer wieder neu ordnet. Dieser Ordnung, so wissen wir heute, wohnt eine unberechenbare Instabilität inne, aus der sich naturgesetzlich neue Ordnung organisiert - etwa wenn unermeßlich heiße und „chaotische" Urmasse nach dem Urknall abkühlt und in Bruchteilen einer Sekunde „Ordnung" in Form von Gravitation und Elektromagnetismus „absondert", um dann erneut in „Unordnung" zu fallen, aus der dann neuerliche Ordnung, etwa Materieteilchen, hervorgeht.

Met zijn gedetailleerde erosielittekens ziet de "Fractal Mountain" er bijna als een echte bergketen uit, hoewel hij ontstaan is op het computerscherm. Met de "fractale geometrie" ontdekten wiskundigen in alle vormen van de natuur regels en ontcijferden daarmee zelfs complexe "chaosprocessen" zoals deze in de kosmos of in het menselijk lichaam kunnen verlopen. "Chaos" representeert in zekere zin het creatieve principe in de natuur, want het zorgt ervoor dat er in de wereld steeds weer een nieuwe orde ontstaat. Zo weten wij tegenwoordig dat er in deze orde een onberekenbare instabiliteit aanwezig is, waaruit volgens de wetten der natuur nieuwe orde wordt georganiseerd - bijvoorbeeld wanneer onmetelijk hete en "chaotische" oermassa na de oerknal afkoelt, in fracties van een seconde "orde" in vorm van gravitatie en elektromagnetisme "afscheidt" om dan opnieuw in "wanorde" te vallen, waar dan nogmaals orde uit ontstaat, bijvoorbeeld materiedeeltjes.

With its detailed erosion scars, the fractal mountain is almost like a real mountain chain although it is a product of computer technology. With fractal geometry, mathematicians were able to discover rules in all the forms of nature and even to decode complex "chaos processes" such as can take place in the cosmos or in the human body. "Chaos" represents, as it were, the creative principle in nature for it enables the world to constantly rearrange its order. This order, we now know, involves an incalculable instability, out of which the laws of nature organize a new order - such as when immeasurably hot and "chaotic" mass cools down following the Big Bang and within fractions of a second "abstracts" "order" in the form of gravitation and electromagnetism, only to fall into "disorder" again, out of which a renewed order, for example in the shape of particles of matter, arises. The universe is continually altering its structure. It has been permanently in flux since the moment of its inception.

Con le sue particolari 'cicatrici' da erosione la fractal mountain assomiglia ad una catena montuosa vera e propria anche se è nata sullo schermo del computer. Grazie alla geometria frattale, i matematici hanno scoperto regole in tutte le forme della natura e sono riusciti a decodificare persino "processi caotici" complessi come quelli che possono svolgersi nel cosmo o all'interno dell'organismo umano. Nella natura il "caos" rappresenta il principio creativo, perchè fa sì che il mondo si dia continuamente un nuovo ordine. Oggi sappiamo che in tale ordine risiede un'instabilità imprevedibile dalla quale, per via delle leggi naturali, si costituisce un nuovo ordine come quando la materia iniziale, smisuratamente calda e caotica, si raffredda dopo la grande esplosione ed entro frazioni di secondo "genera ordine" sottoforma di gravitazione e magnetismo per poi ricadere nuovamente nel "disordine" dal quale a sua volta nasce l'ordine sottoforma di nuove particelle di materia.

Avec ses cicatrices dues à l'érosion, la "Fractal Mountain" fait presque l'effet d'une vraie chaîne de montagnes bien qu'elle soit née sur un écran d'ordinateur. Avec la "géométrie fractale", les mathématiciens ont découvert des règles dans toutes les formes de la nature, décodant même les "processus chaotiques" complexes, pouvant se dérouler dans le cosmos ou le corps humain. Le "chaos" représente en quelque sorte le principe créatif de la nature, puisqu'il entraîne une réorganisation toujours nouvelle du monde. Cet ordre, nous le savons au-jourd'hui, renferme une instabilité imprévisible qui est la source d'un ordre nouveau, conforme aux lois de la nature, comme, p. ex. quand une masse primordiale extrêmement brûlante et "chaotique" se refroidit après le big bang et "secrète" en quelques fractions de seconde un "ordre" composé de gravitation et d'électromagnétisme pour finalement retomber dans un "désordre" d'où naîtra un nouvel ordre, des particules de matière p.ex.

Следы эрозии придают «фрактальной горе» почти натуральный вид, хотя она родилась на экране компьютера. «Фрактальная геометрия» позволила математикам открыть законы во всех формах природы и расшифровать комплексные процессы, могущие протекать в космосе и человеческом теле. «Хаос» представляет собой творческий принцип природы, ибо ведет ко все новой организации мира. Этому порядку присуща непредсказуемая неустойчивость, из которой возникает новый порядок, например, когда сверхгорячая и хаотичная «первичная масса» после Большого взрыва остывает и в доли секунды «выделяет» порядок в форме гравитации и электромагнетизма, чтобы потом снова впасть в беспорядочность, из которой вновь выходит порядок материальных частиц.

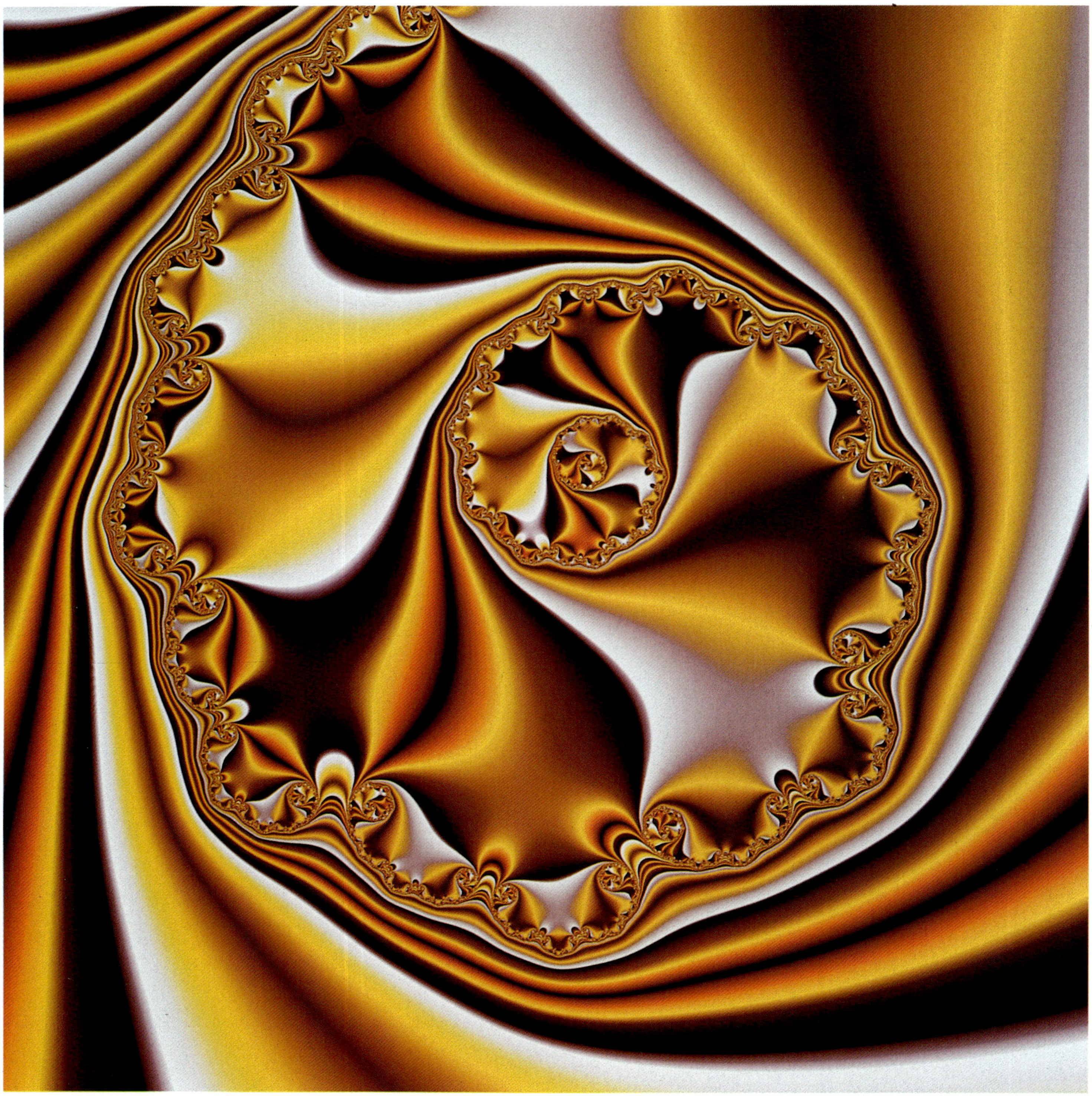

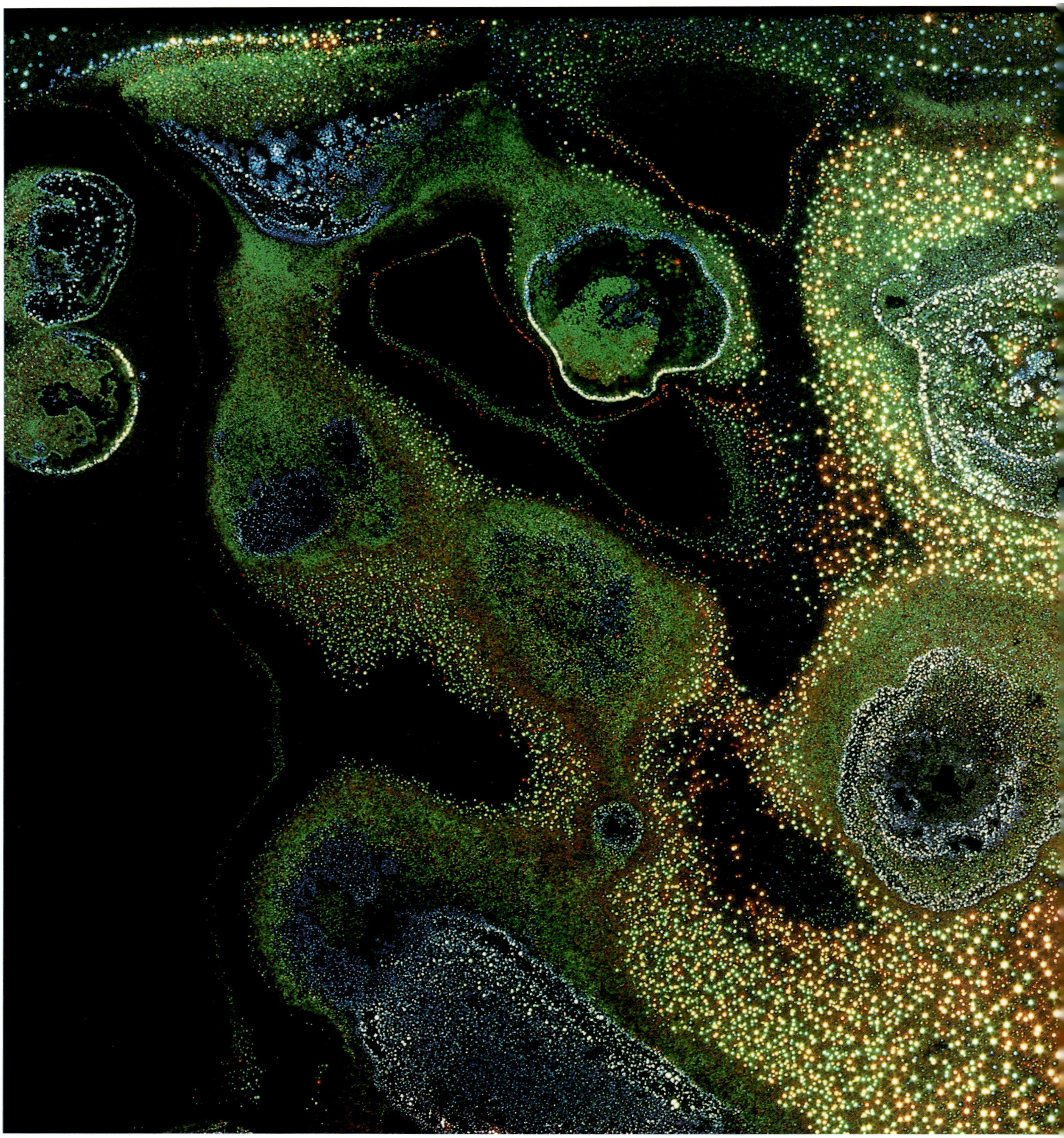

III/9

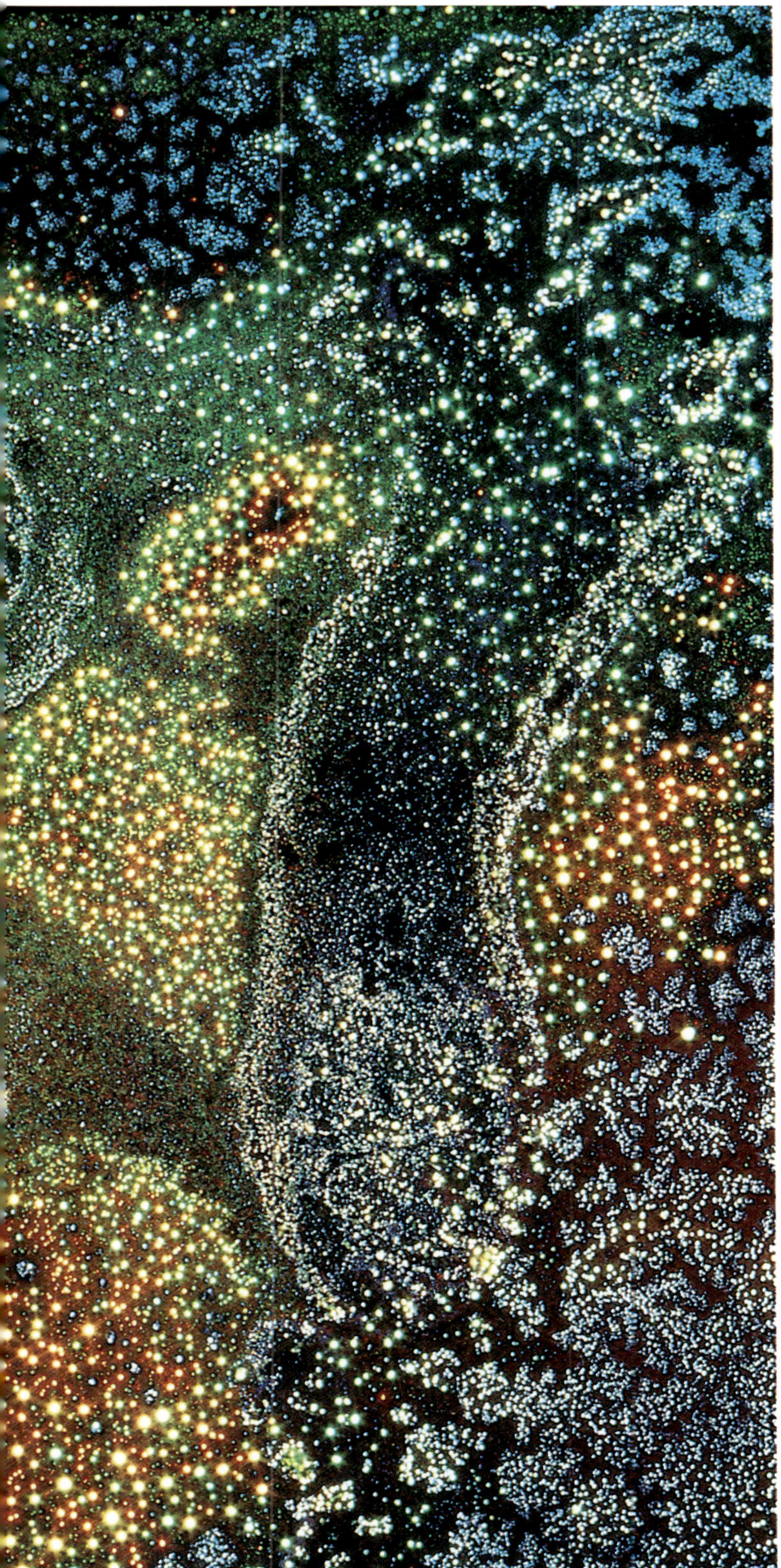

III/11

Die neue Komplexitätsforschung ist taoistisch, das Universum wird als etwas sich Veränderndes angesehen. Die Elemente der Welt bleiben stets die gleichen, sie arrangieren sich bloß neu, wie bei einem Kaleidoskop. Alle Dinge haben jeweils ihr eigenes Ordnungsprinzip, durch gegenseitige Resonanz und wechselseitige Abhängigkeit stehen sie in Einklang miteinander. Komplexität in der Natur entwickelt sich nicht uhrwerkartig, sondern ist vergleichbar mit dem Wachstum einer Pflanze.

Het nieuwe complexiteitsonderzoek is taoistisch. De elementen van de wereld blijven altijd dezelfde, ze worden alleen opnieuw gerangschikt net als bij een caleidoscoop. Alle dingen hebben telkens hun eigen ordeningsprincipe, door onderlinge resonantie en wederzijdse afhankelijkheid zijn ze met elkaar in overeenstemming. Complexiteit in de natuur ontwikkelt zich niet als een uurwerk, maar is te vergelijken met de groei van een plant uit een zaadje, of met erosie door water.

The new complexity research is Taoist, the universe something that is constantly changing. The elements of the world always remain the same; it is just that they rearrange themselves anew like a kaleidoscope. Everything has its principle of order. As a result of mutual resonance and dependence all things are in harmony. Complexity in nature is not something clockwork but can be compared with the growth of a plant out of a seedling or with the rut formations caused by water erosion.

La nuova ricerca della complessità è taoista, l'universo è considerato come qualcosa che muta incessantemente. Gli elementi del mondo rimangono sempre gli stessi, ma si organizzano continuamente, come in un caleidoscopio. Ogni cosa ha un suo principio ordinativo, la reciproca risonanza e la mutua dipendenza creano armonia. Nella natura, la complessità non somiglia al meccanismo di un orologio ma alla crescita di una pianta da un seme o all'erosione dell'acqua.

Les recherches actuelles sur la complexité sont taoïstes; elles considèrent que l'Univers se modifie constamment. Les éléments du monde restent toujours les mêmes; ils se réorganisent uniquement, comme dans un caléidoscope. Toutes les choses ont leur propre principe d'ordre; une résonnance réciproque et une indépendance alternée créent l'harmonie. Dans la nature, la complexité n'obéit pas à un mouvement d'horlogerie; elle est plutôt comparable à la croissance d'une plante.

Изучение комплексности идет по даосскому принципу, Вселенная считается постоянно изменяющейся. Все элементы остаются неизменными, но организуются по-новому, как в калейдоскопе. Все вещи имеют свой принцип организации и находятся в гармонии. Комплексность в природе развивается подобно тому, как растет из семени растение. Простые правила, раскрывающиеся изнутри, дают комплексные структуры.

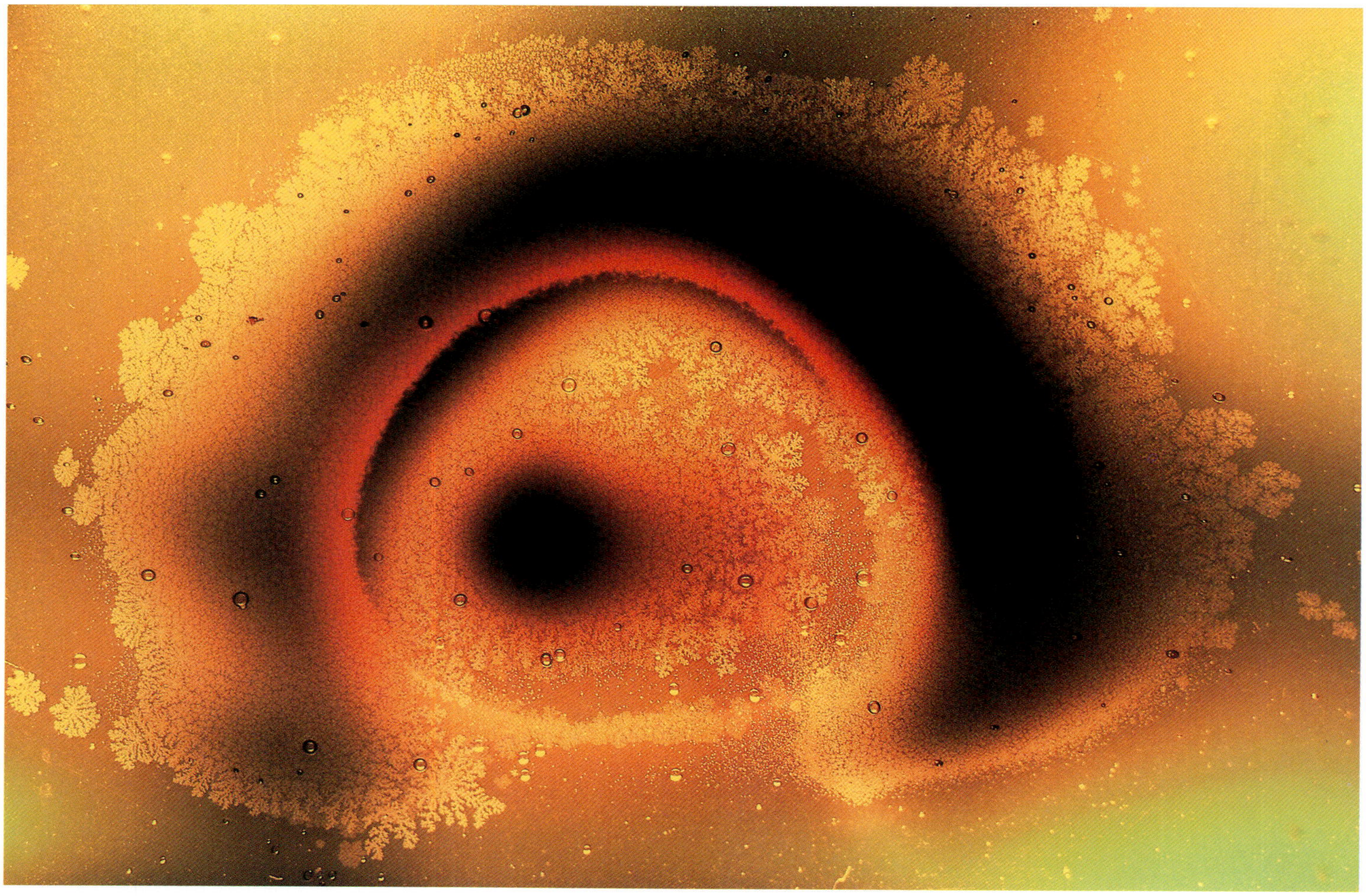

III/13

Das Tao in der Natur gleicht einem Bogen, der gespannt wird. Das Obere wird heruntergezogen, das Untere wird emporgehoben. Was übermäßig ist, wird verringert. Was ungenügend ist, wird ergänzt. Das Tao in der Natur verringert das Übermäßige und ergänzt das Ungenügende. Nicht so das Tao des Menschen: Er verringert das Ungenügende, weil er dem Übermäßigen dient.

Het tao in de natuur lijkt op een boog die gespannen wordt. Het bovenste wordt omlaag, het onderste omhoog getrokken. Wat overmatig is, wordt verkleind. Wat ontoereikend is, wordt aangevuld. Het tao in de natuur verkleint het overmatige en vult het ontoereikende aan. Dit is niet het geval als het gaat om het tao van de mens: hij verkleint het ontoereikende, omdat hij het overmatige dient.

Tao in nature can be compared to a drawn bow. The upper part is pulled down, the lower part is pulled up. What is excessive is reduced, what is insufficient is completed. Tao in nature reduces excesses and completes what is insufficient.
This is not at all the case where Tao and Man is concerned: here it lessens what is insufficient because it serves the excessive.

Il Tao nella natura assomiglia ad un arco quando lo si tende. Ciò che prima stava in alto va giù, mentre ciò che stava in basso va sù. Ciò che esiste in quantità eccessiva diminuisce. Ciò che è presente in quantità insufficiente viene completato. Il Tao nella natura diminuisce gli eccessi e compensa le insufficienze. Diverso il Tao dell'uomo: esso diminuisce le deficienze perchè serve alla dismisura.

Le Tao dans la nature ressemble à un arc que l'on tend. La partie supérieure descend alors que la partie inférieure monte. Le démesuré est amoindri. L'insuffisant est complété. Le Tao dans la nature amoindrit la démesure et complète l'insuffisance. Il en est autrement pour le Tao de l'homme: il amoindrit l'insuffisant parce qu'il est au service de la démesure.

Дао в природе похоже на натягиваемый лук. Верх опускается, низ поднимается. Все избыточное умаляется. Все недостаточное дополняется. Дао в природе умаляет излишнее и дополняет недостаточное. Не так дао человека: он умаляет недостаточное, потому что служит избыточному.

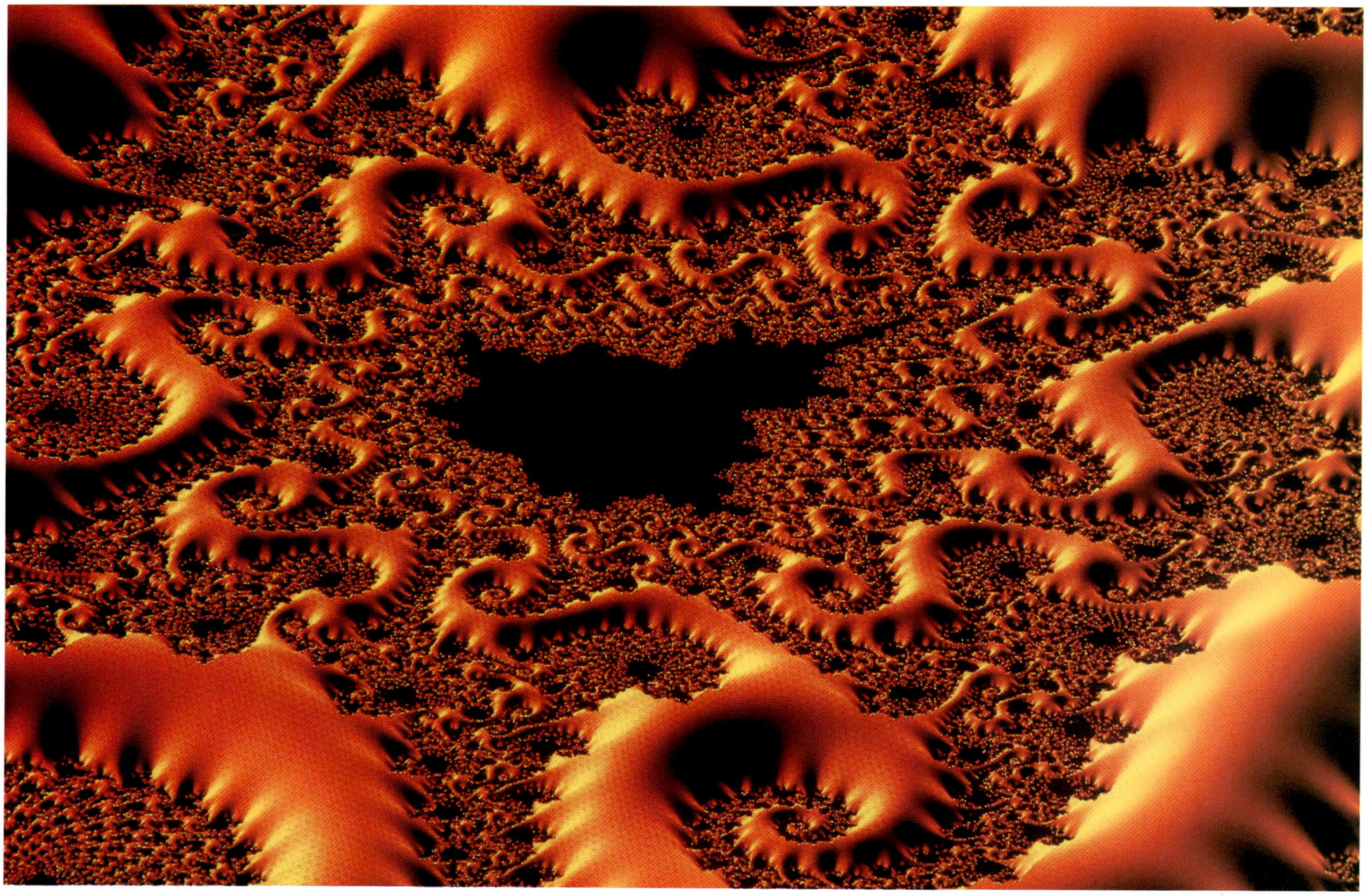

Kreatives Chaos: Immer wieder neu ordnet sich die Welt

Ein bißchen Chaos ist schnell in die Welt gezaubert: Eine Zigarette entzündet, und es steigen Rauchkringel auf von nicht berechenbarer Komplexität. Etwas Sahne in den heißen Kaffee gekippt, und für ein paar Sekunden rotieren Spiralarme in der Tasse - unvorhersehbar in der Entfaltung und im Vergehen ihrer Formen, aber doch universellen Gesetzen folgend. Solch einfache und gleichermaßen hochkomplizierte Vorgänge beschäftigen seit gut 20 Jahren Tausende von Mathematikern, Physikern und Biologen in aller Welt.

Welches essentielle Ordnungsprinzip ließ aus einem explodierenden Feuerball einen penibel geordneten Kosmos mit Planeten, Sonnen und Galaxien entstehen? Wie kam Leben in die Welt, das Auge, das Bewußtsein? Sind wir Menschen reine Zufallsprodukte oder zwangsläufig entstandene Wesen? Das sind einige der Themen, mit denen sich Wissenschaftler beschäftigen, wobei sie sich derzeit nicht nur auf die „chaotischen" Phasen dieser Schöpfungsprozesse konzentrieren, sondern auch auf die frappierende Stabilität sowie die gleichzeitige Simplizität und Komplexität der Ergebnisse. „Komplexitätstheorie" heißt denn heute auch oft, was noch vor kurzem als Chaostheorie bezeichnet wurde. „Chaos und Komplexität jagen einander im Kreis und versuchen herauszufinden, ob sie gleich oder verschieden sind", wie es ein Astrophysiker ausdrückt. Was die Forscher dabei in Bewegung und bei Laune hält, ist die Überzeugung, daß ein fundamentales Naturgesetz darauf wartet, entdeckt zu werden: Ein „Selbstorganisationsgesetz", das exakt beschreibt, unter welchen Umständen bestimmte „Strukturen" der Natur in eine höhere Organisationsform „springen". Leblose Luft zum Beispiel, kann durch wachsende Störung plötzlich zum Wirbelwind, Tornado oder Hurrikan werden: eine komplexere Struktur. Wassermoleküle wiederum „kippen" in eine malstromartige Dynamik, wenn sie von unten erhitzt werden. Komplexes aus Einfachem kann freilich genauso gut beim Abkühlen von Materie entstehen, wie wir inzwischen alle aus der Urknall-Saga wissen: Reine Energie wandelte sich zunächst in simple Teilchen, die Quarks und Elektronen; eine einzige „Superkraft" zerfiel in ihre bis heute wirksamen Teilkräfte - in Gravitation, Elektromagnetismus sowie starke und schwache Kernkraft. Schließlich, nach weiterer Abkühlung des Milliarden Grad heißen Urchaos', entstand jene „komplexe" materielle Ordnung, die bis heute Bestand hat: Quarks gruppierten sich zu Protonen, wurden zu Atomkernen des Wasserstoffs, der einen Gutteil auch unseres Körpers ausmacht. Alle wesentlichen Bausteine des Kosmos' entstanden auf diese Weise innerhalb der ersten drei Minuten nach dem „Big Bang". Bis sich dann die ersten Galaxien bildeten, verging fast eine Milliarde Jahre - ein Indiz dafür, welch komplexe „Entwicklungsschritte" notwendig waren, bis sich Materieteilchen zu Sonnen und Sonnenhaufen zusammenfinden konnten. Ob diese Art von Forschung nun mit „Chaos" umschrieben wird, mit „Komplexität" oder gar mit „Emergenz", wie in jüngster Zeit in Mode gekommen: Tatsache bleibt, daß Neues gesetzmäßig, aber unvorhersagbar entsteht, wie zum Beispiel ein Mensch aus wiederholter Zellteilung oder eine Mandelbrot-Fraktale durch wiederholte Anwendung einer simplen mathematischen Formel. Solche komplexen Systeme, wie es etwa auch das Gehirn ist, die Börse oder ein ökologisches Biotop, weisen Muster auf, die man nur am ganzen System erkennen kann, aber nicht an seinen Teilen. Bislang fehlen noch viel zu viele Bausteine vom Puzzle, um einen seriösen Versuch machen zu können, es zu einem Bild zusammenzusetzen. Aussagen von Fachleuten, wie: „Niemand weiß, was Chaos wirklich ist" oder: „In der Natur ist das Chaos nicht zu fassen", lassen darauf schließen, daß sich das Gebiet noch im Frühstadium seiner Entwicklung befindet.

Wie etwas entsteht in dieser Welt ist nicht zuletzt auch deshalb so schwer zu verstehen, weil auch diese Prozesse den widersprüchlichen „für den menschlichen Geist nicht verstehbaren" Theorien der Quantenmechanik unterworfen sind. Dadurch ist nichts so recht bestimmmbar, der Zufall regiert, alles schwingt in taoistischer Resonanz in einem „Fluß": Die Elemente der Welt bleiben die gleichen, sie arrangieren sich bloß immer wieder neu. Oder, wie es der Kosmologe und Nobelpreisträger Murray Gell-Mann ausdrückt: „Die Entwicklung des Universums ist bestimmt von einem kleinen Anteil fundamentaler Gesetze und einem sehr großen Anteil von Quantenzufällen im Laufe der Jahrmilliarden."

Creatieve Chaos: De wereld herschikt voortdurend zijn orde

Een beetje chaos is snel te toveren in de wereld: een sigaret ontbrandt, en rookwolkjes van niet te berekenen complexiteit stijgen omhoog. Een beetje slagroom in de hete koffie geschonken, en een paar seconden lang roteren spiraalarmen in het kopje - de ontwikkeling en oplossing van de vormen is niet te voorspellen, maar volgt toch universele wetten. Zulke eenvoudige en eveneens uiterst ingewikkelde processen houden duizenden wiskundigen, natuurkundigen en biologen in de hele wereld sinds ruim 20 jaar bezig.

Welk essentieel ordeningsprincipe liet uit een exploderende vuurbal een penibel geordende kosmos met planeten, zonnen en galactische stelsels ontstaan? Hoe kwam er leven in de wereld, het oog, het bewustzijn? Zijn wij mensen zuivere toevalsprodukten of wezens die onvermijdelijk moesten ontstaan? Dat zijn een paar van de onderwerpen waarmee wetenschappers zich bezighouden. Hierbij concentreren zij zich thans niet op de "chaotische" fasen van deze scheppingsprocessen, maar ook op de frapperende stabiliteit en de gelijktijdige simpliciteit en complexiteit van de resultaten. Wat nog niet zo lang geleden chaostheorie werd genoemd, heet tegenwoordig dan ook vaak "complexiteitstheorie". "Chaos en complexiteit achtervolgen elkaar in een kring en trachten er achter te komen of zij gelijk of verschillend zijn", zoals een astrofysicus het uitdrukt. Wat de onderzoekers daarbij in beweging en in een goed humeur houdt, is de overtuiging dat een fundamentele natuurwet erop zit te wachten dat hij wordt ontdekt, een "zelforganisatiewet" die exact beschrijft onder welke omstandigheden bepaalde "structuren" van de natuur in een hogere organisatievorm "springen". Bijvoorbeeld kan levenloze lucht door een toenemende storing plotseling een cycloon, tornado of wervelstorm worden: een complexere structuur.

Watermoleculen "vallen" wederom in een maalstroomachtige dynamiek, wanneer deze van onderen worden verhit. Een complexe vorm uit iets eenvoudigs kan uiteraard net zo goed ontstaan bij het afkoelen van materie, zoals wij allemaal inmiddels uit de oerknalsaga weten: pure energie veranderde eerst in simpele deeltjes, de quarks en elektronen; één enkele "superkracht" viel uiteen in zijn vandaag nog actieve deelkrachten - in zwaartekracht, elektromagnetisme en sterke en zwakke atoomenergie. Tot slot ontstond na verdere afkoeling van de miljarden graden hete oerchaos de "complexe" materiële orde die tot nu is blijven bestaan: quarks groepeerden zich tot protonen, werden atoomkernen van de waterstof waaruit voor een groot deel ook ons lichaam bestaat. Alle essentiële

Creative chaos: The world continually rearranges its order

bouwstenen van de kosmos ontstonden op deze wijze binnen de eerste drie minuten na de "big bang". Het duurde bijna een miljard jaren voordat de eerste galactische stelsels ontstonden - een aanwijzing voor de complexiteit van de ontwikkelingstrappen die nodig waren totdat materiedeeltjes konden samenkomen voor zonnen en zonnenhopen.
Of dit soort onderzoek nu met "chaos" wordt omschreven of met "complexiteit" of zelfs met "emergentie" zoals recentelijk in de mode gekomen, het blijft een feit dat iets nieuws wetmatig, maar onvoorspelbaar ontstaat, net als bijvoorbeeld een mens uit herhaalde celdeling of een fractal van Mandelbrot door de herhaalde toepassing van een eenvoudige wiskundige formule. Zulke complexe systemen, bijvoorbeeld ook de hersenen, de effectenbeurs of een ecologisch biotoop, vertonen patronen die alleen aan het hele systeem, maar niet aan delen ervan te zien zijn. Tot nu ontbreken nog veel te veel stukjes van de puzzel om een serieuze poging te ondernemen er één beeld van te maken. Uitingen van deskundigen in de trant van "niemand weet wat chaos werkelijk is" of "in de natuur is de chaos niet te vatten" leiden tot de conclusie dat het gebied zich nog in het beginstadium van zijn ontwikkeling bevindt.
Hoe iets ontstaat in deze wereld is niet in de laatste plaats ook daarom zo moeilijk te begrijpen, omdat ook deze processen onderworpen zijn aan de tegenstrijdige "voor de menselijke geest niet te begrijpen" grillen van de quantummechanica. Daardoor is niets zo goed te bepalen, het toeval regeert, alles schommelt in taoistische resonantie in een "stroom". De elementen van de wereld blijven dezelfde, ze worden alleen voortdurend opnieuw geordend. Of, zoals de kosmoloog Murray Gell-Mann het uitdrukt: "De ontwikkeling van het universum is voor een klein deel bepaald door fundamentele wetten en voor een heel groot deel door quantumtoevallen in de loop van miljarden jaren."

A bit of chaos is easily produced in this world: light a cigarette and wisps of smoke of incalculable complexity rise into the air. Pour a little cream into hot coffee and, for a few seconds, spiralling arms will rotate in the cup - unpredictable in their development and in the disappearance of their forms, and yet following universal laws. Such simple and yet at the same time highly complex processes have been the concern of thousands of mathematicians, physicists and biologists all over the world for a good twenty years now.
What key principle of order led to the birth of a scrupulously designed cosmos with planets, suns and galaxies, out of a great exploding ball of fire? How did life come about, the eye, consciousness? Are we pure products of chance or beings that have come into existence according to fixed lines? These are some of the subjects with which scientists concern themselves, whereby they do not only concentrate on the "chaotic" phases of these processes of creation but also on the striking stability and the parallel degrees of simplicity and complexity reflected in events.
"Complexity theory" is the name given today to what, a short while ago, was still termed "chaos theory". "Chaos and complexity are chasing each other round in circles and are trying to find out if they are the same or if they are different," was how one astro-physicist put it. What is keeping the research scientist on the move and in good spirits is the conviction that a fundamental law of nature is waiting to be discovered, a "law of self-organization", one which describes exactly under what circumstances certain "structures" in nature "jump" to a higher form of organization. Lifeless air can, for example, as a result of increasing disturbance, develop to suddenly become a whirlwind, a tornado or a hurricane a more complex structure. Molecules of water, on the other hand, "switch" into a maelstrom-like dynamism when they are heated from beneath. Complex results can also result out of something simple when matter cools down, as we now all know from the Big Bang saga: pure energy is transformed first of all into simple particles, quarks and electrons; one single "super force" split up into the partial forces that are still effective today - into gravitation, electro-magnetism and strong and weak nuclear energy. Finally, after the further cooling down of the thousands-of-millions-degrees-hot original chaos, that "complex" order of matter came into being which still exists today: quarks grouped to form protons, became the hydrogen nuclei which also constitute a good part of our bodies. All the essential components of the cosmos came into being in this fashion within the first three minutes following the Big Bang. Almost a thousand million years passed until the first galaxies were then formed - an indication of what complex steps were necessary before the particles of matter could come together to form suns and clusters of suns.
Whether this type of research is now described by the word "chaos" or by "complexity" or even by the term "emergence" which has come into fashion recently, the fact remains that new things come into being according to set principles, but in an unpredictable fashion, just as a human being does as a result of repeated cell division, or a Mandelbrot fractal through the repeated employment of a simple mathematical system. Highly complex systems, such as the brain, or the stock market, or an ecological biotope, produce patterns which one can only recognize in relation to the whole system but not by its parts. Too many elements of the puzzle are still missing to be able to make a serious attempt to form a complete picture. Statements made by experts such as: "No-one knows what chaos really is" or: "Chaos cannot be fathomed in nature" lead one to conclude that the entire area is still at an early stage of its development.
How something comes into being in this world is difficult to understand, not least because the processes involved are all subject to the contradictory whims of quantum mechanics which "are not comprehensible for the human mind". As a result, nothing is clearly definable, chance reigns, everything swings and flows in Taoist resonance: the elements that make up the world remain the same; it is just that they continually rearrange themselves anew. Or as the cosmologist and Nobel Prize winner Murray Gell-Mann puts it: "The development of the universe is determined by a small share of fundamental laws and a very large share of quantum coincidences in the course of thousands of millions of years."

La Fiat sulla ruota dentata, il cielo del Corano, la targa su Giove, le Pleiadi non erano vergini

Si pensi ad una ruota dentata con i denti appuntiti, ma non pericolosi, tipo raggi solari, tutti rivolti all'interno in modo da lasciare al di fuori le estensioni infinite delle tenebre: questa l'immagine della creazione data nel 1619 ad Oppenheim, dallo studioso Robert Fludd. Questa ruota rovesciata, molto suggestiva in verità, è aperta verso l'alto ed è sovrastata da una nube, simbolo del Padre, dalla quale esce, come proiettata, la parola "Fiat", cioè creazione, mentre dall'altra coda vola lo Spirito Santo, ossia una colomba che sembra aver fatto il giro del cosmo. Vale a dire che ne ha circoscritto la parte finalmente ordinata, mentre il buio del caos ne è escluso. Un segno di distensione, perciò, rispetto al magma originario.
Ma come ricostruire, verificare questa preistoria, da sempre mancante di prove? Solo prestando fede alle rivelazioni del suo Creatore o dei suoi profeti, rendendo umano il sovrumano. "In verità è Dio che spacca un seme ed un nocciolo e fa uscire il vivo dal morto ...", si legge sul Corano. Ed ancora: "È lui che spacca il cielo dell'aurora ed ha fatto della notte il momento del riposo e del sole e della luna i mezzi per conteggiare il tempo. Così ha decretato Dio che è potente e saggio. Ed è lui che ha fatto per voi le stelle, affinché con esse poteste orizzontarvi sulla terra e nel mare ..."
Identificazione scientifica è, invece, quella del filosofo greco Proclo, vissuto nel IV secolo d. C. che, commentando i "Dialoghi" di Platone a proposito dei corpi celesti, precisava come il Creatore assegnasse ai medesimi "un duplice movimento, l'uno uniforme e senza spostamenti, l'altro invece con uno spostamento continuo". Ed aggiungeva, quasi a complicare le cose, "che ogni pianeta è un mondo a sé stante che alberga numerose specie divine a noi ignote". Sta di fatto, in piena fiducia con tali concetti, ribaditi dalla scienza successiva che tentava il reperimento di tracce vitali al di fuori della terra, che nel 1972 la sonda americana "Pioneer F" lanciata verso il pianeta Giove, recava a bordo una targa in alluminio e oro da offrire alla lettura, o meglio alla decifrazione di possibili civiltà extraterrestri. Documento suggestivo, sia per la speranza che esistano esseri intelligenti al di fuori dei confini terrestri, sia per la fiducia nel "Codice binario" studiato appositamente per redigere la scritta, sia infine per i dati tecnico-informativi sul nostro pianeta, la sua composizione e collocazione nel sistema solare, nonché un elementare disegnino di un uomo ed una donna.
Tutto questo sorvolando sulla dottrina della molteplicità e contiguità degli universi che talune teorie attribuirebbero alla creazione di entità aliene che avrebbero usato uno schema uniforme per tutti. Prospettive, dunque, fantascientifiche di affidabile probabilità, del tutto contrarie a quella mitologia astrale che mai ebbe fondamento astronomico credibile.
Diamo, ad esempio, un'occhiata alla costellazione delle Pleiadi la cui simbologia è sinonimo di verginità. La loro creazione, tanto per precisare, è dovuta ad un fatto contingente. Esse erano delle ragazze nate da Atlante e da Pleione, amiche di Artemide. La quale, per salvarle dalle grinfie di Orione che le braccava da oltre cinque anni per attentare alla loro purezza, le trasformò nell'omonima costellazione, delle Pleiadi appunto. Ma qui arriva la smentita: quelle ragazze erano tutto altro che vergini! È una seconda leggenda a stabilirlo crudelmente: Zeus aveva fornicato ripetutamente con tre di esse, Maia, Elettra e Taigete; mentre Alcione e Celeno avevano ceduto allegramente la loro illibatezza a Poseidone, emulata da Sterope che si era data ad Ares. Merope, la settima, era invece ripiegata su un mortale di nome Sisifo. Questo la escluse dal brillare in cielo, essendone visibili ad occhio nudo soltanto sei. Il simbolismo delle figure celesti ed i nomi delle costellazioni sono state sempre di fondamentale significato nelle varie credenze e religioni, con larga conferma nei testi sacri per i quali esse sono guida e punizione.
A proposito di quest'ultima sono i demoni a farne le spese. Secondo la tradizione islamica, infatti, essi hanno il discutibile gusto di elevarsi al cielo per ascoltare le conversazioni degli angeli, ma ne vengono ricacciati. Come? Così precisa il Corano: "E ornammo il cielo più basso con lampade (stelle) che destinammo ad essere lanciate contro i demoni, per i quali abbiamo anche preparato il castigo del fuoco."
Il riferimento era alle stelle cadenti; mentre la stella dei Magi è un avvertimento ed una indicazione divina, una cometa, che si vuole divenne nel tempo segnale di evenienze infauste anche per civiltà lontane, come quella che avrebbe annunciato a Montezuma ed all'Inca Huaina Capc, l'arrivo degli spagnoli e la caduta dell'Impero.

Giuseppe Brunamontini

Chaos créatif: une réorganisation toujours nouvelle du monde

Un petit chaos est vite produit comme par enchantement: on allume une cigarette et des ronds de fumée d'une complexité impondérable s'élèvent. Un peu de crème dans le café chaud et pour quelques secondes une hélice tourne dans la tasse, imprévisible dans le développement et l'évanouissement de ses formes, mais obéissant cependant à des lois universelles. Ces réactions à la fois simples et hautement complexes préoccupent depuis une vingtaine d'années des milliers de mathématiciens, physiciens et biologistes du monde entier.
Quel principe d'ordre essentiel fit surgir de l'explosion d'une boule de feu un cosmos méticuleusement ordonné avec des planètes, des soleils et des galaxies? Comment naquit la vie dans le monde, l'œil, la conscience? Les humains ne sont-ils que des produits purement accidentels ou des êtres venus au monde par la force des choses? Voilà quelques-uns des sujets dont s'occupent les scientifiques; il convient cependant de souligner que pour le moment ils ne se concentrent pas seulement sur les phases "chaotiques" des processus de la Genèse mais aussi sur la frappante stabilité ainsi que sur la simplicité et la complexité simultanées des résultats. On appelle souvent "théorie de la complexité" ce qu'on nommait encore récemment la théorie du chaos. "Chaos et complexité tournent en rond, se chassent mutuellement et essaient de trouver si elles sont semblables ou différentes", comme l'exprime un astrophysicien. Ce qui pousse les chercheurs vers l'avant et préserve leur bonne humeur, c'est la conviction qu'une loi naturelle fondamentale attend d'être découverte, une "loi d'autoorganisation" qui décrit exactement dans quelles conditions certaines "structures" de la nature "sautent" dans une forme dont l'organisation est plus complexe. L'air inerte par exemple peut soudain, en cas d'accroissement des perturbations, se transformer en tourbillon, en tornade ou en ouragan, bref en une structure plus complexe. Les molécules d'eau, elles, "basculent" dans une dynamique semblable à un malstrom quand on les chauffe par le bas. Bien entendu, la complexité peut tout aussi bien naître de la simplicité lors du refroidissement de la matière, comme nous le savons tous maintenant, nous qui connaissons la saga de l'explosion primordiale: de l'énergie pure se transforma d'abord en particules simples, les quarks et les électrons; une "superforce" unique se décomposa en forces partielles encore actives de nos jours dans la gravitation, l'électromagnétisme ainsi que l'interaction faible ou forte. Finalement, après un refroidissement plus important du chaos

Творческий хаос: мир постоянно перестраивает себя

premier ayant une température de plusieurs milliards de degrés, naquit cet ordre matériel "complexe" existant encore aujourd'hui: les quarks se regroupèrent en protons pour former les noyaux atomiques de l'hydrogène dont notre corps est en grande partie composé. Tous les éléments essentiels du cosmos naquirent de cette manière pendant les trois premières minutes qui suivirent le "Big Bang". Presque un milliard d'années passa jusqu'à la formation des premières galaxies, ce qui montre combien d'étapes complexes de d'évolution furent nécessaires jusqu'à ce que les particules de matière puissent se réunir en soleils et en amas solaires. Que ce genre de recherches soit appelé "chaos", "complexité" ou même "émergence", comme c'est la mode actuellement, le fait est que toute "nouveauté" naît conformément aux lois mais sans qu'on puisse la prévoir, comme par exemple un être humain à la suite d'une division cellulaire répétée ou un fractal de Mandelbrot par l'utilisation répétée d'une simple formule mathématique. Ce genre de systèmes complexes dont fait partie le cerveau, la bourse ou un biotope écologique, présentent des motifs qu'on ne peut reconnaître que dans l'ensemble du système mais pas dans ses parties. Jusqu'à présent, il manque beaucoup trop de pièces du puzzle pour pouvoir faire une tentative sérieuse de les assembler en vue d'obtenir un tableau. Les déclarations des experts telles que: "Personne ne sait ce qu'est véritablement le chaos" ou "Dans la nature, on ne peut saisir le chaos", permettent de conclure que ce domaine en est encore au premier stade de son développement.
Il est également difficile de comprendre comment quelque chose naît dans ce monde, parce que ces processus sont soumis aux théories, contradictoires et "incompréhensibles pour l'esprit humain", de la mécanique quantique. Ainsi, rien ne peut vraiment être déterminé, le hasard est maître, tout oscille dans un "fleuve" de résonnance taoïste: les éléments du monde restent les mêmes, ils se réorganisent sans cesse, tout simplement. Ou bien, comme l'exprime le cosmologue et lauréat du prix Nobel Murray Gell-Mann: "Le développement de l'Univers est déterminé au cours de milliards d'années par une petite part de lois fondamentales et une très grande part de hasards quantiques".

Модель хаоса сделать нетрудно: дымок от сигареты поднимается причудливыми извивами непредсказуемой формы; если капнуть сливок в горячий кофе, в чашке на секунду появятся «спиральные рукава», также непредсказуемой формы и поведения, но подчиняющиеся всеобщим законам. Эти простые и в то же время в высшей степени сложные процессы вот уже двадцать лет занимают многих математиков, физиков и биологов во всем мире.
Какой высший принцип превратил взорвавшийся огненный шар в детально упорядоченный космос с его планетами, солнцами и галактиками? Как в мире появились жизнь, глаз, сознание? Являемся ли мы чисто случайными продуктами или наше появление было неизбежным? Это только некоторые темы из числа тех, которыми занимаются ученые; причем они сейчас усиленно изучают не только «хаотические фазы» этого процесса творения, но и поразительную устойчивость, а также одновременную простоту и сложность результатов. «Теорией комплексности» называют теперь часто то, что недавно именовалось «теорией хаоса». Как сказал один астрофизик, «хаос и комплексность гоняют друг друга по кругу, пытаясь выяснить, одинаковы они или различны». Исследователей подстегивает и вдохновляет уверенность, что есть один фундаментальный закон, который ждет, чтобы его открыли, «закон самоорганизации», точно описывающий, при каких условиях определенные структуры природы «перескакивают» на более высокую ступень организации. Например, неподвижный воздух с ростом возмущений может превратиться в вихрь, смерч или ураган, то есть в комплексную структуру. Молекулы воды тоже внезапно приобретают динамику водоворота, если снизу поступает тепло. Комплексные структуры могут возникать из простых и при охлаждении, как мы это знаем из «саги о Большом взрыве»: сначала чистая энергия превратилась в элементарные частицы — кварки и электроны; одна единственная сверхсила распалась на единичные силы, которые продолжают действовать до сих пор: гравитация, электромагнетизм, сильное и слабое взаимодействие. В ходе остывания сверхгорячего первичного хаоса возник тот комплексный материальный порядок, который существует по сегодня: кварки сгруппировались в протоны, стали ядерными атомами водорода, из которого в немалой доли состоит и наше тело. Все важнейшие «кирпичики» космоса возникли именно таким образом в первые три минуты после Большого взрыва. А для образования первых галактик потребовался почти один миллиард лет — столь комплексным и сложным было развитие от частиц к солнцам и скоплениям солнц. Как назвать это — «хаосом», «комплексностью» или, как сейчас модно, «эмергенцией» — все равно. Факт то, что новое возникает закономерно, но непредсказуемо, как человек из делящихся клеток или «фракталы Мандельброта» путем многократного применения простых математических формул. Эти комплексные системы, например, мозг, биржа или экологический биотоп, имеют мозаичность, распознаваемую только в целой системе, но не в ее частях. Пока еще не хватает слишком многих деталей этой головоломки, чтобы всерьез попытаться сложить ее в единое целое. Такие замечания специалистов, как: «Никто не знает, что такое хаос на самом деле» или: «В природе хаос понять нельзя» позволяют сделать вывод, что эта область науки находится сейчас на ранней стадии развития.
Как в этом мире возникает новое, очень трудно понять еще и потому, что и эти процессы подчинены противоречивым и «для человеческого разума недоступным» правилам квантовой механики. Нет ничего достоверно определенного, всем управляет случай, все колеблется в резонансе и находится в непрерывном течении: составные части мира остаются прежними, но постоянно организуются по-новому. Или, как выразился космолог и лауреат Нобелевской премии М. Геллмен, «развитие Вселенной определяется в очень малой степени фундаментальными законами и в очень большой — квантовыми случайностями на протяжении миллиардов лет».

MYTHICAL

COSMOS

Ein Schöpfergott, der den Kosmos und das Leben schafft - diese biblische Version des Anfangs aller Dinge hat es heute schwer in Konkurrenz mit „Big-Bang"-Theorie und „Weltformel".

Un dio creatore del cosmo e della vita: la narrazione biblica che spiega l'origine delle cose trova oggi aspra concorrenza con la teoria del "big bang" e della "formula del mondo".

Een scheppende God die kosmos en leven schept - deze versie van de beginne aller dingen heeft het nu in concurrentie met de "big bang" en "wereldformule" theorie niet gemakkelijk.

Un Dieu de la Genèse qui crée le Cosmos et la Vie; cette version biblique du commencement de toute chose a du mal à s'affirmer face à la théorie du "big bang" et à la "formule universelle".

A creator god, who created the cosmos and life - this Biblical version of the beginning of all things must today enter into tough competition with the "Big Bang" theory and the "world formula".

Библейская идея Бога-Творца, создающего космос и жизнь как начало всего, не выдерживает спора с теорией Большого взрыва и «универсальной формулой».

Einige Narren erklären, ein Schöpfer habe die Welt gemacht. Die Lehre von der Erschaffung der Welt aber ist unüberlegt und sollte verworfen werden. Wenn Gott nämlich die Welt erschaffen hat, wo war er vor der Schöpfung? Wie hätte Gott die Welt ohne Rohstoff machen können? Antwortet man darauf aber, er habe erst diesen und dann die Welt gemacht, läßt man sich auf unendliches Weiterfragen ein.

Sommige dwazen verklaren dat een schepper de wereld heeft gemaakt. De leer van de schepping van de wereld is echter ondoordacht. Als God de wereld heeft geschapen, waar was hij dan vóór de schepping? Hoe had God de wereld kunnen maken zonder grondstof? Maar geeft men erop het antwoord dat hij dit eerst en daarna de wereld heeft gemaakt, moet men op eindeloze verdere vragen ingaan.

Some fools explain a creator made the world. This teaching is, however, imprudent. If God created the world, where was he before creation? How could God have made the world without raw material. If one answers that he first made the latter and then the world, then one exposes oneself to an endless chain of questions. Know then that the world is uncreated, like time itself; it is without beginning and without end.

Jinasena (India, 9th Century)

Alcuni folli spiegano che un creatore ha fatto il mondo. La dottrina della creazione del mondo è però sconsiderata e va respinta. Infatti se Dio avesse creato il mondo, dove era prima della creazione? Come avrebbe potuto fare il mondo senza avere materie prime? Ma se si risponde a questa domanda affermando che prima egli ha creato la materia e poi il mondo, si potrebbe continuare all'infinito a chiedere.

Quelques fous déclarent qu'un Créateur a fait le monde. La théorie de la création du monde est cependant irréfléchie et devrait être rejetée. Car si Dieu a créé le monde, où était-il avant la Création? Comment Dieu aurait-il pu fabriquer le monde sans matière première? Si l'on suppose qu'il l'a créée avant le monde, on s'engage dans une série de questions sans fin. Sache donc que le monde n'a pas été créé.

Некоторые глупцы заявляют, будто Творец создал мир. Учение о сотворении неразумно. Если Бог создал мир, где он был до творения? Как он мог сделать мир без материала? Если ответить, что он сначала сделал материал, а потом мир, появятся новые вопросы. Знай, мир не сотворен; как время, не имеет начала и конца и основан на принципах.

Astronomicum Cæsareum:
Historicus, diuina gerens, sophiæque perit⁹,
Hic sua cognoscet, si bona nosse volet.
Namque vetustatis mirator tempora rebus
Distribuet, verè dum canet historias,
Ipse facri præsses noctes æquare diebus
Discet, & hinc serie festa locare sua,
Ipseque naturæ rimator mira commoda
Percipiet, nulli dicta vel acta patens,
Sed caueant animis adsint liuore perustis,
Hæc etenim labes cernere vera negat.

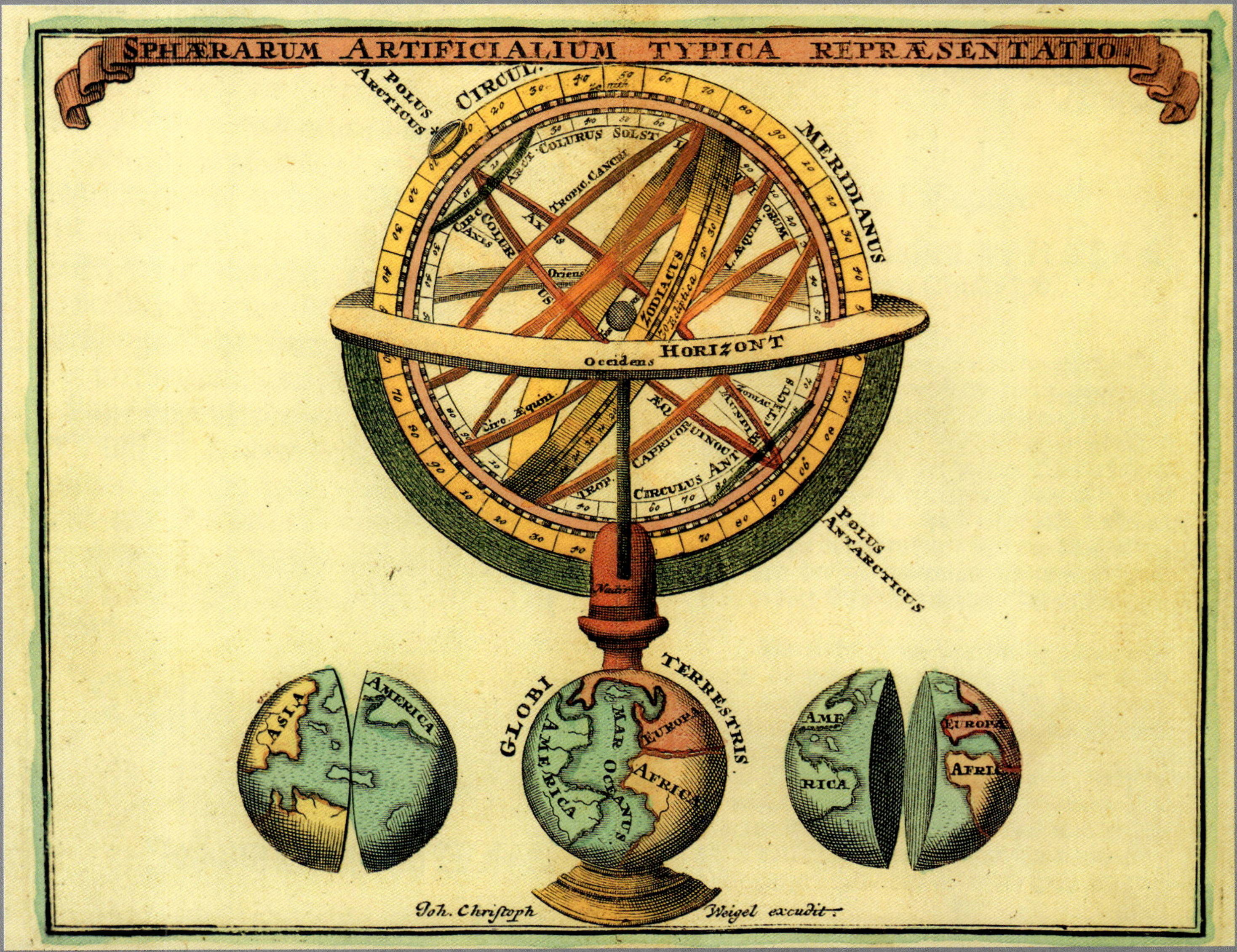

Am siebten Tag des Januars 1610, als ich in der ersten Nachtstunde die Konstellationen durch ein Fernrohr betrachtete, präsentierte sich mir Jupiter. Und da ich mir ein superlativ gutes Instrument gebaut hatte, sah ich drei kleine Sterne in seiner Nähe. Als ich am achten Tag die Beobachtung wiederholte, fand ich sie an anderer Stelle. Alle drei waren nun westlich von Jupiter.

Op de 7de dag van januari in 1610, toen ik in het eerste uur van de nacht de constellaties door een telescoop bekeek, presenteerde Jupiter zich aan mij. En omdat ik voor mij een superlatief goed instrument had gebouwd, zag ik drie kleine sterren in zijn nabijheid - klein, maar heel helder. Toen ik op de achtste dag de waarneming herhaalde, vond ik ze op een andere plaats.

On the 7th day of January in the year 1610 when I went to look at the constellation through the telescope in the first hour of darkness, Jupiter presented itself to me. And as I had built myself a superlatively good instrument, I saw three little stars in its vicinity - small but very bright. When I repected the observation on the eighth day I found them in a different place.

Il settimo giorno di gennaio dell'anno 1610, quando presi ad osservare le costellazioni nella prima ora della notte attraverso un telescopio, si presentò ai miei occhi Giove. E poichè mi ero costruito un ottimo strumento, vidi anche tre piccole stelle vicino a lui, piccole ma straordinariamente lucenti. Quando, nell'ottavo giorno, ripresi ad osservare, le ritrovai in un altro luogo.

Le septième jour du mois de janvier 1610, alors que je regardais les constellations à travers une lunette à la première heure de la nuit, je découvris Jupiter. Etant donné que j'avais construit un instrument exceptionnel, je vis trois petites étoiles à côté de lui, petites mais très brillantes. Lorsque, le lendemain, je repris mes observations, elles avaient changé de place.

В седьмой день января 1610 года, в первом часу ночи, когда я в трубу рассматривал конфигурации, я увидел Юпитер. Поскольку мой инструмент был исключительно хорош, я увидел рядом с планетой три маленькие, но очень яркие звезды. Когда я на восьмой день повторил наблюдения, они оказались в другом месте.

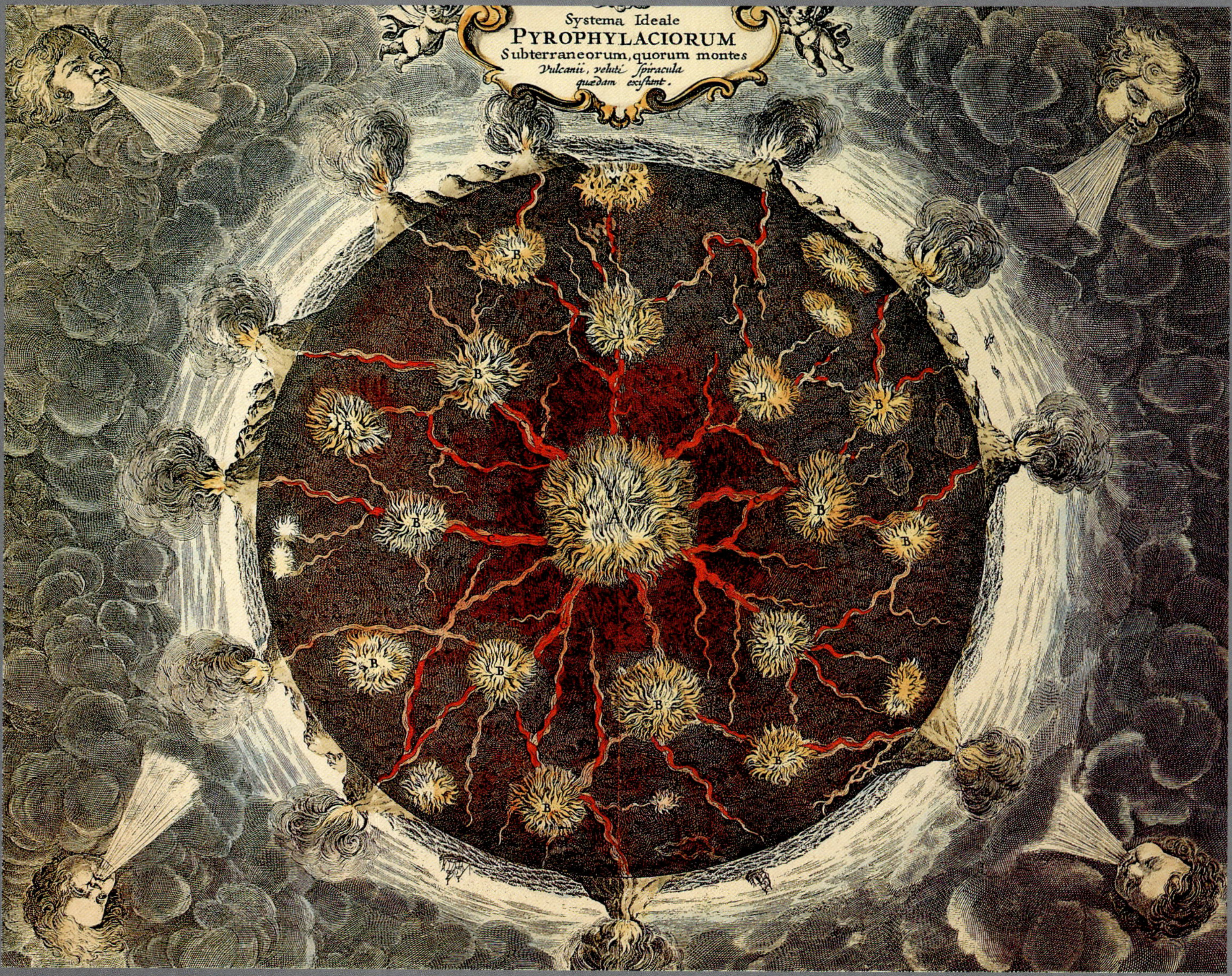

Christiaan Huygens, 1690

Wir mögen aufsteigen von dieser Erde und, von oben auf sie herabblickend, bedenken, ob die Natur all ihre Pracht und Herrlichkeit nur auf sie verschwendet hat. So werden wir, wie Reisende, die in anderen, fernen Ländern weilen, ein besseres Urteil über die Vorgänge zu Hause gewinnen und ein jeglich Ding nach seinem wahren Wert schätzen.

Wij mogen opstijgen van deze beperkte aarde en, op haar neerziend, bedenken of de natuur al haar pracht alleen aan haar heeft verspild. Zo zullen wij die in verre landen verwijlen, een beter oordeel over thuis krijgen en ieder ding naar zijn ware waarde schatten. Hetgeen de wereld groot noemt, zullen wij minder bewonderen, en de nietigheden verachten.

We may rise up and leave this limited Earth and from above gaze down on her and wonder if nature has wasted all her magnificence on her alone. And so, like travellers who visit other faraway countries, we will acquire a much better judgement of things that happen at home and will be able to assess everything according to its true value.

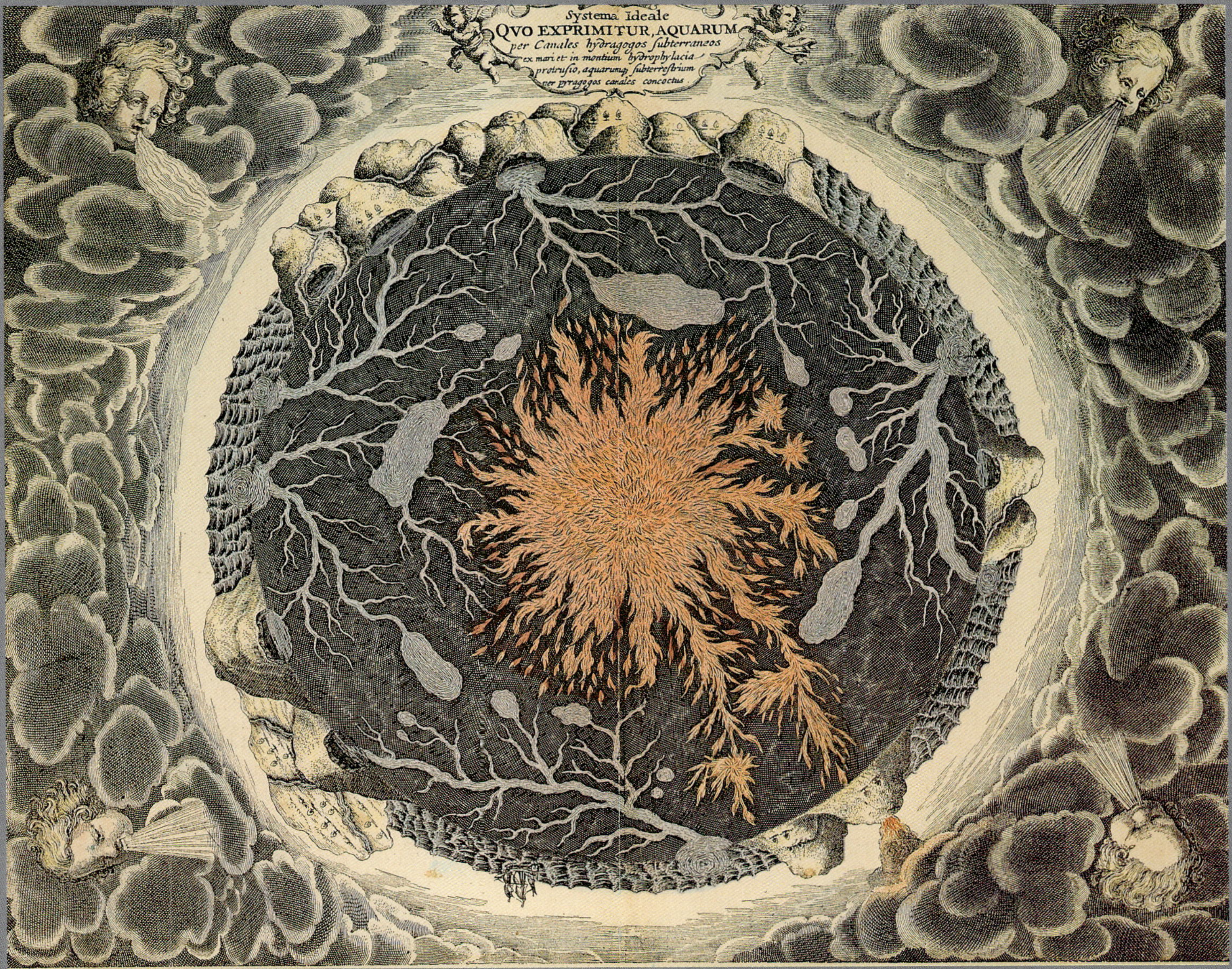

Possiamo innalzarci da questa Terra e, mirandola dall'alto, chiederci se la natura ha sprecato tutto il suo splendore e la sua magnificenza solo per lei. Così noi come viaggiatori, che sostano in paesi lontani, potremo giudicare meglio quanto accade a casa nostra e impareremo a stimare le cose secondo il loro vero valore e ad evitare gli eccessi.

Nous pouvons nous élever au-dessus des limites de cette Terre et, la regardant d'en haut, réfléchir si la nature n'a gaspillé qu'ici son faste et sa splendeur. Ainsi, comme des voyageurs séjournant dans d'autres pays lointains, nous jugerons mieux de ce qui se passe chez nous et estimerons chaque chose à sa véritable valeur et mépriseront les futilités.

Можно оторваться от этой Земли и, глядя на нее сверху, подумать, потратила ли природа всэ свое великолепие лишь на нее. Так путешественник в дальней стране лучше понимает, что происходит дома и верно оценивает все вещи. Тем, что зовется великим, не будем восхищаться и будем презирать ничтожность.

EFFIGIES TYCHONIS BRAHE O.F.
ÆDIFICII ET INSTRUMENTORVM
ASTRONOMICORVM STRVC TORIS.
Aº. DOMINI 1587 ÆTATIS SVÆ 40

Astrologia
ptolomeus

PLANIGLOBIU
Taurus
Aries
Pisces
Andromeda
Orion
Perseus
Pegasus
Cassiopeia
Cepheus
Cygnus
Canis minor
Gemini
Delphinus
Cancer
Draco
Lira
Aquila
Leo
Bootes
Coma Berenices
Virgo
Serpens
Iupiter

COELESTE.
Cetus
Eridanus
Lepus
Phenix
Aquarius
Grus
Notius
Toucan
Rhomb.
Dorado
Columba
Monoceros
Indus
Capricornus
Pavo
Arco Navis
Hydra
Crux
Ara
Lupus
Sagittarius
Centaurus
Scorpion
Libra
Corvus
Crater
Saturnus observatus a celeb. Dn. Hevelio

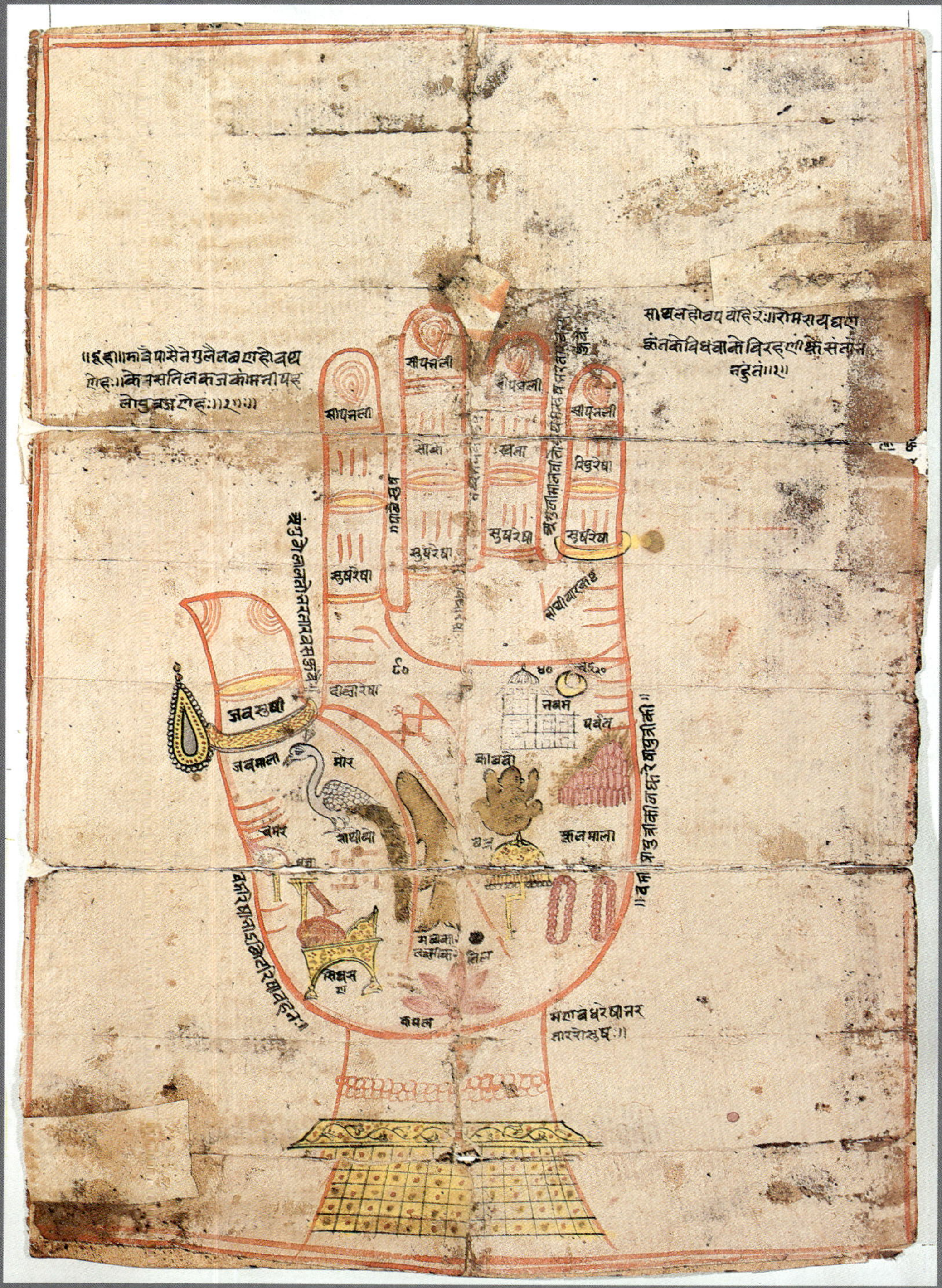

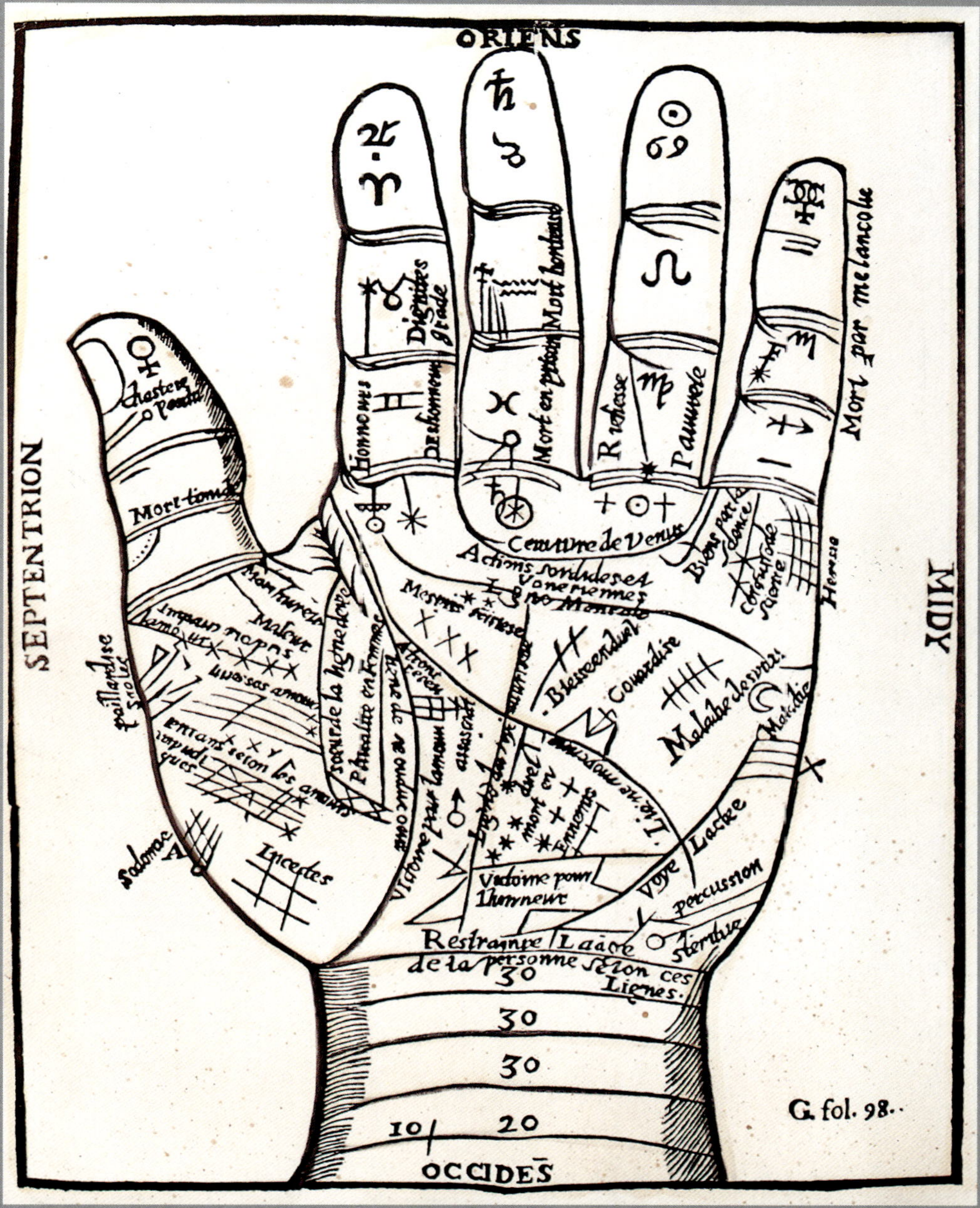
ORIENS
SEPTENTRION
MIDY
OCCIDĒS
Mort per melancolie
Ceinture de Venus
Couardise
Voye Lactee
percussion
Incestes
Restrainte Laâce
de la personne selon ces
Lignes.
30
30
30
10
20
G. fol. 98..

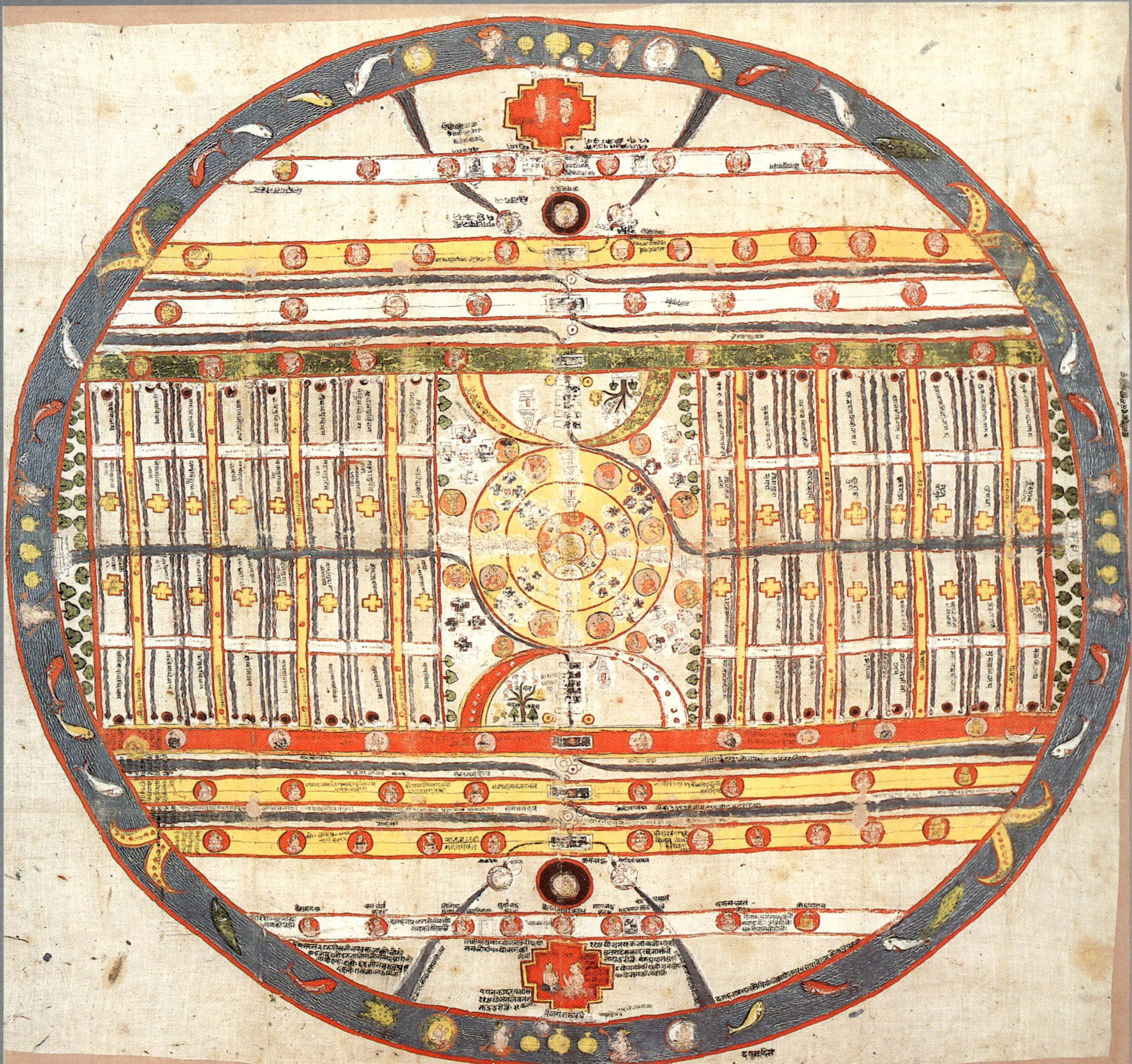

Wenn rechts der Venus ein Stern sichtbar ist, wird es gute Ernten im Land geben. Wenn kein Stern sichtbar ist, bedeutet es Unglück. Ist Venus groß und ein Stern neben ihr, verheißt es Kraft und Macht für die Menschen von Elam.

Als rechts van de Venus een ster zichtbaar is, zullen er in het land goede oogsten zijn. Als er geen ster zichtbaar is, betekent dat ongeluk. Is Venus groot en er een ster naast haar, belooft dat kracht en macht voor de mensen van Elam.

When a star is visible to the right of Venus, then there will be good harvests in the land. If no star is to be seen, this means ill fortune. If Venus is big and with a star alongside her, then this means strength and power for the people of Elam.

Astrological rule
of the Sumerians,
2000 B.C.

Se a destra di Venere si può vedere una stella, il raccolto sarà buono. Altrimenti, se la stella non si vede, significa che incombe disgrazia. Se Venere è grande e c'è una stella al suo fianco, la gente di Elam avrà forza e potere.

Quand une étoile est visible à droite de Vénus, la récolte sera bonne. Si aucune étoile n'est visible, c'est signe de malheur. Si Vénus est grande et une étoile à ses côtés, c'est un présage de force et de puissance pour les habitants d'Elam.

Если справа от Венеры видна звезда — будет хороший урожай. Если звезды не видно, это означает беду. Если Венера большая и рядом с ней стоит звезда, то значит будет сила и власть у людей Элама.

Schöpfer des Alls: „Götterpartikel" oder „Vater im Himmel"

Schepper van het Heelal: "Goddelijk Deeltje" of "Vader in de Hemel"

Der verkrüppelte Mann im Rollstuhl mit dem Gesicht eines Kindes ist Millionen Menschen in aller Welt vertraut. Vielleicht gerade wegen seiner schweren Krankheit ist er so etwas wie ein Weiser des globalen Dorfes geworden, der uns abends am Lagerfeuer den Himmel erklärt. Wie kein Wissenschaftler seit Albert Einstein hat der britische Mathematiker Stephen Hawking heutige Vorstellungen vom Universum geprägt. „Babykosmen", „Imaginäre Zeit" oder „Parallele Welten" sind durch ihn zu geläufigen Begriffen geworden.
Mehr Aufsehen als mit seiner Wissenschaft erregte Hawking freilich vor einiger Zeit mit der beiläufig ausgedrückten „Hoffnung", die Vereinigung der Urkräfte in einer „Weltformel" werde eines Tages das „Denken Gottes" enthüllen. Hawking hatte aber nicht Gott gemeint, sondern den Begriff „Naturgesetz" bildhaft umschreiben wollen, wie er später klarstellte. Zu einem ähnlichen Dementi sah sich unlängst der Nobelpreisträger Leon Lederman veranlaßt, der ein Atomteilchen „Gottespartikel" genannt hatte: Das Wort, ließ er verlauten, sei als Metapher zu verstehen „für alles, was wir noch nicht wisssen."
Im Grenzbereich zwischen Wissenschaft und Religion, das zeigen die zwei Beispiele, liegt explosiver Stoff verborgen in einer Zeit spiritueller Verunsicherung, in der viele Menschen nach wissenschaftlich fundierter Hilfestellung in religiösen Fragen suchen. Sie glauben „irgendwie" an eine „höhere Macht" oder etwas ähnlich Unbestimmtes und fühlen sich enttäuscht, wenn etwa ein Mann wie Hawking auf ihre Frage die übliche Forscherantwort gibt: „Warum das Universum existiert? Was sein Sinn ist? Ich weiß es nicht." Andere Physiker haben sich indes herausgewagt aus dem Kokon des Spekulationstabus und damit begonnen, etwas freier mit Ideen zu jonglieren. Warum es eine Welt gibt, was vor dem Urknall war und ähnlich harte Nüsse meiden auch diese Autoren. Aber sie beschreiben Gottheiten in einer fernen Zukunft, wenn - wie sie ausmalen - die Menschheit das Weltall erobert haben wird und es in einem „Omegapunkt" zur „Vollendung" alles Seienden kommt. Der Astronom und Physiker David Darling etwa entwirft eine Welt, in der Homo sapiens das in quantenhafter Unbestimmtheit schwingende Universum durch seine Existenz und sein Schauen erst „wirklich macht". Während die Menschen also über Millionen von Jahren Elektronen zur Existenz verhelfen, werden sie selbst von einer nahezu unendlichen Intelligenz beobachtet. Das Universum, so Darling, wird damit sein eigener Gott geworden sein: „Und dann, wenn der kosmische Geist zu denken beginnt, wird sich alle Substanz des Universums spontan neu organisieren. Unsere Träume und Ideen und Abstraktionen werden sich manifestieren und Wirklichkeit werden." Quantenphysik ermöglicht virtuose Gedankenspiele - so auch das Szenarium des renommierten Kosmologen Frank Tipler: Mit vielen mathematischen Gleichungen - etwa die „universelle Wellenfunktion des Heiligen Geistes" - „beweist" er die Möglichkeit der Auferstehung von den Toten und eines ewigen Lebens für die Menschheit. Der Physiker Paul Davies wiederum leitet aus der „Kreativität" des Kosmos das Vorhandensein eines „Planers" ab. Und der Astrophysiker John Gibbin kommt zu dem Ergebnis, daß der Sinn des Universums schlicht darin liege, immer neue Schwarze Löcher und daraus Kosmen zu produzieren.
Als Leser dieser widersprüchlichen Kosmotopien fragt man sich, ob wir Heutigen mit unseren Vorstellungen des Universums wirklich so viel weiter sind als unsere Vorfahren. Ihre Vision des Kosmos' sah zunächst typischerweise so aus, daß sie in den Gestirnen göttliche Wesen sahen und in deren Bewegungen Omen und Anweisungen für die Menschen. Die himmlische Ordnung wurde als göttlicher Plan angesehen, nach dem auch alles Irdische zu ordnen war. Herrscher wurden zu Sonnenkönigen, Konstellationen zur Zeit der Geburt bestimmten das Schicksal des Menschen. Mit dem Christentum verloren die Gestirne ihre Macht, ein einziger Schöpfergott wurde in einen fernen „Himmel" jenseits der Sterne gehoben. Und schließlich fanden wir heraus, daß die Erde nicht im Zentrum des Universums liegt, sondern am Rand einer unbedeutenden Galaxie durchs All zieht - in einem Kosmos, der möglicherweise nur einer von vielen ist. Glauben spielt freilich auch bei unserer heutigen Suche nach einer „Weltformel" eine Rolle: Wir gehen einfach davon aus, daß es in einem nach festen Regeln geordneten Universum ein solches Gesetz geben muß. Vielleicht gibt es aber gar keine Weltformel. Und vielleicht ist unser Auffassungsvermögen in der gegenwärtigen Evolutionsstufe noch zu klein, um das Universum verstehen zu können. Daß wir auf die Frage nach dem Sinn von allem bislang nur spirituelle Antworten gefunden haben, bedeutet freilich nicht, daß es keine „rationale" Antwort gibt. Eine der Stärken der Wissenschaft liegt darin, daß sie Ignoranz bewußt macht. Das Wissen, nicht zu wissen, wird zweifellos noch vieler Denker Geist und Herz beflügeln, hinter das Geheimnis des Seins kommen zu wollen.

De misvormde man in de rolstoel met het gezicht van een kind is miljoenen mensen in de hele wereld vertrouwd. Wellicht is hij juist wegens zijn zware ziekte zoiets als een wijze van het globale dorp geworden die ons 's avonds bij het kampvuur de hemel verklaart. De Britse wiskundige Stephen Hawking heeft als geen andere wetenschapper sinds Albert Einstein zijn stempel op de hedendaagse voorstellingen van het universum gedrukt. "Babykosmossen", "imaginaire tijd" of "evenwijdige werelden" zijn door hem gangbare begrippen geworden.
Meer opzien dan met zijn wetenschap baarde Hawking enige tijd geleden uiteraard met de bijkomstig uitgesproken "hoop" dat het verenigen van de oerkrachten in een "wereldformule" op een dag het "denken van God" zou onthullen. Hawking had echter niet God bedoeld, maar had, zoals hij later duidelijk maakte, het begrip "natuurwet" aanschouwelijk willen omschrijven. Ook de Nobelprijswinnaar Leon Lederman zag zich onlangs, toen hij een atoomdeeltje "goddelijk deeltje" had genoemd, tot een soortgelijke rectificatie genoodzaakt: Hij liet mededelen dat het woord als een metafoor "voor alles wat wij nog niet weten" moest worden opgevat.
In het grensgebied tussen wetenschap en religie ligt, zoals deze twee voorbeelden tonen, een explosieve stof verborgen in een tijd van spirituele onzekerheid waarin veel mensen naar wetenschappelijk gefundeerde hulp bij religieuze vragen zoeken. Zij geloven "op de een of andere manier" in een "hogere macht" of iets dat net zo onbepaald is, en voelen zich teleurgesteld, wanneer bijvoorbeeld een man als Hawking op hun vraag het gebruikelijke antwoord van een onderzoeker geeft: "Waarom het universum bestaat? Wat de zin ervan is? Ik weet het niet." Andere natuurkundigen hebben zich intussen uit de cocon van het speculatietaboe gewaagd en zijn ermee begonnen iets vrijer met ideeën te jongleren. Waarom er een wereld is, wat er vóór de oerknal bestond en soortgelijke moeilijke onderwerpen gaan ook deze auteurs uit de weg. Maar zij beschrijven godheden in een verre toekomst, wanneer - zoals zij het afschilderen - de mensheid het heelal veroverd zal hebben en het in een "omega-punt" tot "voleindiging" van al het zijn komt.
De astronoom en natuurkundige David Darling ontwerpt bijvoorbeeld een wereld waarin de homo sapiens het in quantumachtige onbepaaldheid slingerende universum door zijn existentie en kijken pas "werkelijk maakt". Terwijl de mensen dus miljoenen jaren lang voor de existentie van elektronen zorgen, worden zij zelf door een nagenoeg oneindige

The creator of the universe: "God particles" or "our father in heaven"

intelligentie geobserveerd. Volgens Darling zal het universum daardoor zijn eigen god geworden zijn: "En wanneer de kosmische geest dan begint te denken, zal de gehele substantie van het universum zich spontaan opnieuw organiseren. Onze dromen en ideeën en abstracties zullen zichtbaar en werkelijkheid worden." Quantumfysica maakt virtuoze gedachtenspelletjes mogelijk - zoals ook het scenario van de gerenommeerde kosmoloog Frank Tipler. Met talloze wiskundige vergelijkingen - bijvoorbeeld de "universele golffunctie van de Heilige Geest" - "bewijst" hij de mogelijkheid van de opstanding van de doden en van een eeuwig leven voor de mensheid. De natuurkundige Paul Davies leidt daarentegen uit de "creativiteit" van de kosmos de aanwezigheid van een "planner" af. En de astrofysicus John Gibbin komt tot de conclusie dat de zin van het universum is, steeds nieuwe zwarte gaten en hieruit kosmossen te produceren. Als lezer van deze tegenstrijdige kosmotopieën vraagt men zich af of wij die vandaag leven, met onze voorstellingen van het universum nu echt wel zoveel verder zijn dan onze voorouders. Hun visie op de kosmos zag er aanvankelijk typisch genoeg zo uit dat zij in de hemellichamen goddelijke wezens zagen en in de bewegingen ervan voortekens en aanwijzingen voor de mensen. De hemelse orde werd gezien als een goddelijk plan en volgens dit plan was ook al het aardse te ordenen. Heersers werden zonnekoningen, constellaties ten tijde van de geboorte bepaalden het lot van de mens. Met het christendom verloren de hemellichamen hun macht, één enkele scheppende god werd in een verre "hemel" aan gene zijde der sterren "opgehemeld". En tot slot ontdekten wij dat de aarde niet in het centrum van het universum ligt, maar aan de rand van een onbelangrijk galactisch stelsel door het heelal trekt - in een kosmos die eventueel slechts een van vele is. Geloof speelt uiteraard ook bij onze huidige speurtocht naar een "wereldformule" een rol. Wij gaan er gewoonweg van uit dat er in een universum dat volgens vaststaande regels geordend is, zo'n wet moet bestaan. Misschien is ons bevattingsvermogen in de tegenwoordige evolutiefase nog te klein om het universum te kunnen begrijpen. Dat wij op de vraag naar de zin van alles tot nu alleen maar spirituele antwoorden hebben gevonden, wil uiteraard niet zeggen dat er geen "rationeel" antwoord bestaat. Een van de sterke zijden van de wetenschap is dat zij onwetendheid bewust maakt. Te weten dat men het niet weet, dat zal ongetwijfeld nog talloze denkers hart en geest bevleugelen om achter het geheim van het zijn te willen komen.

The crippled man in the wheelchair with the face of a child is familiar to millions of people all over the world. Perhaps it is precisely because of his serious disability that he has become something like a wise man in this global village, one who at the end of the day round the campfire explains the mysteries of the heavens. Like no other scientist since Albert Einstein, the British mathematician Stephen Hawking has stamped today's ideas of the universe. "Baby cosmoses" "imaginary time" or "parallel worlds" are terms which have become known through his work.
Some time ago, however, Hawking caused more of a stir than he did with his work when he expressed in passing the hope that the unification of the elementary forces in a "world formula" might one day reveal "God's thinking". Hawking did not however literally mean God, as he later made clear; he had only wanted to graphically paraphrase the idea of a "natural law". The Nobel prize winner Leon Lederman also found himself compelled to make a similar denial just recently. He had called an atomic particle the "god particle", but his terminology, he declared, was only to be understood as a metaphor "for everything that we do not yet know".
In the borderland between science and religion, the two examples show, there remains hidden a lot of dynamite in this a time of spiritual disquiet, in which many people are looking for scientifically-founded help in religious matters.
They believe "somehow" in a "higher power" or some other undefined thing and are disappointed when a man like Hawking answers their questions in typical research scientist style: "Why does the universe exist? What is its purpose? I don't know." Other physicists have dared to depart the cocoon of the speculation taboo and have begun to juggle somewhat more freely with ideas. Why the world exists, what there was before the Big Bang and other hard nuts to crack are admittedly also avoided by these authors. However they describe deities in a distant future when - as they depict it - mankind will have conquered space and we will witness an "Omega point" of "perfection" of all things existing. The astronomer and physicist David Darling, for example, describes a world in which homo sapiens by his mere existence and his way of seeing things makes the universe and its quantum indeterminacy "real". Just as over the course of millions of years Man has helped electrons to come into their right, so Man himself is being observed by a virtually infinite intelligence. The universe, according to Darling, will thus become its own god: "And then, when the Cosmic Spirit begins to think, all the substance in the universe will begin to organize itself anew on a spontaneous basis. Our dreams and ideas and abstractions will then manifest themselves and become reality." Quantum physics makes possible such virtuose worlds of thought and this applies, too, to the scenario of the renowned cosmologist Frank Tipler: with the help of four mathematical equations - such as the "universal wave function of the Holy Spirit" - he "proves" the possibility of the resurrection of the dead and of eternal life for mankind. The physicist Paul Davies, on the other hand, derives from the idea of the "creativity" of the cosmos the existence of a "planer". And the astro-physicist John Gibbin comes to the conclusion that the purpose of the universe lies simply in discovering ever new Black Holes and producing cosmoses from them.
When one reads all these contradictory cosmo-utopias, one asks oneself if, today, we and our ideas on the universe, are really so much further on than our ancestors. Their typical vision of the cosmos was at first that they saw in the stars divine beings and in the movement of the stars omens and directions for mankind. The heavenly order was looked upon as a divine plan, to which everything earthly had to subordinate itself. Rulers became sun-gods, constellations at the time of birth determined individual fates. With the coming of Christianity, the stars lost their power; a sole Creator God was elevated into a far "Heaven" beyond the stars. Then we found out that the Earth is not the centre of the universe but was moving through space on the fringe of an insignificant galaxy - in a cosmos which might possibly be just one of many. Admittedly, faith still plays a role in our search for a "world formula" today: we simply assume that, in a universe organized according to set rules, there must be such a law. But perhaps there is no universal formula at all. And perhaps our powers of perception at this present stage of evolution are still too limited for us to understand the universe. That, up to now, in our questions relating to the purpose of all things we have found only spiritual answers, does not mean that there is not a "rational" answer. One of the strengths of science lies in the fact that it makes ignorance obvious, and the knowledge that Man lacks in knowledge will doubtless encourage the heart and the spirit of many thinkers to attempt to find out the secret of being.

La sabbia di Archimede, la gloria di colui che tutto muove, l'energia di Plank, il Big-Bang ed il bombardamento di Giove

È innegabile il mistero, il fascino del caos. Quel caos dal quale divinità siderali hanno ritagliato l'ordine, tanto più comprensibile ed affidabile per la mentalità umana. Infatti, lo informe è terrore, insicurezza, energia potenziale e terribile, eppure primordiale origine della sicurezza del cerchio, via dallo smarrimento, con l'uomo e la donna modellati e sicuri nell'allegria del Paradiso terrestre.
Tutto questo ha concesso ad Archimede, il più grande matematico dell'antichità e probabilmente di tutti i tempi, il più rilevante fisico nel senso attuale della parola, di divagarsi un giorno con il calcolo del numero dei granellini di sabbia che coprivano il mondo allora conosciuto. Per quanto piccolissimi, avevano forma definitiva ed autonoma gli uni dagli altri. Questo suo calcolo lo portò a stabilire che tale numero era enorme: non infinito come sostenevano gli altri. Analogo calcolo si sarebbe potuto fare per i corpi celesti vaganti nello spazio se avesse potuto disporre di qualche mezzo tecnico adeguato. Già questo, comunque, era ordine. Rassicurante, indubbiamente, che aveva spinto Ipparco di Nicea, il maggiore astronomo della antichità intera, tanto ammirato dai suoi colleghi moderni, a scoprire gli enigmi della trigonometria per mezzo della quale misurò le distanze che separavano il sole dalla terra. E soprattutto a catalogare 850 stelle fisse, fornendo su di esse notizie circostanziate.
Era un corale contributo all'armonia, all'assetto definitivo che Dante Alighieri, nel primo canto del Paradiso attribuiva a "La gloria di Colui che tutto muove / Per l'universo penetra e risplende / In una parte più, e meno altrove.", subito spiegando: "... Le cose tutte quante / Hann'ordine tra loro; e questo è forma / Che l'universo a Dio fa somigliante."
Oggi, molto crudelmente, ma molto più realmente questo motore si chiama "energia". Il Prof. Antonino Zichichi nelle sue "Cronache scientifiche", informa: "Volendo capire come è fatto il mondo bisogna andare ancora più in alto con i livelli di energia, fino a toccare il massimo concepibile: e cioè l'energia di Plank. Infatti la nostra esistenza materiale ha inizio con il cosiddetto grande botto, meglio noto come Big-Bang. Allora tutto era terribilmente caldo. Da allora sono passati venti miliardi di anni ed il Cosmo si è raffreddato: oggi la temperatura (energia media) dello spazio intergalattico è di appena duecentosettanta gradi sotto lo zero dei nostri termometri."
Ecco, dunque, i due vertici: "Si va dalla bassissima energia media (temperatura) del cosmo a quella altissima del Big-Bang, passando per il fenomeno vita, per le stelle, per i collassi stellari..."; collassi che si ripetono nel tempo come l'attuale bombardamento di meteoriti giganti sul pianeta Giove: masse incandescenti di chilometri di diametro con impatti a catena. Un finimondo che neppure Fetonte, quando ottenne di guidare il carro del Sole, riuscì a combinare. Questa la cronaca che ne fa Ovidio: "Ecco che grandi città van distrutte con le loro mura e gli incendi riducono in cenere intere regioni con le loro popolazioni. Bruciano i boschi coi monti. Ardono l'Ato ed il Tauro in Cilicia e lo Tmolo e l'Eta, e l'Ida che prima pullulava di sorgenti, prosciugato, e l'Elicona ... Un rogo immenso è l'Etna, aggiunto fuoco a fuoco; ardono il Parnaso dalle due cime e l'Erice e il Cinto ... Neppure la Scizia si salva, malgrado il suo freddo: il Caucaso brucia, come l'Ossa e il Pindo e l'Olimpo più grande di entrambi, e le Alpi sublimi e l'Appennino rannuvolato ..."
Meglio non trovarsi nella traiettoria di questi grappoli di incommensurabili atomiche! E a che servirebbe prevederle?
Quanto più tranquillo, dunque, è l'ordine! Su di esso, sì, che si possono intessere favole e leggende accovacciati attorno ai fuochi e raccontarsi, come si crede in Giappone, che la terra sia trasportata da un pesce gigantesco; da una tartaruga in India; da un serpente presso gli Amerindi; da un elefante nel Sud-Est asiatico; da uno scarabeo in Egitto: l'essenziale è che compia regolarmente i suoi giri lasciando gli uomini ad immaginare la matrice magica della sua esistenza.
Tanto da circondarsi di templi, di spazi sacri, per un ritorno necessario e continuo alla meditazione sul provvisorio abbisognevole di certezze, oppure al silenzio assoluto sul caos interiore e universale. Un tunnel che esigerebbe un'uscita, difficile da trovare però. E la scienza chiarisce e complica.
Perciò è tanto più facile dire: "... Non capita mai che il fine a cui aspiriamo sia il disordine", come ha fatto l'architetto Heinrich Tessenow all'inizio del secolo. È una massima che già era nel cervello di tutti e ripeterla è superfluo: a meno che non si sia fra i pochi della altra parte, anarchici visionari o illuminati. L'universo, frattanto, sembra apparentemente immutabile!

Giuseppe Brunamontini

Créateur du Cosmos: "Particules divines" ou "Notre Père qui es aux Cieux"

Des millions de personnes dans le monde entier connaissent la photo de cet homme au visage d'enfant recroquevillé dans son fauteuil roulant. Peut-être justement à cause de sa grave maladie, il est devenu une sorte de sage du village global qui nous explique le ciel, le soir, autour du feu de camp. Aucun scientifique depuis Albert Einstein n'a autant marqué la représentation actuelle de l'Univers que le mathématicien britannique Stephen Hawking. C'est lui qui nous a familiarisés avec les notions de "Bébécosmes", "Temps imaginaire" ou "Mondes parallèles".
Il y a quelque temps, certes, Hawking a fait encore plus sensation en exprimant incidemment "l'espoir" que la réunion des forces originelles en une "formule universelle" dévoilerait un jour la "pensée de Dieu". Hawking précisa ultérieurement qu'il ne s'agissait pas véritablement de Dieu mais d'une circonlocution imageant la notion de "loi de la nature". Dernièrement, Leon Lederman, lauréat du Prix Nobel, qui avait appelé une particule atomique "particule divine", se vit également contraint de démentir l'expression; ce mot, affirma-t-il, doit être compris comme une métaphore "pour tout ce que nous ne savons pas encore".
Ces deux exemples montrent bien que le domaine limitoire situé entre la science et la religion renferme une matière explosive à une époque d'incertitude spirituelle où beaucoup de gens cherchent de l'aide dans les fondements de la science pour résoudre les questions religieuses. Ils croient "d'une certaine manière" à une "puissance supérieure" ou à quelque chose de similaire mais indéfini et sont déçus quand un homme comme Hawking répond à leur question comme les scientifiques le font d'habitude: "Pourquoi est-ce que l'Univers existe? Quel est son sens? Je ne sais pas." D'autres physiciens ont toutefois osé sortir du cocon formé autour du tabou spéculatif et commencé à jongler plus librement avec leurs idées. Mais ils évitent eux aussi de se prononcer sur les questions délicates telles que "pourquoi le monde existe-t-il", "qu'y avait-il avant l'explosion originelle". Ils décrivent cependant les divinités d'un avenir lointain quand l'humanité aura conquis l'Univers et qu'on arrivera en un "point Oméga" à "l'accomplissement" de tout ce qui existe. L'astronome et physicien David Darling, par exemple, conçoit un monde dans lequel l'Homo sapiens, par son existence et ses observations, "rend réel" l'Univers oscillant dans une indétermination quantique. Pendant des millions d'années, l'humanité permet donc aux électrons de vivre tout en étant elle-même observée par une intelligence quasiment

Творцы Вселенной: «Божии частицы» или «отец Небесный»

infinie. Ainsi, selon Darling, l'Univers deviendra un jour son propre Dieu: "Alors, quand l'esprit cosmique commencera à penser, toute la substance de l'Univers se réorganisera spontanément. Nos rêves, idées et abstractions se manifesteront et deviendront réalité". La physique quantique permet de faire jouer son esprit avec virtuosité, comme le prouve le scénario du célèbre cosmologue Frank Tipler: à l'aide de nombreuses équations mathématiques (comme la "fonction ondulatoire universelle de l'Esprit Saint"), il "prouve" la possibilité d'une résurrection des morts et d'une vie éternelle pour l'humanité. Le physicien Paul Davies, de son côté, déduit de la "créativité" du cosmos l'existence d'un "planificateur". Et l'astrophysicien John Gibbin conclut que la raison de l'existence de l'Univers est simplement de produire sans cesse des trous noirs dont naissent les cosmos.
En tant que lecteur de ces cosmogonies contradictoires, on se demande si, avec l'Univers tel que nous le concevons, nous avons tant progressé par rapport à nos ancêtres. Notre vision première et typique du cosmos considérait les astres comme des êtres divins dont les mouvements représentaient des présages et instructions à l'intention des humains.
L'ordre céleste était regardé comme un plan divin commandant également tout élément terrestre. Les souverains devinrent des "rois-soleils", les constellations en présence à la naissance déterminèrent le destin de l'homme. Avec la chrétienté, les astres perdirent leur puissance, un seul Dieu créateur fut placé dans un "Ciel" éloigné au-delà des étoiles. Finalement, on s'aperçut que la Terre n'était pas au centre de l'Univers mais qu'elle flottait dans l'espace à la limite d'une galaxie insignifiante, dans un cosmos qui n'était peut-être pas le seul. De nos jours, la foi joue certes un rôle dans la quête d'une "formule universelle": nous supposons tout simplement qu' une telle loi doit exister dans un Univers réagissant selon des lois déterminées.
Mais il n'existe peut-être pas de loi universelle. La phase évolutive actuelle de notre intelligence est peut-être encore trop infime pour pouvoir comprendre l'Univers. Le fait que nous n'ayons jusqu'à présent trouvé que des réponses spirituelles à la question du "pourquoi" de l'Univers ne signifie pas qu'il n'existe pas de réponse "rationnelle". L'une des forces de la science est qu'elle nous fait prendre conscience de notre ignorance; le fait de savoir qu'on ne sait rien donnera sans doute encore des ailes à l'esprit et au coeur de bien des penseurs à la recherche du secret de l'existence.

Этот человек с детским лицом, сидящий в кресле-каталке, знаком миллионам людей. Может, именно из-за тяжелого недуга он стал как бы мудрецом всей мировой деревни, который вечером сидит у костра и объясняет звездное небо. Никто после Эйнштейна не оказал такого влияния на наши представления о Вселенной, как британский математик Стивен Хаукинг. Благодаря ему, такие понятия, как «бэби-космос», «воображаемое время» или «параллельные миры» стали привычными.
Еще большее внимание он привлек к себе некоторое время назад, высказав как бы между прочим свою надежду на то, что соединение всех первичных сил в одной «универсальной формуле» когда-нибудь раскроет «замысел Божий». Позднее он пояснил, что имел в виду не Бога, а хотел сказать «закон природы». Подобное опровержение недавно пришлось дать и небелевскому лауреату Л. Ледерману, назвавшему одну атомную частицу «Божией частицей». Он сказал, что это — метафора «для всего, чего мы еще не знаем».
Как показывают эти два примера, пограничная область между наукой и религией взрывоопасна в наше время всеобщей духовной неуверенности, когда многие ищут научно обоснованные ответы на религиозные вопросы. Они «где-то» верят в «высшую силу» или еще во что-то такое неясное и чувствуют разочарование, когда такие люди, как Хаукинг, дают на их вопрос стандартный ответ ученого: «Почему существует Вселенная? Не знаю.» Другие физики рискнули покинуть кокон умозрительных табу и начали свободно жонглировать идеями. Однако таких вопросов, как «почему существует мир?» или «что было до Большого взрыва?» избегают даже эти авторы. Но они описывают божества отдаленного будущего, когда — по их представлениям — человечество завоюет Вселенную, и в какой-то «точке Омега» все сущее придет к своему «завершению». Так, астроном и физик Д. Дарлинг изображает мир, в котором «хомо сапиенс» одним своим существованием и созерцанием «делает истинной» Вселенную, раскачивающуюся в квантовой неопределенности. Итак, пока люди в течение миллионов лет помогают электронам существовать, за ними самими наблюдает бесконечный Разум. Вселенная, по Дарлингу, станет тем самым своим собственным богом: «И тогда, когда космический дух начнет думать, вся материя Вселенной спонтанно реорганизуется. Наши мечты, идеи и абстракции станут реальной действительностью». Квантовая физика позволяет играть в виртуозные умственные игры. Таков сценарий известного космолога Ф. Типлера: с помощью многих уравнений — вроде «универсальной волновой функции Св. Духа» — он «доказывает» возможность воскресения из мертвых и вечной жизни всего человечества. Физик П. Дэвис в свою очередь выводит из «творческого характера» космоса наличие «Того, Кто планирует». А астрофизик Дж. Гиббин делает заключение, что смысл Вселенной всего-навсего в том, чтобы производить все новые «черные дыры», а из них — новые космосы.
Когда читаешь эти противоречивые космо-утопии, спрашиваешь себя, действительно ли мы с нашими представлениями о Вселенной так далеко ушли от предков. Их представление о космосе вначале было очень характерным: они видели в созвездиях богов, а в их движении — указания и благословение людям. Небесный порядок рассматривался как Божий промысел, по которому следовало организовать и все земное. Властители становились «королями-солнцами», расположение планет при рождении определяло судьбу человека. В христианстве созвездия потеряли свою власть, единственный Бог-Творец был вознесен на удаленное «небо» по ту сторону звезд. И вот выяснилось, что Земля лежит не в центре Вселенной, а путешествует по ней на краю заурядной галактики, в космосе, который, возможно, является лишь одним из многих. Вера играет, конечно, свою роль и сейчас, при поисках «универсальной формулы»: мы надеемся, что в упорядоченной по твердым правилам Вселенной такой закон должен быть. А может, нет никакой «универсальной формулы». А может, наша способность понимания на нынешней ступени эволюции недостаточна, чтобы постичь Вселенную. То, что на вопрос о смысле всего сущего мы пока получали ответ из духовной области, не означает, что нет «рационального» ответа. Наука сильна тем, что она раскрывает незнание; знание собственного незнания вдохновит сердца и умы тех, кто стремится проникнуть в тайну бытия.

COSMOS

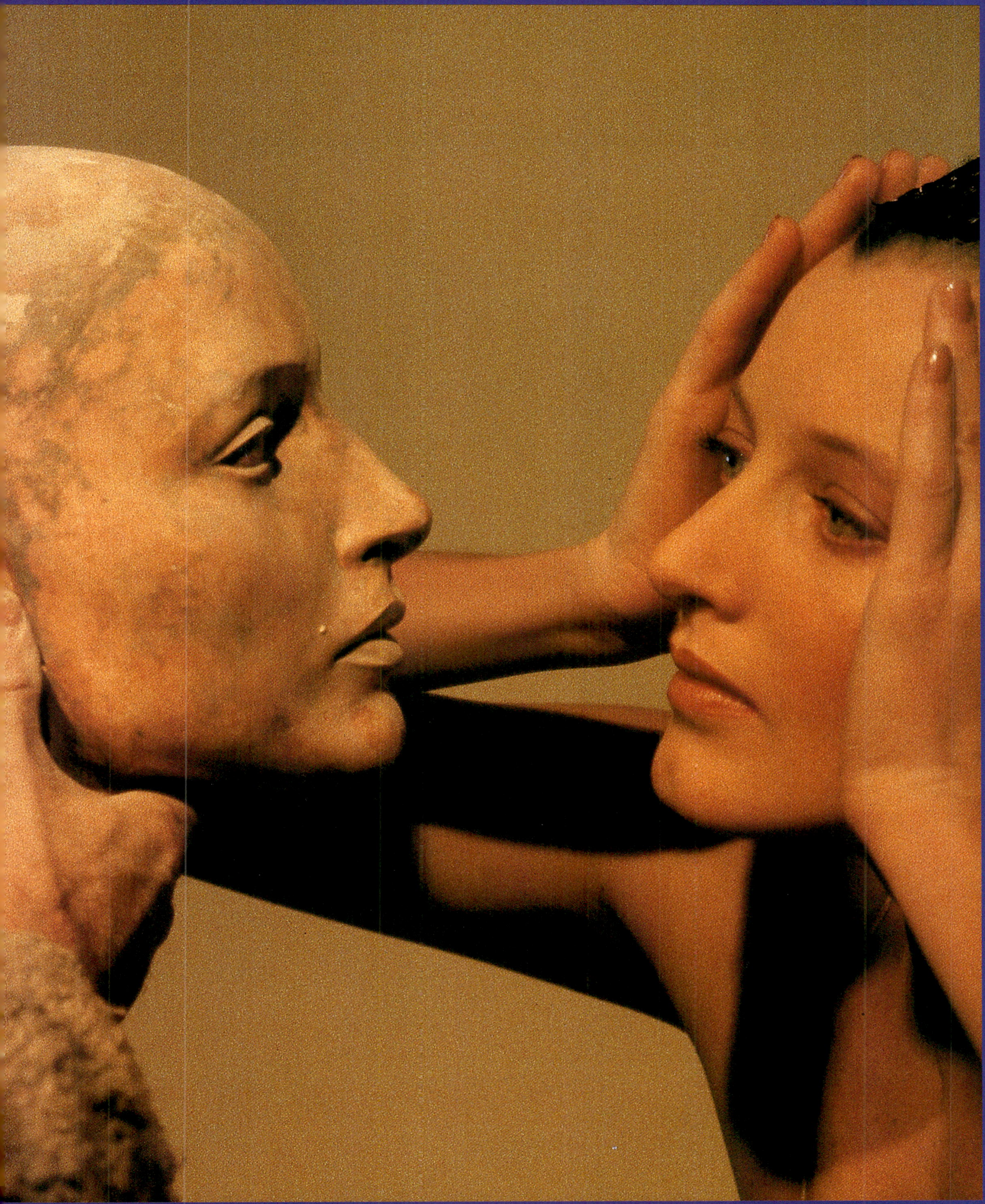

MAN

Der Archetyp des „kosmischen Menschen", der sich als göttlich definiert, weil er sich als Teil eines göttlichen Kosmos' weiß, ist in zahlreichen Mythen und religiösen Lehren der Welt vertreten. In einer frühen Form tauchte der „Mensch, der ein Gott ist" in der altägyptischen Hochkultur auf, zweitausend Jahre später sagte der Menschensohn Jesus Christus zu seinen Jüngern: „Ich und der Vater sind eins" und „Wißt ihr nicht, daß ihr Götter seid?" Die buddhistische Version des „kosmischen Menschen" unterscheidet sich von den westlichen Erscheinungsformen in vielen Punkten: Es gibt keinen Gott, sondern einen Kosmos, der von einer Art göttlichem Prinzip durchsetzt ist - ein Prinzip, das dem Menschen, als Teil des Ganzen, Göttlichkeit verleiht. Ziel des Buddhismus ist es, das Wesen des Selbst zu erkennen - was notwendig ist, um der „Illusion der Getrenntheit" entfliehen zu können, um eins zu werden mit dem Kosmos. Nach dem Tod löst sich für ihn alle Individualität auf.

Het archetype van de "kosmische mens" die zich als goddelijk definieert, omdat hij in de wetenschap leeft een deel van een goddelijke kosmos te zijn, is in talrijke mythen en religies van de wereld vertegenwoordigd. In een vroege vorm doemde de "mens die een god is" op in de Osiris Horus mythe van de oud-Egyptische hoge cultuur. Tweeduizend jaar later zei de mensenzoon Jezus Christus tegen zijn discipelen: "Ik en de Vader zijn één" en "Weet gij niet dat gij goden zijt?" De Boeddhistische versie van de "kosmische mens" verschilt op veel punten van de westerse verschijningsvormen: er is geen god, maar een kosmos die van een soort goddelijk principe doordrongen is - een principe dat de mens als deel van het geheel, goddelijkheid verleent. Doel van het boeddhisme: inzicht in het wezen van het eigen ik - nodig om aan de "illusie van het gescheiden zijn" te ontkomen om één te worden met de kosmos. Na de dood verdwijnt alle individualiteit, alles wordt "brahman", er is geen leven meer.

The archetype of "cosmic man" who defines himself as divine, because he knows himself to be a part of a divine cosmos, is to be found in numerous myths and religious teachings. In an early form, the "man who is a god" appeared in the Osiris Horus myth of the very advanced ancient Egyptian civilization. Two thousand years later, the Son of Man, Jesus Christ, said to his disciples: "I and the father are one" and "Do you not know that you are gods?" The Buddhist version of the "cosmic man" differs from Western ideas in many respects: there is no god but a cosmos instead, a cosmos imbued with a kind of divine principle, one which imparts divinity to man as a part of the Whole. The goal of Buddhism is to recognize the nature of the Self, a necessary step towards getting rid of the "illusion of separation", towards becoming one with cosmos. After death, all individuality disintegrates, every thing becomes "Brahman"; there is no more life.

L'archetipo dell'uomo cosmico che si definisce divino perchè sa di essere parte di un cosmo divino è presente in numerosi miti e ricorre nelle dottrine religiose di tutto il mondo. Una delle prime forme di "uomo-dio", la troviamo nella civiltà dello antico Egitto, poi Gesù Cristo, figlio dell'uomo, disse ai suoi discepoli: "Io e il Padre siamo uno" e "Non sapete che voi siete degli dèi?". La versione buddista dello "uomo cosmico" si distingue dalle manfestazioni occidentali sotto numerosi aspetti: qui infatti non troviamo un dio, bensì un cosmo, permeato da una specie di principio divino, un principio che conferisce la divinità all'uomo perchè lo vede quale parte del tutto. L'obbiettivo primario del buddismo consiste nel riconoscere se stessi. Questo processo è necessario per poter sfuggire successivamente "allillusione della separazione" e per poter divenire finalmente uno con tutto il cosmo. Dopo la morte si dissolve per esso ogni individualità, e la vita cessa di esistere.

L'archétype de l'"homme cosmique", qui se veut divin parce qu'il est conscient de son appartenance au cosmos divin, est représenté dans de nombreux mythes et dogmes religieux. L'"homme qui est un Dieu" apparut autrefois dans le mythe d'Osiris de la haute culture de l'Egypte ancienne; deux mille ans plus tard, Jésus Christ, fils d'un humain, disait à ses disciples: "Moi et le Père ne faisons qu'un" et "Ne savez-vous pas que vous êtes des Dieux?" La version bouddhiste de l'"homme cosmique" se distingue des visions occidentales en de nombreux points: "Il n'y a pas de Dieu mais un Cosmos pénétré d'une sorte de principe divin, un principe qui confère à l'homme en tant que partie du Tout un caractère divin. Le but primaire du bouddhisme est de reconnaître la nature du Moi, ce qui est nécessaire pour pouvoir échapper à l'"illusion d'une séparation", pour ne faire qu'un avec le Cosmos. Après la mort, toute individualité se dissout; tout devient "Brahman"; il n'y a plus de vie.

Архетип «космического человека», определяющего себя как божественый, считающего себя частью божественного космоса, представлен во многих мифах и религиозных учениях. В ранней форме «человек-бог» появился как Озирис у древних египтян. Через две тысячи лет Сын Человеческий Иисус сказал ученикам: «Я и Отец едины» и: «Разве вы не знаете, что вы боги?» Буддистское понимание «космического человека» отличается от западного: Бога нет, есть космос, пронизанный божественным принципом, делающим человека как часть его божественным. Цель буддизма — понять свое Я, чтобы избежать «иллюзии разделенности» и слиться с космосом. После смерти жизни нет, индивидуальность окончательно распадается, все становится «брахман».

„Und das Wort ward Fleisch und wohnte unter uns, und wir sahen seine Herrlichkeit als des eingeborenen Sohnes vom Vater..." Jesus spricht zu ihm: „Ich bin der Weg und die Wahrheit und das Leben; niemand kommt zum Vater denn durch mich. Wenn ihr mich erkannt habt, werdet ihr auch meinen Vater erkennen. Glaubst du nicht, daß ich im Vater bin und der Vater in mir? Die Worte, die ich zu euch rede, die rede ich nicht von mir selbst aus. Und der Vater, der in mir wohnt, der tut seine Werke..."

"Het woord is vlees geworden en het heeft onder ons gewoond en wij hebben zijn heerlijkheid aanschouwd, een heerlijkheid als van de eniggeborene des Vaders... Jezus zeide tot hem: Ik ben de weg, de waarheid en het leven; niemand komt tot de Vader dan door Mij. Indien gij Mij kendet, zoudt gij ook mijn Vader gekend hebben. Gelooft gij niet, dat Ik in de Vader ben en de Vader in Mij is? De woorden die Ik tot u spreek, zeg Ik uit Mijzelf niet; maar de Vader die in Mij blijft, doet zijn werken..."

"And the word became flesh and lived for a while among us. We have seen his glory, the glory of the one and only Son who came from the Father"... Jesus answered: "I am the way and the truth and the life. No-one comes to the Father except through me. If you really knew me, you would know my Father as well...Don't you believe that I am in the Father and that the Father is in me. The words I say to you are not just my own. Rather, it is the Father, living in me, who is doing his work."

"Colui che è il Verbo è diventato un uomo e ha vissuto in mezzo a noi uomini. Noi abbiamo contemplato il suo splendore divino. È lo splendore del Figlio unico di Dio Padre..." Gesù gli disse: "Io sono la via, io sono la verità e la vita, nessuno arriva al Padre se non per mezzo di me. Se mi conoscete, conoscerete anche mio Padre. Dunque non credi che io vivo nel Padre e che il Padre vive in me? Le parole che vi dico non vengono da me; il Padre abita in me, ed è lui che compie le sue opere...".

"Et le Verbe s'est fait chair, et il a habité parmi nous - et nous avons contemplé sa gloire, gloire que tient de son Père un Fils unique... Jésus lui dit: C'est moi la Voie, la Vérité et la Vie; personne ne va au Père que par moi. Quand vous m'aurez reconnu, vous aurez aussi reconnu mon Père. Ne crois-tu pas que je suis dans le Père et que le Père est en moi? Les paroles que je vous dis, je ne les dis pas de mon propre chef. Mais c'est le Père qui est en moi à demeure, qui lui-même accomplit ses œuvres."

И Слово стало плотью, и обитало с нами, и мы видели славу Его, как Единородного от Отца... Иисус сказал ему: Я есть путь и истина и жизнь; никто не приходит к Отцу, как только через Меня. Если бы вы знали Меня, то знали бы и Отца Моего. Разве ты не веришь, что Я в Отце и Отец во Мне? Слова, которые говорю Я вам, говорю не от Себя; Отец, пребывающий во Мне, Он творит дела.

Die „Anima“ ist eine Personifizierung aller weiblichen Tendenzen in der Psyche eines Mannes - einschließlich intuitiver Gefühle des Einsseins mit Natur und Kosmos. Am häufigsten manifestiert sich diese „innere Frau“ in

Form erotischer Phantasien, die typischerweise von Anzeigen, Plakaten oder Filmen angeregt werden. Die Anima kann dem Mann aber auch in Gestalt einer Frau entgegentreten, in die er sein feminines Selbst hineinprojiziert: Er empfindet es als ihr zugehörig und entbrennt in „kosmischer“ Liebe. Diese Illusionen produzierende „dunkle Anima“ ist in zahlreichen Mythen personifiziert worden, etwa in Gestalt von Hexen, Sirenen oder Flußgeistern, die mit ihren Reizen Männer in den Tod ziehen.

De "Anima" is een personificatie van alle vrouwelijke neigingen in de psyche van een man, met inbegrip van intuïtieve gevoelens van het één zijn met de natuur en kosmos. Het meest deze "inwendige vrouw" in de vorm van erotische fantasieën te voorschijn, die typisch genoeg door advertenties, affiches of films worden geanimeerd. De "Anima" kan de man echter ook in de gestalte van een bepaalde vrouw tegemoet treden waar hij zijn feminiene ik op projecteert. Hij ervaart het als een stuk van haar en ontbrandt in een hartstochtelijke "kosmische" liefde. Deze illusies producerende "donkere Anima" is in talrijke mythen gepersonifieerd, heksen, sirenen of riviergeesten die de mannen met hun bekoring de dood insturen.

The "anima" is a personification of all female tendencies in the psyche of a man - including intuitive feelings of unity with nature and the cosmos. This "woman inside" manifests itself in the form of markedly erotic fantasies such as are stimulated by advertising or films. The anima can however also confront a man in the form of a certain woman into whom he then projects his feminine self: he feels that all this is part of her and is inflamed with a passionate "cosmic" love. The "dark anima" which produces these illusions has been personified in countless myths in the form for example of witches, sirens or river spirits who, with the help of their charms, lure unwitting men to their death.

Marie-Louise von Franz

L'anima è una personificazione di tutte le tendenze femminili esistenti nella psiche di un uomo, compresi i sentimenti intuitivi dell'essere tutt'uno con la natura ed il cosmo. Spesso questa "donna interiore" si manifesta sottoforma di fantasie erotiche stimolate da inserzioni pubblicitarie, manifesti o film. L'anima può manifestarsi all'uomo anche sottoforma di una donna nella quale egli proietta il suo ego femminile: egli allora si sente di appartenere a lei e si infiamma per lei di un appassionato amore "cosmico". La anima oscura che produce queste illusioni è stata personificata in numerosi miti sotto le sembianze di streghe, sirene o spiriti dei fiumi che, con la loro bellezza, attirano gli uomini in trappole mortali.

L'"anima" est une personnification de toutes les tendances féminines dans le psychisme d'un homme, y compris l'intuition de ne faire qu'un avec la nature et le cosmos. Cette "femme intérieure" se manifeste le plus souvent sous forme de visions érotiques suggérées par des annonces ou panneaux publicitaires ou par des films. L'anima peut aussi se présenter à l'homme sous les traits d'une femme précise sur laquelle il projette son Moi féminin: il a le sentiment de lui appartenir et s'enflamme d'un amour "cosmique" passionnel. Cette "anima ténébreuse" génératrice d'illusions a été personnifiée dans de nombreux mythes: ondines, sirènes ou esprits des fleuves dont le charme entraîne les hommes dans la mort.

«Анима» — это воплощение женского начала в психике мужчины, включая интуитивное сознание единства с природой и космосом. Чаще всего «внутренняя женщина» проявляется в эротических фантазиях, вызываемых рекламой или кинофильмами. «Анима» может встретиться в виде определенной женщины, на которую будет спроецировано у мужчины «женское Я»; тогда он будет испытывать к ней «космическую любовь». «Темная анима» встречается во многих мифах в образах ведьм или сирен, губящих мужчин своими прелестями.

Der Animus, der „innere Mann“ in der Psyche einer jeden Frau, ist eine wandelhafte Gestalt, die archetypische Formen bis zum „kosmischen Menschen“ annehmen kann. In Träumen erscheint das Männliche mitunter als Athlet oder Abenteurer. Weit häufiger jedoch taucht es als „Träger des Worts“ auf: als Dichter und Denker, als Politiker. Häufige Archetypen bei reiferen Frauen sind „weise Begleiter auf dem Weg zu spiritueller Wahrheit“.

De "Animus", de "inwendige man" in de psyche van iedere vrouw, is een veranderlijke gestalte die archetypische vormen, tot een "kosmische mens" toe, kan aannemen. In dromen verschijnt het mannelijke af en toe als atleet of avonturier. Maar veel vaker duikt het als "drager van het woord" op: als dichter en denker, als politicus. Archetypen die vaak bij rijpere vrouwen optreden, zijn "wijze begeleiders op de weg naar spirituele waarheid".

The "animus", the "man insude" in the psyche of every woman, is a changeable figure who can assume archetypal forms up to that of the "cosmic man". In dreams, the masculine element sometimes appears as an athlete or an adventurer. Much more frequently, however, he is a "bearer of words" - a poet and thinker, a politician. Frequent archetypes where more mature women are concerned are "wise companions on the way to spiritual truth".

L'animo, "l'uomo interiore" nella psiche di ogni donna, è una figura mutevole, che può assumere forme archetipiche fino ad arrivare "all'individuo cosmico". Nei sogni, la mascolinità appare talvolta come un atleta o un avventuriero. Molto più spesso però essa compare quale "portatore della parola": come poeta e pensatore, come politico. Gli archetipi frequenti nel caso di donne più mature sono "saggi accompagnatori sulla via che conduce alla verità spirituale".

L'animus, "l'homme intérieur" dans le psychisme de la femme, est d'aspect changeant; il peut adopter des formes archétypales allant jusqu'à "l'homme cosmique". Dans les rêves, le masculin a entre autres les traits d'un athlète ou d'un aventurier. Cependant, il apparaît plus souvent comme "porteur du Verbe": poète, penseur ou politicien. Les archétypes fréquents chez les femmes mûres sont les "sages compagnons sur le chemin de la vérité spirituelle".

«Анимус», «внутренний мужчина» в психике любой женщины может принимать архетипические формы вплоть до «космического человека». В мечтах мужское начало представлено атлетом, героем. Но чаще он предстает как «носитель слова» — поэт, мыслитель, политик. Частые архетипы у зрелых женщин — «руководители на пути к духовной мудрости»: Ганди, Будда, Лао-цзы.

DREYHAUPT
CHRONIK
DES
SAAL-CREYSES

Stets besitzen die Helden, die sich in den Mythen aller Völker tummeln gottähnliche Eigenschaften - ein Indiz für die Neigung von Homo sapiens, einem „Übermenschen" Autorität zu übertragen, selbst um den Preis der Versklavung. Ganz anders die Haltung des „kosmischen Menschen": Er sucht Freiheit im Erkennen des „Göttlichen in allem", im Erkennen des „Göttlichen, das ist": im Kosmos, in der Natur, in sich selbst. „Göttliche" Helden haben keinen Platz in diesem Weltverständnis.

In de mythen van alle volkeren zijn er helden die het kwaad overwinnen en mensen voor de ondergang behoeden. Steeds hebben deze redders godgelijke eigenschappen - een aanwijzing voor de neiging van de homo sapiens om autoriteit aan een "supermens" over te dragen. Heel anders de "kosmische mens". Deze zoekt vrijheid in het zien van het "goddelijke in alles", in het zien van het "goddelijke dat is", in de kosmos, in de natuur, in zichzelf. Voor "goddelijke" helden is er geen plaats.

In the myths of all the peoples of this world there are heroes who overcome evil and who save mankind from death and destruction. These saviours always have god-like qualities - an indication of the readiness of homo sapiens to transfer authority to a "superman", even at the price of enslavement. The attitude of "cosmic man" is different: he seeks freedom by recognizing "the divine" everywhere: in the cosmos, in nature, in himself. There is no room for "divine" heroes in this view of the world.

I miti di tutti i popoli sono pieni di eroi che salvano gli uomini dalla distruzione e dalla morte. Questi salvatori hanno sempre qualità simili a quelle degli dèi, un indizio per la tendenza dell'homo sapiens a trasferire l'autorità ad un "super uomo", anche se il prezzo è la schiavitù. La posizione "dell'uomo cosmico" è invece del tutto diversa: egli cerca libertà nel riconoscere il "divino" in tutto, il "divino che esiste" nel cosmo, nella natura, in se stesso. In tale concezione del mondo non c'è posto per eroi "divini".

Les mythes de tous les peuples fourmillent de héros qui possèdent des attributs semblables à ceux de Dieu, ce qui traduit le penchant de l'Homo sapiens à confier l'autorité à un "surhomme", même au prix de l'asservissement. Le comportement de l'"homme cosmique" est tout autre: il cherche la liberté dans la reconnaissance du "divin en toute chose", dans la reconnaissance du "divin qui existe": dans le cosmos, dans la nature, en soi-même. Les héros n'ont pas leur place dans un tel monde.

В мифах всех народов есть герои, побеждающие зло и спасающие от смерти. Они всегда имеют божественные черты. Люди признают авторитет «сверх-человека» даже ценой порабощения. Совсем иное у «космического человека»: он ищет свободы в познании «божественного во всем — в космосе, природе, в себе». «Божественным» героям нет места в таком восприятии мира.

FRIEDERICH
PRINZ V. HESSEN HOMB
D. CAV. OESTRR.
GR. YORK V. WARTENBURG
GEN. D. INFANTERIE
PREUSSEN
GR. KLENAU
GEN. D. CAVALLERIE
OESTERREICH
GEN. D. INFANTERIE
PREUSSEN
GRAF GYULAI
FELDZEUGMEISTER
OESTERREICH

GRAF COLLOREDO
FELDZEUGMEISTER
OESTERREICH

Kosmischer Mensch: Eine Welle auf der Suche nach dem Ozean

Es ist schon seltsam mit uns Menschen, wie wir über Millionen Lichtjahre entfernte Galaxien weit besser Bescheid wissen als über uns selbst. Wer wir eigentlich sind und wohin die Reise führen soll, wußten unsere Vorfahren in vorwissenschaftlicher Zeit noch mit einer gewissen Glaubensüberzeugung zu beantworten. Mit dem Verschwinden kosmischer Mythen und dem Verblassen des Christentums bröckelten dann aber wesentliche Identitätsstützen - mit dem Ergebnis, daß sich viele Menschen heute reichlich verloren fühlen in einem Kosmos, in dem sie sich nicht mehr als Geschöpfe Gottes fühlen können, sondern nur als Zufallsprodukte eiskalter, unpersönlicher Naturgesetze. Je mehr wir über das Universum herausfinden, sagt der Schriftsteller Walker Percy, desto verlorener fühlen wir uns darin: „Jeder Fortschritt beim objektiven Verstehen des Kosmos' und bei seiner technologischen Kontrolle distanziert das Selbst ein Stück mehr davon, so daß am Ende das Ich zu einem dahintreibenden Gespenst in jenem Kosmos wird, den es so genau erforscht hat." Wie also dem zerflatternden Selbst eine Identität geben, damit es sich heimisch fühlen kann im Universum, zu dem es ja schließlich gehört? Manche versuchen es in ihrer Not mit der ältesten aller Methoden der Ich-Verankerung im Hier und Jetzt: Sie lernen in einem Workshop, auf schamanistische Art „Bär" zu werden, „Adler" oder „Kojote", was aber natürlich nur sehr kurzfristig gelingt, weil der rechte Glaube an das Totemtier fehlt. Ihre Tücken kann auch eine ekstatische Wiedergeburts-Erfahrung „in Jesus Christus" haben, wenn einem später klar wird, daß sich an dem paternalistischen, aus weiter Ferne regierenden Gott seit Kindertagen nicht viel geändert hat. Und schließlich ist da zum Beispiel immer noch das gute alte „immanente Selbst" des Konsumenten in einer wissenschaftlich definierten Welt, das sich vielleicht ein bißchen „transzendieren" läßt als Therapie gegen das Verlorenheitsgefühl im Kosmos. Etwa über die Identifizierung mit „göttlichen" Superstars und Urmutter-Divas, Staatsrettern und „Wüstensturm"-Superhelden - oder gar mit „begnadeten" Physik-Nobelpreisträgern und deren neueste Weltformel-Theorie. Solche Ersatz-Spiritualität, sagt der Religionsphilosoph Ken Wilber, verdeckt nur das tief und unbewußt in allen Menschen vorhandene Bedürfnis nach Einheit mit einem göttlichen Kosmos, der sie ein aktiver Teil dieser Göttlichkeit sein läßt. „Dringt dieser Wunsch eines Tages spontan hervor, so ist das vergleichbar mit einer Meereswelle, die ihrer selbst plötzlich bewußt wird und entdeckt, daß sie eins ist mit dem Ozean - und dadurch eins mit allen anderen Wellen." Dieser Archetyp des göttlichen „kosmischen Menschen" ist in zahlreichen Mythen und religiösen Lehren der Welt vertreten. Er erscheint als Adam, als der persische Gayomart oder als hinduistischer Purusha. Nach Carl Jung war es keineswegs Jesus Christus, der den Mythos des göttlichen Menschen schuf (Wißt ihr denn nicht, daß ihr Götter seid?) Die symbolische Idee existierte vielmehr Jahrhunderte vor ihm „und er selbst wurde von der Idee des kosmischen Menschen erfaßt und aus dem Leben eines Handwerkers in Nazareth gehoben." Während sich viele Elemente des Christentums - vor allem die Existenz eines persönlichen Gottes - mit dem archetypischen Ideal des „kosmischen Menschen" nicht in Übereinstimmung bringen lassen, „paßt" das buddhistische Weltbild geradezu perfekt dazu. In der fernöstlichen Lehre gibt es keinerlei Gott, Schöpfer oder andere allmächtige Gestalt. Es gibt nur einen Kosmos, der von einer Art göttlichem Prinzip - dem „Wesen von allem, was ist" - durchsetzt ist. Als Teil dieses Kosmos' ist auch jeder Mensch göttlich; seine Aufgabe auf Erden besteht darin, das Wesen des Selbst zu suchen - was zur Erkenntnis führen soll, daß es kein Selbst gibt, und daß alle Wesen und Objekte des Universums Ausdruck derselben göttlichen Realität sind. Auf die Frage „Wer bist du?" könnte ein Buddhist antworten: „Ich bin Atman, bin Buddha-Natur, das heißt, ich bin göttlich in mir, aber ich bin auch Brahman, das Göttliche des Kosmos'." Da es in seinem Universum keinen großen Vater Gott gibt, kann und will ein Buddhist auch keine Gottheit um Hilfe, Gnade oder Schutz anflehen. Sein Ziel ist, Ganzheit zu erreichen, der Illusion der Getrenntheit zu entfliehen, das heißt, eins zu werden mit dem Kosmos, wie es dem Ideal des kosmischen Menschen entspricht. Eine große Hürde auf dem Weg dorthin ist die Überwindung des Ich mit seinen Eitelkeiten und Begierden. Nach dem Tod löst sich dann alle Individualität auf; es gibt keine Seele, die weiterbesteht. Zeitlose Ganzheit sei ewigem Leben vorzuziehen, sagt der Buddhist. Statt den Kosmos besitzen zu wollen, solle man versuchen, eins mit ihm zu werden. Und statt Göttlichkeit zu spielen, solle man danach streben, göttlich zu sein: „Nur wer weiß, daß er selbst Gott ist, sucht ihn nicht anderswo."

Kosmische mens: Een golf op zoek naar de oceaan

Het is wel iets merkwaardigs met ons mensen dat wij meer weten over galactische stelsels die miljoenen lichtjaren van ons verwijderd zijn dan over onszelf. Wie wij eigenlijk zijn en waar de reis naar toe moet gaan, daarop hebben onze voorouders in voorwetenschappelijke tijd nog met een zekere geloofsovertuiging weten te antwoorden. Met het verdwijnen van kosmische mythen en het verbleken van het christendom zijn dan evenwel essentiële identiteitspilaren brokkelig geworden - met het gevolg dat veel mensen zich tegenwoordig rijkelijk verloren voelen in een kosmos waarin zij zich niet meer als schepselen van God kunnen voelen, maar slechts als toevalsprodukten van ijskoude, onpersoonlijke natuurwetten. Hoe meer wij over het universum te weten komen, zegt de auteur Walker Percy, des te meer voelen wij ons erin verloren: "Door elke vooruitgang bij het objectief begrijpen van de kosmos en bij de technische controle ervan, raakt het eigen ik er een stuk verder van verwijderd, zodat het eigen wezen tenslotte in de kosmos die hij juist zo nauwkeurig heeft bestudeerd, een rondwarend spook wordt."

Hoe kan men nu het uiteenfladderende eigen ik een identiteit geven zodat het zich in het universum, waar het per slot van rekening bij hoort, thuisvoelen kan? Sommigen proberen het in hun nood met de oudste van alle methoden der ik-verankering in het Hier en Nu: In een workshop leren zij op sjamanistische wijze een beer, adelaar of prairiewolf te worden, hetgeen echter natuurlijk voor slechts korte tijd lukt, omdat er niet echt in het totemdier geloofd wordt. Ook een extatische rebirthing-ervaring "in Jezus Christus" kan verraderlijk zijn, als men later tot het inzicht komt dat er sinds de kindertijd niets veranderd is aan de paternalistische, van verre regerende god. En tot slot is er bijvoorbeeld altijd nog het goede oude "immanente eigen ik" van de consument in een wetenschappelijk gedefinieerde wereld dat wellicht een beetje te "transcenderen" is, als therapie voor het verlorenheidsgevoel in de kosmos. Bijvoorbeeld door zich te identificeren met "goddelijke" superstars en oermoeder-diva's, bevrijders en superhelden uit de "woestijnstorm". Of zelfs met "begenadigde" Nobelprijswinnaars in de natuurkunde en hun nieuwe theorie voor de wereldformule.

Zo'n imitatiespiritualiteit, zegt de godsdienstfilosoof Ken Wilber, overdekt slechts het diep en onbewust in alle mensen aanwezige verlangen naar eenheid met een goddelijke kosmos die hen een actief deel van deze goddelijkheid laat zijn. "Als deze wens op een dag spontaan naar buiten dringt, dan kan men

dat vergelijken met een oceaangolf die zich plotseling van zichzelf bewust wordt en ontdekt dat hij één is met de oceaan - en daardoor één met alle andere golven." Dit archetype van de goddelijke "kosmische mens" komt in de wereld in talrijke mythen en menige geloofsleer voor. Hij verschijnt als Adam, als de Perzische Gayomart of als hindoeïstische Purusha. Volgens Carl Jung was het geenszins Jezus Christus die de mythe van de goddelijke mens schiep (Weet gij dan niet dat gij goden zijt?. Deze symbolische idee bestond veeleer eeuwen vóór hem "en hijzelf werd door de idee van de kosmische mens gegrepen en uit het leven van een ambachtsman in Nazareth geheven." Terwijl veel elementen van het Christendom - vooral de existentie van een persoonlijke God - niet in overeenstemming te brengen zijn met het archetypische ideaal van de "kosmische mens", "past" het Boeddhistische wereldbeeld er welhaast perfect op. In de leer uit het Verre Oosten is er generlei god, schepper of almachtige figuur. Er is alleen een kosmos die door een soort goddelijk principe - het wezen van alles wat is - is doordrenkt. Als deel van deze kosmos is ook iedere mens goddelijk; zijn taak op aarde is het wezen van zijn eigen ik te zoeken - hetgeen tot het inzicht moet leiden dat er geen eigen ik is en dat alle wezens en objecten van het universum uitdrukking van dezelfde goddelijke realiteit zijn. Op de vraag "Wie ben jij?" kan een boeddhist antwoorden: "Ik ben Atman, ben Boeddha-natuur, dat wil zeggen, ik ben goddelijk in mij, maar ik ben ook Brahman, het goddelijke van de kosmos."
Omdat er in zijn universum geen grote vader God bestaat, kan en wil een boeddhist ook geen godheid om hulp, genade of bescherming smeken. Zijn doel is volkomenheid te bereiken, de illusie van het gescheiden zijn te ontvluchten, dat wil zeggen: één te worden met de kosmos, hetgeen in overeenstemming is met het ideaal van de kosmische mens. Een grote hindernis op de weg daarheen is de overwinning van het eigen wezen met zijn ijdelheden en begeerten. Na de dood wordt alle individualiteit opgeheven; er is geen ziel die verder bestaat. Tijdloze volkomenheid is boven eeuwig leven te verkiezen, zegt de boeddhist. In plaats van de kosmos te willen bezitten, moet men trachten één met hem te worden. En in plaats van goddelijkheid te spelen, moet men er naar streven goddelijk te zijn: "Alleen degene die weet dat hij zelf god is, zoekt hem niet elders."

Cosmic Man: A wave searching for the ocean

It is a strange thing that we humans know more about galaxies that are millions of light years away than we do about ourselves. Who we really are and where the journey is taking us; these were questions that our ancestors were, before the advent of science, able to answer with a certain degree of conviction. As the cosmic myths disappeared and Christianity paled, essential supports in our attempt to find our identity crumbled, with the result that today many people feel themselves to be very much lost in a cosmos in which they can no longer feel themselves to be God's creatures, viewing themselves instead as but the chance products of ice-cold, impersonal natural laws. The more we find out about the universe, says the writer Walker Percy, the more lost we feel in it: "Every advance in our objective understanding of the cosmos and in its technological control alienates the self a little more from it, so that in the end the ego becomes but a drifting ghost in that very cosmos that it has so very precisely researched. How therefore can one give the shaky ego an identity so that it can feel at home in the universe of which it is after all a part?
In their distress, many turn to the very oldest of methods of anchoring their ego in the here and now: they learn in a workshop, in the manner of the shamans, how to become a "bear", an "eagle" or a "coyote", something which is of course only temporarily possible because real faith in the totem animal is lacking. The ecstatic "born again in Jesus Christ" experience can also involve problems, when one recognizes later that not much has changed since childhood days in the paternalistic God ruling from a great distance. And finally there is also the good, old "immanent self" of the consumer in a scientifically defined world which can perhaps be "upgraded" a bit as a therapy against the feeling of being lost in the cosmos, for example by identifying with "divine" superstars and "first mother" divas, saviours of the state and "Desert Storm" super heroes or even with "gifted" Nobel prize winners for physics and their latest universal formulas. Such substitute-spirituality, states the religious philosopher Ken Wilber, only conceals the deep, unconscious need of all men to be united with a divine cosmos, a cosmos which enables them to be an active part of this divinity. "If this desire one day spontaneously comes to the fore then it can be compared in its effect to a wave in the ocean which has suddenly become aware of itself and discovers that it is at one with this ocean - and, as a result, united with all the other waves." This archetype of the divine "Cosmic Man" is to be found in numerous myths and religious teachings. It appears as Adam, as the Persian Gayomart or the Hindu figure of Purusha. According to Carl Jung, it was by no means Jesus Christ who created the myth of divine man ("Do you not know then that you are Gods?"). The symbolic idea existed thousands of years before him "and he himself was seized by the idea of Cosmic Man and elevated above the life of a craftsman in Nazareth." While many elements of Christianity - above all the existence of a personal god - cannot be reconciled with the archetypal ideal of Cosmic Man, the Buddhist world-view "suits" virtually perfectly. In this Far Eastern teaching there is no god, or creator or other all-powerful figure of any kind. There is only a cosmos which is imbued with a kind of divine principle - the "being of all that is". As a part of this cosmos, each individual is also divine; his task on Earth consists in seeking the nature of the Self - a search which should lead to the recognition that there is no Self and that all the beings and objects in the universe are but an expression of the same divine reality. In answer to the question "Who are you?" a Buddhist might answer: "I am Atman, I am of the nature of Buddha, that is, I am divine in my own right, but I am also Brahman, the divine element of the cosmos." Since there is no big father-god in his universe, the Buddhist cannot and does not wish to implore a god for help, mercy or protection. His goal is to attain to the Whole, to leave behind him the illusion of separation, that is to be one with the cosmos as befits the ideal of Cosmic Man. A big hurdle on the way is the overcoming of the ego and all its vanities and desires. After death all individuality ceases to exist; there is no goal that lives on.
A timeless whole is to be preferred to eternal life, the Buddhist says. Instead of wanting to possess the cosmos, one should try to become one with it. And instead of playing god, one should strive to be divine: "Only he who knows that he himself is God, does not seek him elsewhere."

L'interpretazione dei sogni, la cabala e la smorfia, la tartaruga assassina, E.T. e la Luna

Nell'intera opera "L'interpretazione dei Sogni" di Sigmund Freud non c'è uno straccio di sogno, uno che è uno, che abbia a veleggiare per il cosmo, ad ambientarsi fra i pianeti e costellazioni. E siccome la fonte è ineccepibile e la casistica analizzata è vasta, se ne possono trarre gli estremi per una indagine conoscitiva i cui risultati stabiliscono che l'umanità, nella sua vita onirica, lascia ben poco spazio alle stelle. Evidentemente preferisce contemplarle ad occhi aperti con piena soddisfazione di ogni stimolo, di ogni carenza, di ogni scontento interiore. Ciò non toglie che i vari libri sulla "Cabala", la "Enciclopedia popolare del Lotto", "La Smorfia" e simili, non prevedessero nei loro dizionari ad indicare i numeri corrispondenti alle varie voci riguardanti il cosmo: gli astri per primi, accompagnati da astronomi con telescopio ed astrologhi, astronomia ed astrologia; pianeti e stelle. E questo con testi editi prima ancora che l'uomo prendesse dimestichezza con lo spazio attraverso le iniziali imprese astronautiche. Evidentemente la fantasia popolare aveva più sbocchi rispetto alle persone sottoposte ad analisi.
Due situazioni differenziate, dunque, nelle quali evitiamo di inoltrarci non essendo quello dei sogni, l'argomento del nostro lavoro. È attraverso l'interesse diretto collegato alla astrologia che l'uomo ha avuto da sempre più vicina la sensazione del suo cosmo. Di essa, J. Collin De Plancy, nel "Dizionario Infernale" pubblicato nel 1818 con l'approvazione dell'Arcivescovo di Parigi, dice testualmente: "Arte di dire la buona ventura, di estrarre gli oroscopi, di predire gli avvenimenti, dall'aspetto, la posizione e l'influenza dei corpi celesti ... Il commissario Lamarre, nel suo "Trattato di Polizia" (tit. VII, cap. I) non respinge l'opinione che stabilisce che sia stata insegnata dal diavolo"; mentre degli astrologhi dice: "Si fanno vanto delle due o tre predizioni avveratesi, ancorché sovente in modo approssimativo, tra le mille che non hanno avuto successo. L'oroscopo del poeta Eschilo prediceva che egli sarebbe rimasto schiacciato dalla caduta di una casa: per sottrarsi al suo destino, si dice, il poeta si installò in mezzo ad un campo. Ma un'aquila che aveva rapito una tartaruga, gliela lasciò cadere sulla testa e fu ucciso."
Eppure l'astrologia non ha tanto prosperato come in questi ultimi decenni. Più della metà degli italiani conosce il proprio segno ed almeno una volta alla settimana legge l'oroscopo che gli appartiene: è persuaso che gli astri influiscano sui caratteri e sui temperamenti, la carriera, la salute e l'amore. Ne è scaturita un'industria sulla quale tenta di veder chiaro, anche a tutela dei numerosi utenti, il Comitato italiano per il controllo delle affermazioni sul paranormale (Cicap), ostacolato, però, nel suo lavoro da scogliere di convinzioni radicate. Quali per esempio? La solitudine, la nullità dell'uomo rispetto al cosmo. Il misticismo che lo porta a cercare corrispondenze importanti sopra la sua testa. "Viviamo tutti in una foresta di simboli", diceva infatti Baudelaire; e Shakespeare, per bocca di un suo personaggio: "Orazio, vi sono molte più cose nel cielo e nella terra che non ne può immaginare tutta la nostra filosofia". Sono queste che cercano gli ufologi? Indubbiamente, perché gli "oggetti non identificati" potrebbero essere i ricognitori, le staffette di altre comunità viventi nel cosmo. L'Aeronautica militare italiana ha dato una grossa mano in questa ricerca ed i frutti sono riassunti in questo schematico dato: 111 Ufo segnalati nel periodo compreso tra il 1979 ed il 1990. Si tratta di avvistamenti per così dire ufficiali, con data, ora e forma. La quale era quadrata, triangolare, cilindrica, irrispettosa cioè di quella classica del vero e proprio disco volante. Questo, il cosmo che popola il cielo e la distratta fantasia dell'uomo moderno, pronta però a risvegliarsi di fronte all'evenienza, al fatto eccezionale. Così che nell'anno 1980 l'Aeronautica ha incamerato 32 segnalazioni, mentre gli studiosi del CISU, Centro Italiano di Studi Ufologi 381. Un record irripetuto, quasi a proporre l'incantevole presenza di tanti E.T. fra noi. Meno romantico, tuttavia, di quello della Luna con Endimione, raccontato da Luciano nei "Dialoghi". Ad Afrodite che le domandava se era vero che fermasse la sua quadriga per scendere ad ammirare il ragazzo, la Luna rispondeva: "Non c'è cosa al mondo più bella di lui. Soprattutto quando, fattosi letto della clamide sulla pietra, dorme, e appena trattiene le frecce che sembra gli stiano per sfuggir di mano: la destra mollemente appoggiata sul suo capo, inquadra graziosamente il suo volto; e in questo abbandono del sonno respira con un dolcissimo alito d'ambrosia. E allora io scendo senza far rumore, camminando sulla punta dei piedi per non turbarlo nel sonno.

Giuseppe Brunamontini

L'homme cosmique: une vague à la recherche d'un océan

Il est quand même curieux que nous en sachions beaucoup plus sur des galaxies éloignées de millions d'années-lumière que sur nous-mêmes. Nos ancêtres des temps préscientifiques savaient encore répondre avec une certaine conviction religieuse à la question de savoir qui nous sommes et où ce voyage nous conduit. Quand les mythes cosmiques disparurent et que le christianisme s'estompa, d'essentiels piliers de l'identité humaine s'effritèrent; en conséquence, beaucoup de nos contemporains se sentent absolument perdus dans un cosmos où ils ne peuvent plus se considérer comme des créatures de Dieu mais uniquement comme des produits accidentels de lois de la nature froides et impersonnelles. Plus nous en découvrons sur l'Univers, dit l'écrivain Walker Percy, plus nous nous y sentons perdus: "Chaque progrès dans la compréhension objective du cosmos et dans son contrôle technique le sépare encore plus du Moi, de sorte qu'à la fin, le Moi se transforme en spectre errant dans chaque cosmos qu'il a exploré avec exactitude". Comment donc donner une identité au moi se dispersant au vent pour qu'il puisse se sentir chez lui dans un Univers auquel il appartient bien? Dans leur détresse, certains essaient la plus ancienne de toutes les méthodes, l'ancrage du Moi ici et maintenant; ils apprennent en groupe à devenir "ours", "aigle" ou "coyote" à la manière chamaniste, ce qui, bien sûr, ne réussit que pour une très courte durée puisque la vraie foi dans l'animal totémique leur manque. L'expérience d'une renaissance extatique "en Jésus Christ" peut aussi se révéler perfide quand on s'aperçoit plus tard que le Dieu paternaliste qui règne de si loin n'a guère changé depuis notre enfance. Et puis finalement, il y a toujours le bon vieux "Moi immanent" du consommateur dans un monde défini par la science, qui se laisse peut-être un peu "transcender" pour servir de thérapie contre le sentiment d'abandon dans le cosmos. Par exemple, par la voie de l'identification avec des superstars "divines", des divas nommées "Eve", des sauveurs nationaux et des superhéros "tempêtes du désert" ou même avec des lauréats "surdoués" du Prix Nobel de physique et leur dernière théorie de la formule universelle. Selon le philosophe des religions Ken Wilber, une telle spiritualité de substitution masque uniquement le besoin, inconsciemment présent au plus profond de l'homme, d'unité avec un cosmos divin qui fait d'eux une partie active de cette nature divine. "Quand ce désir surgit un jour spontanément, il est comparable à une vague marine qui prend soudain conscience d'elle-même et découvre qu'elle ne fait qu'un avec l'océan et donc

Космический человек: волна, которая ищет океан

aussi un avec toutes les autres vagues". Cet archétype de l'"homme cosmique" divin est présent dans de nombreux mythes et doctrines religieuses. Il apparaît en la personne d'Adam, du Gayomart perse ou du Purusha hindou. Selon Carl Jung, ce ne fut pas Jésus Christ qui créa le mythe de l'homme divin (Ne savez-vous pas que vous êtes des Dieux?). L'idée symbolique existait déjà bien des siècles avant "et lui-même fut saisi par l'idée de l'homme cosmique et quitta la vie d'artisan de Nazareth". Alors que de nombreux éléments du Christia-nisme, en particulier l'existence d'un Dieu per-sonnel, ne peuvent concorder avec l'idéal archétypal de l'"homme cosmique", la vision bouddhiste du monde y correspond parfaitement. Dans la doctrine d'Extrême-Orient, il n'existe aucun Dieu, Créateur ou autre personnage tout-puissant. Il n'y a qu'un cosmos imprégné d'une sorte de principe divin, de l' "être de tout ce qui est". En tant que partie du cosmos, chaque être humain est aussi divin; sa tâche sur Terre est de chercher la nature du Moi, ce qui doit le conduire à la reconnaissance qu'il n'y a pas de Moi et que tous les êtres et objets de l'Univers sont l'expression de la même réalité divine. A la question "Qui es-tu?", un bouddhiste pourrait répondre: "Je suis Atman, je suis la nature de Bouddha, autrement dit ce qui est en moi est divin; mais je suis aussi Brahman, le Divin du Cosmos". Puisqu'il n'y a pas de grand Dieu le Père dans son Univers, un Bouddhiste ne veut pas non plus implorer l'aide, la grâce ou la protection d'une divinité. Son but est d'atteindre l'intégralité, d'échapper à l'illusion de la séparation, bref, de ne faire qu'un avec le cosmos, ce qui correspond à l'idéal de l'homme cosmique. Surmonter le Moi, sa vanité et ses désirs, est un grand obstacle sur le chemin qui y mène. Après la mort, toute individualité se désagrège; il n'y a pas d'âme qui survive. L'intégralité intemporelle est préférable à la vie éternelle, dit le Bouddhiste. Au lieu de vouloir posséder le cosmos, il faut essayer de ne plus faire qu'un avec lui.

Et au lieu de jouer à la divinité, il faut s'efforcer d'être divin: "Seul celui qui sait qu'il est lui-même Dieu ne le cherche pas autre part".

Странные все же мы, люди, существа: о галактиках, отстоящих от нас на миллионы световых лет, мы знаем больше, чем о самих себе. На вопрос о том, кто мы такие и куда ведет наш путь, предки в донаучную эпоху еще могли ответить по своей вере. Однако с исчезновением космических мифов и ослаблением христианства рухнула важнейшая опора нашего самосознания; в результате многие чувствуют себя затерянными в космосе, не творениями Божиими, а случайными продуктами бездушных и безликих законов природы. Писатель У. Перси говорит, что чем больше мы о Вселенной знаем, тем потеряннее себя в ней чувствуем: «Любой прогресс в объективном понимании космоса и в технических средствах наблюдения удаляет личность на шаг от него, так что в конце концов Я превращается в блуждающее привидение в том самом космосе, который оно так досконально изучило». Как же вернуть личности самосознание, чтобы она почувствовала себя уютно во Вселенной, которой, собственно, и принадлежит? Некоторые пытаются помочь беде древнейшим методом конкретного фиксирования своего «Я»: в каком-нибудь кружке они учатся на шаманский манер «превращаться» в медведя, орла или койота, что удается, конечно, на очень короткий срок, потому что подлинная вера в тотем отсутствует. Коварным может оказаться и экстатический опыт «возрождения в Иисусе Христе», когда человек убеждается, что образ Бога, отечески управляющего миром из далекого далека, со времен собственного детства мало изменился. И, наконец, есть еще доброе старое «внутреннее Я» потребителя в научно определенном мире, которое можно попытаться слегка «трансцендировать», чтобы снять чувство потерянности в космосе, например, путем отождествления себя с «божественными супер-звездами» и оперно-театральными дивами, спасителями государства и героями «Бури в пустыне» или даже нобелевскими лауреатами по физике с их новейшей теорией «универсальной формулы». Религиозный философ К. Вайлбер говорит, что такая эрзац-духовность лишь скрывает глубокую и неосознанную потребность единенения с божественным космосом, который дает ей возможность быть активной частью этого божества. «Когда это желание спонтанно прорывается, его можно сравнить с морской волной, которая вдруг осознает, что она принадлежит океану и родственна всем другим волнам.» Этот архетип божественного «космического человека» представлен в многочисленных мифах и религиозных учениях. Он встречается как Адам, персидский Гайомарт, индуистский Пуруша. По мнению К. Юнга, отнюдь не Иисус Христос создал миф о божественном человеке («Разве вы не знаете, что вы боги?»). Эта символическая идея существовала за столетия до него, «им завладела идея космического человека и подняла его над жизнью назаретского ремесленника». Если многие элементы христианства, прежде всего существование личного Бога, нельзя совместить с архетипическим идеалом «космического человека», то, например, буддистская картина мира полностью укладывается в него. В этом учении нет Бога, Творца или иного всесильного существа. Есть только один космос, который пронизан божественным принципом, «сущностью всего, что есть». Будучи частью этого космоса, божественен каждый человек, а его задача на земле — поиски своего «я», которые должны привести к пониманию того, что никакого «я» нет и что все существа и объекты Вселенной являются выражением одной и той же божественной реальности. На вопрос: «Кто ты?» буддист мог бы ответить: «Я атман, от природы Будды, это значит, я божественен в себе самом, но я еще и брахман, божественность космоса.» Поскольку в его космосе нет Бога Отца, буддист не может и не хочет молить никакое божество о помощи, милости и защите. Его цель — достичь целостности, уйти от иллюзии разделенности, то есть слиться с космосом, что отвечает идеалу космического человека. Важнейшая задача на этом пути — преодоление своего «я» с его суетностью и страстями. После смерти всякая индивидуальность исчезает; нет никакой души, которая продолжала бы жить. Буддист скажет, что следует предпочесть безвременную целостность вечной жизни. Надо стремиться не к обладанию космосом, а к слиянию с ним. И не изображать божество, а стараться быть божественным: «Только тот, кто знает, что он сам бог, не ищет его нигде.»

SACRAL

COSMOS

Zu jener Zeit, als Moses und sein Volk aus Ägypten auszogen und dann Gott auf dem Berg Sinai seine Botschaft verkündete, galten noch unzählige Punkte der Landschaft als heilig und als Orte spiritueller Kraft. Beseelte Materie gehörte zu Raum und Zeit; man redete mit den Sternen und lauschte den Planeten.

Destijds toen Mozes met zijn volk uit Egypte ging en God op de berg Sinaï zijn boodschap verkondigde, werden nog talloze punten van het landschap voor heilige plaatsen van spirituele kracht gehouden. Bezielde materie behoorde tot ruimte en tijd; er werd met de sterren gepraat en naar de planeten geluisterd.

At the time when Moses and his People left Egypt and God made known his message on Mount Sinai many points in the landscape were considered to be holy and locations of special spiritual power. Spiritual matter was a part of space and time; people talked with the stars and listened carefully to the planets.

Quando Mosè ed il suo popolo abbandonarono l'Egitto e Dio, sul monte Sinai, proclamò il suo messaggio, innumerevoli punti della terra erano considerati sacri ed erano luoghi di forza spirituale. La materia animata apparteneva allo spazio ed al tempo, si parlava con le stelle e si ascoltava il suono dei pianeti.

En ces temps lointains où Moïse et son peuple quittèrent l'Egypte et où Dieu dicta ses lois sur le Mont Sinaï, beaucoup d'endroits de cette région étaient considérés comme sacrés, renfermant une force spirituelle. La matière animée faisait partie de l'espace et du temps: on parlait avec les étoiles et écoutait les planètes.

К тому времени, когда Моисей и его народ вышли из Египта и Бог на Синае дал заповеди, многие места считались священными, исполненными духовной силы. Одухотворенная материя жила во времени и пространстве, люди говорили со звездами и слушали язык планет.

Die schönste Emotion ist die mystische. Zu wissen, daß das, was für uns unergründlich ist, wirklich existiert und sich als die höchste Weisheit und die strahlendste Schönheit offenbart, - dieses Wissen, dieses Gefühl ist der Kernpunkt wahrer Religiosität.

De mooiste emotie die wij kunnen ervaren, is de mystieke. Te weten dat hetgeen voor ons ondoorgrondelijk is, werkelijk bestaat en zich openbaart als de hoogste wijsheid en de stralendste schoonheid - dit gevoel is de kern van ware religiositeit.

The most beautiful emotion we can experience is of a mystical nature. To know that what is impenetrable to us really exists, manifesting itself as the highest wisdom and the most radiant beauty - this feeling, is at the center of true religiousness.

L'emozione più bella che ci è concesso è quella mistica. Sapere che ciò che per noi è impenetrabile esiste realmente e che si manifesta come la saggezza più alta e la bellezza più splendente - questa nozione, questo sentimento è al centro della vera religiosità.

La plus belle émotion dont nous puissions faire l'expérience est de nature mystique. Savoir que l'impénétrable existe vraiment et se manifeste sous forme de sagesse suprême et de beauté radieuse - ce savoir, ce sentiment est le centre de la véritable religiosité.

Самая приятная эмоция — мистическая. Знание, что нечто для нас непознаваемое действительно существует и проявляется как высшая мудрость и красота, — это знание, это чувство есть зерно истинной религиозности.

Albert Einstein

Stuart Piggott

„Es ist Mode geworden, unsere Vorfahren des Neolithikums und der Bronzezeit als geradezu wissenschaftlich orientierte Himmelsbeobachter zu interpretieren, die Steinkreise, Megalithen und andere Steinobjekte in raffinierter Weise als Observatorien benutzten. Ich halte das für Wunschdenken."

"Het is mode geworden onze voorouders van het Neolithicum en bronzen tijdperk te interpreteren als wetenschappelijk georiënteerde hemelobservatoren die stenen cirkels, megalieten en andere stenen voorwerpen op geraffineerde wijze als observatoria gebruikten. Ik vind dat wishful thinking.

"It has become fashion to interpret our Neolithic and Bronze Age ancestors as scientifically-orientated observers of the heavens who used stone circles, megaliths and other stone objects as observatories. This is wishful thinking. Everything points to the fact that ecstatic rituals played a major role."

È di moda pensare che i nostri antenati del Neolitico e della Età del Bronzo fossero degli osservatori del cielo che seguivano un orientamento scientifico usando cromlech, megaliti ed altri oggetti di pietra come sofisticati osservatori. Questa visione appartiene più ad un sogno che alla realtà.

On aime considérer nos ancêtres du néolithique et de l'âge du bronze comme de savants observateurs du ciel auxquels des cercles en pierre, des mégalithes et autres objets de pierre servaient d'observatoires. Une chimère à mon avis. Tout semble indiquer le rôle essentiel de leurs rituels extatiques.

«Ныне модно считать наших предков прямо-таки учеными-астрономами, использовавшими каменные круги, мегалиты и прочее как обсерватории. Я называю это выдавать желаемое за действительное. Все же самым главным были ритуалы.

Bis vor kurzem für astronomisch relevant gehalten, gelten die Figuren und Linien von Nazca, Peru, heute als „heilige Pfade".

Vandaag beschouwt men de figuren en lijnen van Nazca, die men nog niet lang geleden voor astronomisch relevant hield, als "heilige paden".

Considered to be astronomically relevant up until recently, the figures and lines of Nazca in Peru are now thought to be "holy paths".

Ritenute fino a poco tempo fa rilevanti per l'astronomia, oggi le figure e linee di Nazca, in Peru, non sono considerate che "sentieri sacri".

Autrefois étudiés par les astronomes, les dessins et lignes de Nazca au Pérou sont aujourd'hui considérés comme des "sentiers sacrés".

Фигуры и линии в Наске/Перу, считавшиеся астрономическими объектами, теперь считают культовыми местами.

NORTH
Denton
30
EAST
Texarkana
TO
SOUTH
Industrial Blvd

Heilige Orte: Wie im Himmel, so auf Erden

Schlange-Jaguar schmeckte den salzigen Schweiß, der in seine Mundwinkel sickerte. Trotz der Kühle in der Tiefe des heiligen Berges war ihm heiß von den anstrengenden Ritualen der vergangenen Stunden, und er spürte, daß das dreitägige Fasten stärker an seinen Kräften gezehrt hatte als ihm lieb war. Für einen Moment lehnte er sich zurück gegen die Mauer der Grabkammer und schloß die Augen. Dann riß er sich hoch aus seiner Erschöpfung und gab den Priestern das Zeichen, daß er bereit war zur abschließenden Zeremonie für seinen Vater: Das Blutopfer von fünf Sklaven, um die Götter günstig zu stimmen für seine Wiedergeburt. Schlange-Jaguars Vater, der große Gottkönig Pacal, war wenige Tage zuvor gestorben, und nun lag es an ihm, seinem Sohn, das Maya-Volk von Palenque in Einklang mit dem Willen von Sonne und Venus zu führen - in Einklang mit den Gestirnen, als deren direkte Abkömmlinge sich die Herrscherfamilie betrachtete und deren himmlischen Willen sie von den Astronomen-Schamanen deuten ließen. Der verstorbene Pacal, das wußten alle Mayas, war als Zeichen seiner Göttlichkeit am gleichen Tag zur Welt gekommen wie die Sonne vor 5000 Jahren. Keine Zweifel am Wohlwollen der Götter hatte Schlange-Jaguar denn auch drei Monate nach der Verabschiedung seines Vaters während der Feiern zu seiner Inthronisation. Man schrieb den 20. Januar des Jahres 690, und für jedermann klar zu sehen flammten vor dem sich verdunkelnden Himmel drei Planeten auf, in göttlicher Dreieinigkeit knapp nebeneinander in seltener Konstellation: Saturn im Osten, Jupiter in der Mitte, Mars im Westen. Selbstbewußt und zuversichtlich trat der neue Gottkönig in dieser Nacht hoch oben auf der Sonnenpyramide vor sein Volk und brachte dem Kosmos das größtmögliche und begehrenswerteste Opfer dar: Blut, das in ein Gefäß herabfloß, als er seine Penishaut mehrmals mit einem Rosenstachel durchbohrte. Für die Maya-Indianer wie für andere Kulturen der Vergangenheit war die Welt ein noch viel lebendigerer Ort als er es für uns Heutige ist. Heilige Kraft, spirituelle Präsenz, beseelte Materie waren überall Teil der Landschaft und gehörten selbstverständlich zu Raum und Zeit. Erde und Himmel waren noch nicht voneinander getrennt, die Gestirne kehrten noch in ihre irdische Heimstatt zurück. Und die Menschen redeten mit den Sternen, sie lauschten den Planeten. Vor allen aber beobachteten sie die Gestirne und lasen aus jedem Knick und jeder Wendung in den Planetenbahnen eine Botschaft heraus. Vom Urwalddorf in den Tropen bis zum Städtchen in den Alpen projizierten sie dabei ihre Gesellschaft, ihr Leben, Ihre Hoffnungen und Ängste in das Geschehen des Himmels - und entdeckten dann - wen kann es wundern - eine Welt, die ihnen überraschend vertraut vorkam. Je nach Lebensform, also etwa Jäger, Ackerbauer oder Viehzüchter, konnte dann die Sonne feindlich oder freundlich gesinnt erscheinen. Außerdem verhielten sich dieselben Planeten je nach Betrachter „typisch weiblich", „typisch männlich" oder sogar satanisch bösartig. Himmelskörper schienen oft unberechenbar zu sein, Stimmungen unterworfen, schwankten sie doch einmal hierhin, dann dorthin, hielten an, bewegten sich rückwärts, um dann wieder doppelt so schnell vorwärts zu eilen. Ein Maya-Gottkönig wie Chan-Bahlum (Schlangen-Jaguar) hatte wohl so etwas wie die Funktion eines Vermittlers zwischen diesem Oben und dem „Unten" des Volkes. Seine rituellen Handlungen liessen die Kraft des Übernatürlichen in das Leben der normal Sterblichen überfließen und sie gaben den Widersprüchen des Himmelsgeschehens eine klare Bedeutung. Opfergaben dienten dazu, Verbindungen herzustellen zwischen unten und oben, zwischen mächtiger Natur und den weit unterlegenen Menschen; die Opfer garantierten die Wiederkehr der zyklischen Lebensvorgänge, sie ließen die Menschen aktiv teilhaben an den kreisförmigen Bewegungen der Natur.

Uns heutige Moderne verlangt es ganz gewiß nicht weniger nach Kontakt mit dem restlichen Teil der Schöpfung, wenn nicht sogar nach Einssein mit ihr - ob wir uns das nun bewußt eingestehen oder nicht. Statt eines Tempels, Schreins oder einer Kirche, suchen wir eher ein Stück unberührter Natur auf, um Verbindung herzustellen. Wir wollen Gefühlsbrücken bauen, wollen uns auch irgendwie verpflichten, indem wir vielleicht einem Baum oder einem Berg eine Feder oder einen bunten Stein hinlegen. Eine symbolhafte Geste, sicherlich, bei der es vielleicht darum geht, sich das gemeinsame Schicksal mit dem Kosmos bewußt zu machen. Carl Jung merkte vor vielen Jahren einmal an, daß uns Menschen manche Steine in der Natur vielleicht deshalb so faszinieren, weil sie unserem „innersten Kern auf seltsame Weise verwandt sind." In diesem Sinn symbolisiert der Stein die „einfachste und tiefste Erfahrung - die Erfahrung von etwas Ewigem."

Heilige Plaatsen: In de Hemel, zo ook op Aarde

Slang-Jaguar proefde het zoute zweet wat in zijn mondhoeken sijpelde. Ondanks de koelte in de diepte van de heilige berg had hij het warm gekregen door de inspannende rituelen van de afgelopen uren. Hij voelde dat de drie vastendagen zijn krachten meer gesloopt hadden dan hem lief was. Een moment lang leunde hij tegen de muur van de grafkamer en sloot zijn ogen. Toen verzamelde hij zijn laatste krachten en gaf de priesters het teken dat hij bereid was voor de laatste ceremonie voor zijn vader: het bloedoffer van vijf slaven om de goden gunstig te stemmen voor zijn wedergeboorte. De vader van Slang-Jaguar was enkele dagen tevoren gestorven en nu was het aan hem, zijn zoon, het Mayavolk van Palenque te leiden in overeenstemming met de wil van Zon en Venus - in overeenstemming met de hemellichamen, waarvan de heersende familie zich als directe afstammeling beschouwde en wier hemelse wil zij door astronoom-sjamanen lieten uitleggen. De overleden Pacal - dat wisten alle Maya's - was als teken van zijn goddelijkheid op dezelfde dag ter wereld gekomen als de Zon 5000 jaar geleden. Slang-Jaguar had tijdens de plechtigheden ter gelegenheid van zijn inhuldiging drie maanden na het afscheid van zijn vader dan ook geen twijfel aan de gunstige gezindheid van de goden.

Men schreef 20 januari van het jaar 690 en voor iedereen duidelijk zichtbaar vlamden voor de verduisterende hemel drie planeten op, in goddelijke drieënigheid vlak naast elkaar in een zelden voorkomende constellatie: Saturnus in het oosten, Jupiter in het midden, Mars in het westen. Zelfbewust en vol vertrouwen trad de nieuwe godkoning in deze nacht hoog boven op de zonnepiramide voor zijn volk en bracht de kosmos het grootste mogelijke en meest begerenswaardige offer: bloed dat in een schaal stroomde, toen hij zijn penishuid meermaals met de doorn van een roos doorboorde. Voor de Maya-Indianen was de wereld net als voor andere culturen uit het verleden, nog veel levendiger dan voor ons mensen die vandaag leven. Heilige kracht, spirituele presentie, bezielde materie waren overal deel van het landschap en behoorden vanzelfsprekend bij ruimte en tijd. Aarde en hemel waren nog niet van elkaar gescheiden, de hemellichamen keerden 's nachts nog in hun aardse onderkomen terug. En de mensen spraken met de sterren, zij luisterden naar de planeten. Maar vooral observeerden zij de hemellichamen en lazen in elke knik en iedere wending in de planeetbaan een boodschap. Of het oerwouddorp in de tropen of het stadje in de Alpen, zij projecteerden daarbij hun samenleving, hun leven, hoop en

Holy places: As in Heaven, so on Earth

vrees in het hemelse gebeuren en ontdekten dan - wie kan het verwonderen - een wereld die hun verrassend vertrouwd voorkwam. Al naar levensvorm, dus bijvoorbeeld jagers, akkerbouwers of veefokkers, kon de zon vijandig of vriendelijk gezind verschijnen. En al naar observeerder gedroegen dezelfde planeten zich "typisch vrouwelijk", "typisch mannelijk" of zelfs duivels boosaardig. Hemellichamen leken dikwijls onberekenbaar, aan stemmingen onderhevig te zijn, want soms schommelden zij hierheen en dan daarheen, bleven staan, gingen terug om vervolgens weer dubbel zo snel vooruit te snellen. Een Maya-godkoning als Chan-Bahlum (Slang-Jaguar) had zoiets als de functie van een bemiddelaar tussen dit Boven en het Onder van zijn volk. Zijn rituele handelingen lieten de kracht van het bovennatuurlijke in het leven van de gewone stervelingen overvloeien en zij gaven een duidelijke verklaring voor de tegenstrijdigheden van het hemelse gebeuren. Offergaven dienden voor het tot stand brengen van verbindingen tussen onder en boven, tussen de machtige natuur en de veel zwakkere mensen; de offers garandeerden de terugkeer van de cyclische levensprocessen, zij lieten de mensen actief aan de kringbewegingen van de natuur deelnemen.
Wij hedendaagse, moderne mensen verlangen zeer zeker niet minder naar contact met het overige deel van de schepping, wanneer niet zelfs naar het één-zijn ermee - of wij dat nu bewust toegeven of niet. In plaats van een tempel, schrijn of kerk zoeken wij eerder een stukje ongerepte natuur op om contact tot stand te brengen. Wij willen gevoelsbruggen bouwen, willen ons op de een of andere manier ook verplichten, door misschien voor een boom of een berg een veer of een gekleurde steen neer te leggen. Een symbolisch gebaar, zeker, dat wellicht bedoeld is voor de bewustwording van het lot dat met de kosmos wordt gedeeld. Carl Jung maakte vele jaren geleden eens de opmerking dat wij mensen door sommige stenen in de natuur misschien, omdat zij "met onze diepste kern op een wonderlijke wijze verwant zijn" zo gefascineerd zijn. In deze zin symboliseert de steen de "eenvoudigste en innigste ervaring - de ervaring van iets eeuwigs".

Snake-Jaguar tasted the salty sweat which trickled into the corner of his mouth. In spite of the cool temperature in the depth of the Holy Mountain he was hot, following the strenuous rituals of the last few hours and he noticed that the three days of fasting had sapped his strength more than he would have liked. For a moment he leaned back against the wall of the burial chamber and closed his eyes. Then he jerked himself out of his exhaustion and gave the priests a sign that he was ready for the final ceremony for his father: the blood sacrifice of five slaves in order to win the favour of the gods for his rebirth. Snake-Jaguar's father, the great god-king Pacal, had died just a few days previous and now it was up to his son to lead the Maya people of Palenque in accordance with the will of the Sun and Venus, in accordance with the stars, of which the ruling family looked upon itself as the direct descendants and whose heavenly will they had astronomer-shamans interpret. Pacal, as all the Mayas knew, had, as a sign of his deity, been born on the same day as the Sun, 5,000 years ago. And three months after taking leave of his father, Snake-Jaguar left no doubt about the goodwill of the gods in the celebrations on the occasion of his en-thronement. It was 20 January of the year 690 and as everyone could clearly see three planets lit up against the darkening sky, in a divine triunity, close together in a rare constellation: Saturn in the east, Jupiter in the centre, Mars in the west. Self-confident and assured, the new god-king appeared high up on the Sun Pyramid before his people and made the greatest possible and most desirable sacrifice: blood, blood that flowed into a vessel when he bored a rose thorn through the skin of his penis several times. For the Maya Indians as well as for other civilizations in the past, the world was much more alive than it is for us today. Holy forces, spiritual presence, and psychic matter were part of the landscape and a natural part of space and time. Heaven and Earth were not yet separate from each other; the celestial bodies still returned at night to their earthly home. And people talked with the stars and listened to what the planets had to say. And, above all, they observed the stars and the heavenly bodies and interpreted out of each twist and turn in the movenents of the planets a message. From the village in the tropical forest to the little town in the Alps, people projected their society, their life, their hopes and fear into celestial events and discovered in the process - no wonder - a world which seemed to them to be surprisingly familiar. Depending on their life-style, be it that of the hunter, the farmer or the cattle breeder, the sun could appear to them as friend or foe. And one and the same planet could, depending on the viewpoint involved, behave in a "typically female" or "typically male" or even satanic fashion. Celestial bodies often seemed to be unpredictable, appeared to be moody; they swayed in one direction and then in the other, they stopped, they moved backwards and then hurried forwards at twice their previous speed. A Maya god-king like Chan-Bahlum (Snake-Jaguar) probably functioned as a kind of intermediary between "above" and "below". His ritual acts allowed the power of the super-natural to flow over into the life of the ordinary man and they gave a clear meaning to the celestial contradictions. Sacrificial offerings served to create links between above and below, between the mighty forces of nature and the greatly inferior humans; the offerings guaranteed the return of vital cycles of nature, they enabled people to play an active part in the circular movements of nature.
We, the inhabitants of the modern world, doubtless also yearn just as much for contact with the rest of creation, and even for unity with it, whether we consciously admit this or not. Instead of a temple, a shrine or a church, we tend to seek out a spot of untouched nature in order to establish a link. We want to build emotional bridges, we want to pledge ourselves somehow by perhaps laying a feather or a brightly coloured stone under a tree or at the foot of a mountain. A symbolic gesture certainly where it is a matter of demonstrating one common fate with the cosmos. Many years ago, Carl Jung observed that stones perhaps fascinate us so much because they are "in a strange way related to our innermost core". Seen in this light, the stone symbolizes the "simplest and the deepest experience - the experience of something eternal".

L'interpretazione di Michelangelo, Le Voyage dans la Lune, il pianeta Kripton, musica dalle galassie

Adesso che la Cappella Sistina è stata restaurata, erompe più forte che mai la pittura di Michelangelo. Una pittura che dà conto nella sua collocazione architettonica, costituita di finti archi marmorei e da vani dai diversi sfondi, dei fatti della creazione. Che riguardano il cosmo, soprattutto, e l'universo, da quando il Creatore, come racconta la Genesi, separò la luce dal caos delle tenebre, creò gli astri, per poi occuparsi direttamente del pianeta Terra disciplinandone le acque, creando Adamo e poi Eva, e centrando stupendamente la scena del Peccato e la successiva cacciata dal Paradiso.
Michelangelo, chiamato Michelagnolo nel suo tempo (1475-1560), Buonarroti per l'anagrafe ed anche per la storia dell'arte, iniziò molto giovane quest'opera immensa, poco più di 33 anni e ve ne impiegò quattro per portarla a termine. Ebbene, in questo affresco, sul cui unico valore è superfluo qui soffermarsi, dà consapevole conto di quanto avvenne all'inizio del tempo quasi avesse competenza e tavolozza di preastronautica. E su tali strutture, poi, ha esteso i colori del suo temperamento di immortale artista. Vivissimi, che il restauro ha riportato alla vera luce.
Naturalmente altre esperienze eccezionali ci sono state, ma l'idea che l'uomo può farsi del cosmo nell'arte, dall'origine al diluvio universale, può largamente nascere dalla vista della Cappella Sistina. Questo, senza togliere valore documentario a statuette, affreschi e graffiti di cui è cosparsa la terra. Antichità stupefacenti, dati i temi trattati, che non potevano avere ragionevoli riscontri reali se non nell'immaginario umano, oppure in evenienze inspiegabili. Infatti misteriosissime appaiono le statuette di bronzo "dogu" venute alla luce da scavi archeologici in Giappone. La più alta misura 60 centimetri e tutte stupiscono per la identificazione precisa delle tute spaziali e degli occhialoni curvi della nostra epoca e non risalenti a 600 anni avanti Cristo. Possibile che gli autori scolpissero qualcosa di già visto? O è solo frutto dell'idea magica del cosmo che accende profetiche fantasie?
Sulla stessa tastiera, possono essere collocati gli affreschi del chiostro di Desani, nel Kosovo-Metohija in Jugoslavia. Essi sono stati scoperti nel 1964, ma risalgono al XIV secolo. I soggetti riguardano navi spaziali nel cielo di una scena religiosa, comprendente angeli probabilmente terrorizzati che si coprono occhi e orecchi. La panoramica è necessariamente veloce, ma i temi da toccare non sono pochi poiché il cosmo nell'idea dell'uomo ha seguito le varie piste dell'arte compresa la settima. Il cinema, esattamente, che contemporaneamente alla sua nascita realizzò il primo film di fantascienza. Siamo nell'anno 1900 ed il pubblico poté assistere, non solo alla proiezione di uno dei primi film della storia del cinema, ma a "Le Voyage dans la Lune" e la fortuna vuole che è una delle pochissime pellicole conservate della produzione di allora. Ebbe origine dalla spinta dei palloni che già si levavano nel cielo e da due romanzi famosi di Jules Verne e di H. G. Wells. Ecco come George Sadoul, nella "Storia del Cinema" ne riassume la trama: "Degli astronomi a consesso, vestiti da astrologhi, decidono di fare un viaggio nella luna. Visitano un'officina piena di macchine ingenue e complicate. Dall'alto di un tetto assistono alla fusione del cannone. Delle belle marinaie dalle forme opulente recano il proiettile, nel quale prendono posto gli esploratori, il cannone è caricato. Fuoco!"
E poi c'è l'arrivo e la permanenza sulla luna con tutte le avventure e le visioni che si succedono, fino alla fuga con il paracadute. Un film di trascinante interesse a guardarlo con gli occhi di allora e che oggi si può tranquillamente continuare a giudicare un successo, considerando le elementari tecnologie di ripresa e le attrezzature di una semplicità al limite dell'impossibile. E fu l'inizio di una produzione cinematografica infinita, molto più intensa in questi ultimi decenni, collocata nel cosmo. Cosmo, e siamo nel mondo dei fumetti, dal quale arrivò il mitico Superman. Il quale non è un terrestre, ma approdò qui, ancora fanciullo, dal pianeta Kripton. "Kripton - racconta Umberto Eco - stava per essere distrutto da una catastrofe cosmica ed il padre di Superman, abile scienziato, era riuscito a mettere in salvo il proprio figlio, affidandolo ad un veicolo spaziale."
La musica, infine, per cercare di concludere questo nostro libro con il pieno dell'orchestra; ma di una nuova armonia: quella ascoltata nel Planetario dell'Osservatorio di Los Angeles Griffith Park. Proveniva dalle stelle, per un pubblico rapito. A comporla e dirigerla alla fine del 1991, era l'astrofisica e musicista italiana Fiorenza Terenzi. Titolo: "Musica dalle Galassie". Informa l'autrice: "Nel comporre la musica proveniente dalle emissioni radio della galassia Ugc 6697 ho cercato di rispettarne le tonalità prevalenti ...".

Giuseppe Brunamontini

Endroits sacrés: sur la Terre comme au Ciel

Serpent-Jaguar sentait le goût salé de la sueur qui perlait aux coins de sa bouche. Malgré la fraîcheur dans les profondeurs de la Montagne Sacrée, les rituels harassants des dernières heures lui avaient donné chaud et il sentait que les trois jours de jeûne lui avaient coûté bien des forces. Pour un instant, il s'adossa contre le mur de la chambre mortuaire et ferma les yeux. Puis il s'arracha à son épuisement et fit signe aux prêtres qu'il était prêt à célébrer la cérémonie finale pour son père: le sacrifice du sang de cinq esclaves pour que les Dieux soient favorables à sa renaissance. Le père de Serpent-Jaguar, le grand Dieu-Roi Pacal, était mort il y a quelques jours et c'était maintenant à lui, son fils, de conduire le peuple Maya de Palenque, en accord avec la volonté du Soleil et de Vénus, en accord avec les astres dont la famille régnante se considérait être les descendants directs et dont ils faisaient interpréter la volonté céleste par des chamans-astronomes. Le défunt Pacal, tous les Mayas le savaient, était né le même jour que le soleil 5000 ans auparavant; c'était le signe de sa nature divine. Serpent-Jaguar n'avait donc aucun doute sur la bienveillance des Dieux, trois mois après les adieux faits à son père, au cours des festivités accompagnant son intronisation.
Le 20 janvier de l'année 690, tout le monde put nettement voir trois planètes s'enflammer dans un ciel s'assombrissant et former par leur proximité une trinité divine, une rare constellation: Saturne à l'Est, Jupiter au centre, Mars à l'Ouest. Confiant, sûr de lui, le nouveau Dieu-Roi parut cette nuit-là aux yeux de son peuple tout en haut de la pyramide du soleil et offrit au cosmos le plus grand et le plus désirable sacrifice: du sang qui coula dans un récipient lorsqu'il perça plusieurs fois la peau de son pénis avec une épine de rose. Pour les indiens Maya, comme pour d'autres civilisations du passé, le monde était un endroit encore plus vivant qu'il ne l'est pour nous aujourd'hui. Force sacrée, présence spirituelle, matière animée faisaient en tout lieu partie du paysage et étaient naturellement liées à l'espace et au temps. Ciel et Terre n'étaient pas encore séparés, les astres retournaient encore dans leur foyer terrestre pendant la nuit. Et les hommes parlaient aux étoiles et écoutaient les planètes. Mais surtout, ils observaient les astres et lisaient un message dans chaque courbure et chaque changement de direction de l'orbite des planètes. Du village de la forêt vierge jusqu'à la bourgade des Alpes, ils projetaient leur société, leur vie, leurs espoirs et leurs peurs sur les événements célestes et découvraient alors (qui s'en étonnerait) un monde qui leur semblait familier.

Святые места: как на небе, так и на земле

Selon qu'ils étaient chasseurs, cultivateurs ou éleveurs, le soleil pouvait alors leur apparaître hostile ou bienveillant. Et les mêmes planètes se comportaient, selon l'observateur, de manière "typiquement féminine", "typiquement masculine" ou même satanique. Les corps célestes paraissaient souvent imprévisibles, soumis à des humeurs, puisqu'on pouvait les voir aller de ci, de là, s'arrêter, puis reculer pour finalement avancer deux fois plus vite. La fonction d'un Dieu-Roi Maya comme Chan-Bahlum (Serpent-Jaguar) était en quelque sorte de négocier entre ce Haut et le "Bas" que représentait le peuple. Ses actes rituels faisaient couler la force du surnaturel dans la vie des mortels et donnaient aux contradictions des phénomènes célestes une signification claire et nette. Les sacrifices servaient à établir des relations entre le haut et le bas, entre la puissante nature et les humains conscients de l'ampleur de leur infériorité; les sacrifices garantissaient le retour des événements cycliques qui marquaient l'existence et laissaient participer les hommes de manière active aux mouvements circulaires de la nature.
Le besoin de contact avec le reste de la création, voire même le désir de s'unir à elle, n'est certainement pas moindre à l'époque moderne, que nous en soyons conscients ou non. Au lieu de nous rendre dans un temple, une église ou de nous agenouiller devant une relique, nous cherchons plutôt un morceau de nature vierge pour communiquer. Nous voulons construire des ponts émotionnels, nous voulons aussi nous engager en déposant une plume ou une pierre colorée au pied d'un arbre ou d'une montagne. Un geste symbolique, certes, qui ne sert peut-être qu'à nous faire prendre conscience de notre destin commun avec le cosmos. Carl Jung remarqua il y a bien des années que certaines pierres trouvées dans la nature nous fascinent peut-être parce qu'elles "s'apparentent curieusement à ce que nous avons de plus profond". Ainsi, la pierre symbolise l'"expérience la plus simple et la plus profonde - l'expérience de quelque chose d'éternel".

Змея-Ягуар почувствовал соленый вкус пота в уголках губ. Несмотря на прохладу, царившую в глубине священной горы, ему было жарко от утомительных многочасовых обрядов. Да и трехдневный пост утомил его больше, чем он ожидал. На мгновение он прислонился спиной к стене погребальной камеры и закрыл глаза. Потом он стряхнул с себя усталость и дал жрецам знак, что готов к заключительному обряду по отцу — жертвоприношению пяти рабов для умилостивления богов и новой жизни для него. Отец Змеи-Ягуара, божественный царь Пакаль, умер несколько дней назад, и теперь он, его сын, должен был установить согласие народа майя в Паленке с волей Солнца и Венеры, с созвездиями, прямыми потомками которых считала себя царская семья, чью небесную волю истолковывали по ее повелению звездочеты-шаманы. Покойный Пакаль — это знали все майя — в знак своей божественности родился в тот же день, что и Солнце 5000 лет назад. У Ягуара-Змеи не было сомнений в благосклонности богов и спустя три месяца после похорон отца, во время коронования.
Было 20 января 690 г., и на темном небе, хорошо видимые, ярко светились рядом три планеты в божественном триединстве и редкой конфигурации — Сатурн на востоке, Юпитер в центре и Марс на западе. С достоинством и уверенностью в себе поднялся в эту ночь новый божественный царь на вершину пирамиды Солнца перед лицом своего народа и принес космосу самую большую и вожделенную жертву — кровь, которая стекла в чашу, когда он многократно проколол свою крайнюю плоть розовым шипом. Для индейцев майя, как и для прочих культур прошлого, мир был гораздо более одухотворенным, чем для нас сегодня. Священная сила, духовное присутствие, одухотворенная материя были частью природы, безусловно принадлежали времени и пространству. Земля и Небо еще не были разделены, созвездия еще возвращались в свой земной дом. И люди беседовали со звездами, вслушивались в язык планет. Но с особым вниманием они следили за созвездиями и истолковывали все движения планет как важные вести. И жители тропического леса, и обитатели альпийского городка связывали свою жизнь, надежды и тревоги с событиями на небе, и неудивительно, что они открывали при этом мир, который был им прекрасно знаком. В зависимости от рода занятий, т. е. у охотников, земледельцев или скотоводов, Солнце могло быть настроено враждебно или благосклонно. И одни и те же планеты проявляли себя то «по-мужски», то «по-женски», то вовсе сатанински злобно. Небесные тела вели себя непредсказуемо, были подвержены смене настроений — отклонялись в сторону, останавливались, двигались вспять, чтобы потом еще быстрее устремляться вперед. Божественный царь майя, например, Хан-Бахлум (Змея-Ягуар) был чем-то вроде посредника между высшим миром и низшим, т. е. народом. Через выполняемые им ритуальные действия сверхъестественная сила входила в жизнь простых смертных, и он же придавал ясное значение всем противоречивым событиям на Небе. Жертвы должны были служить связующим звеном между верхним и нижним мирами, между могучей природой и слабым человеком; жертвы гарантировали повторение жизненных циклов и позволяли людям активно участвовать в круговороте природы.
Современность требует от нас в неменьшей степени установления связи с остальным тварным миром, если вообще не соединения с ним, сознаем мы это или нет. Вместо храма, ковчега или церкви мы для этой связи ищем, скорее, нетронутый уголок природы. Мы стремимся передать свои чувства, принять на себя какое-то обязательство, хотя бы положив у дерева или горы перышко или красивый камень. Конечно — чисто символический жест, который, скорее, служит осознанию общности нашей судьбы с космосом. К. Юнг заметил много лет назад, что некоторые камни потому нас так привлекают, что они «странным образом сродни нашему внутреннему зерну». В этом смысле камень символизирует «самый простой и самый глубокий опыт — опыт чего-то вечного».

COSMOS

LIFE

Die Lebensweise der Menschen bestimmt ihr Denken - so auch im Fall dieser bäuerlichen Navajo-Familie in Arizona. Morgendliche Gebete zur Sonne und Heilungszeremonien sind hier noch selbstverständlicher Teil des Jahresrhythmus'.

Il modo di vivere degli uomini determina il loro pensiero anche per questa famiglia di contadini Navajo in Arizona. Qui le preghiere rivolte al sole e le cerimonie di guarigione sono una parte naturale dello avvicendarsi delle stagioni.

De levenswijze van de mensen bepaalt hun denken - ook zo in het geval van deze Navajo boerenfamilie in Arizona. Ochtendgebeden tot de zon en genezingsceremoniën zijn hier vanzelfsprekend nog een deel van het ritme der jaargetijden.

Le mode de vie des humains détermine leur mode de pensée; c'est aussi le cas de ces paysans Navajo en Arizona. Ici, les prières matinales adressées au soleil et les cérémonies de guérison font encore partie du rythme annuel.

How man lives determines his way of thinking - and this is equally true of this big Navajo farming family in Arizona. Morning prayers to the sun and healing ceremonies are here a perfectly natural part of the rhythm of the year.

Образ жизни определяет мышление людей, например, у этих крестьян навахо в Аризоне. Утренняя молитва Солнцу и обряды исцеления являются для них естественной частью годового ритма.

Vogel mit Deinen Flügeln, stiehl mir mein verwundetes Herz. Flieg' damit hoch und höher, bis der Mensch und seine Erde verschwunden sind. Blende meine Augen und mache meine Ohren taub, daß ich gefeit bin gegen Lanzen, die durchdringen, Stimmen, die besänftigen und Hände, die falsch halten. Trage mein Herz in ein friedvolles Land der unberührten Schönheit, wo Seele und Geist zusammen in seliger Liebe ruhen. Dann bringe mein geheiligtes Herz zur Erde zurück.
Bringe mit dir den Samen einer neuen Generation.

Gevleugelde vogel, steel mijn gewond hart, vlieg ermee omhoog, en nog hoger totdat de mens en zijn aarde zijn vervaagd en verdwenen. Draag mijn hart naar een vreedzaam land van ongerepte schoonheid waar geest en ziel tezamen zonder iets anders te storen in zalige liefde rusten.
Breng dan mijn genezen hart naar de aarde terug en breng het zaad van een nieuwe generatie met je mee.
Begin alleen als het veilig is en zaai het verstandig.
Ik zal dit zaad met mijn leven zorgvuldig beschermen.

Winged Bird steal my pierced heart. Fly with it high and higher till man and his earth are blurred and lost. Blind my eyes and deafen my ears to lances that pierce, voices that soothe, and hands that hold falsely. Carry my heart to a peaceful land of untouched beauty where soul and mind together, disturbing no other, lie in blissful love. Child of beauty, are you ready? Then return my healed heart to earth. Bring with you seeds of a new generation. Begin only when it is safe. And plant wisely. I will guard these seeds with my life.

Veronica Riley

Uccello, tu che hai le ali, ruba il mio cuore ferito. Portalo in alto, sempre più in alto, finché non si possa più distinguere l'uomo e la sua terra. Acceca i miei occhi e fai sorde le mie orecchie, che io sia immune contro le lance che trapassano. Porta il mio cuore in un paese di pace dove l'anima e lo spirito possano riposare, senza disturbare tutto il resto, in amore beato. Figlio della bellezza sei pronto? Allora riporta il mio cuore santificato alla terra. Porta con te il seme di una nuova generazione. Inizia quando sarà sicuro e seminalo saggiamente.

Oiseau, avec tes ailes, vole-moi mon cœur blessé. Emmène-le haut, encore plus haut jusqu'à ce que l'homme et sa terre disparaissent. Eblouis mes yeux et rend sourdes mes oreilles pour me protéger des lances qui transpercent, des voix qui apaisent et des mains qui trahissent.Emporte mon cœur dans la région paisible où la beauté est vierge, où l'âme et l'esprit reposent dans un amour serein sans détruire tout le reste. Ramène alors sur terre mon cœur guéri. Apporte la semence d'une nouvelle génération. Plante-la avec sagesse.

Птица, возьми своими крыльями мое израненное сердце. Полети с ним высоко, так высоко, чтобы уже не было видно людей и их земли. Отнеси мое сердце в мирную страну нетронутой красоты, где душа и дух, никому не мешая, покоятся в блаженной любви. А потом принеси исцеленное сердце назад. Принеси с собой семя нового поколения. Посей его мудро, когда оно будет в безопасности. Я буду охранять его своей жизнью.

Das Plakat lädt mich ein, das Museum zu besuchen: Pueblos, Paluten, Apachen in Originalkleidung. Museumsführung. Töpfereien ruhen in Glasvitrinen. Sie konzentrieren in sich einen speziellen Geist, eine Geometrie der Verständigung von einem anderen Ort. Das Ganze erinnerte an die protzigen Reklametafeln am Highway 66 in Richtung Westen: Seht zu, wie richtige Indianer echten indianischen Schmuck herstellen.

Het aanplakbiljet nodigt mij uit het museum te bezoeken. Indianen tentoongesteld: Pueblo, Paiute, Apache in originele kleding. Pottenbakkerskunst rust in vitrines om in musea rond te waren. Een speciale geest is erin, een geometrie van begrip van een andere plaats. Het geheel herinnert aan die reclameborden aan de highway 66 in richting westen: Kijk hoe echte Indianen echte Indiaanse sieraden maken.

The poster invites me to take a trip to the Museum - Native Americans on display: Pueblo - Paiute - Apache in authentic native costumes: Pottery rests in glass cases. They concentrate a formal essence, a geometry of understanding, from another place. The whole thing reminded me of those awful gaudy signs riding West on highway 66 when we were young: See real Indians make authentic Indian jewelry.

Il manifesto invita a visitare il museo. Indigeni americani messi in mostra: Pueblos, paiuti, apache in costume. La guida del museo spiega. Quei cocci di vasellame nelle vetrine portano spiriti nei musei. Vi si concentra un'essenza formale, una geometria della comprensione da un altro luogo. Il tutto ricorda una pubblicità lungo la Highway 66: Guardate come veri indiani producono veri monili indiani.

L'affiche m'invite à visiter le musée: Pueblos, Paoutes, Apaches en costume authentique. Des poteries reposent dans les vitrines en verre. Elles renferment un esprit particulier, une géométrie d'un autre endroit. Tout ressemble aux panneaux publicitaires qui bordaient le Highway 66 en direction de l'Ouest quand nous étions jeunes: Regardez comment de vrais Indiens fabriquent de vrais bijoux indiens.

Плакат приглашает в музей. Выставлены коренные американцы: пуэбло, паюты, апачи — в своих одеждах. Объяснения экскурсовода. Гончарный товар в витрине путешествует из музея в музей. Он несет в себе особый дух, геометрию иных краев. Все это напоминает рекламу: «Поглядите, как настоящие индейцы делают настоящие индейские украшения!»

VII/23

Wenn der Frühling kommt, wenn unsere Erdmutter strömt von lebendigen Wassern, dann pflanzen wir im Boden die verschiedenen Arten Mais, die Quelle unseres Fleisches.

Als de lente komt, wanneer onze aardmoeder stroomt van levendige wateren, dan planten wij in de aarde de verschillende soorten maïs, de bron van ons vlees.

When the first signs of spring appear, when our Earth Mother streams from living waters, then we plant the different sorts of maize, the source of our flesh.

Quando viene la primavera, quando acque vive scorrono per la madre terra, noi piantiamo nel suolo i tipi di mais più diversi. Sono la fonte della nostra carne.

Quand le printemps arrive, quand notre Terre-Mère regorge d'eaux vives, nous plantons alors dans le sol les différentes sortes de maïs, la source de notre chair.

Когда приходит весна, когда по нашей Матери Земле текут живые воды, мы сажаем в землю кукурузу, источник нашей плоти.

Zuni song

Es gibt Gesang, vollkommenen Gesang, eine einfache Melodie; weit schallend durch den Himmel und in dem Raum zwischen Himmel und Erde wächst du und bist vibrierend und einmalig. Mit der kleinsten Blume fing deine Existenz an, ein Sämling, lebendig und wachsend am richtigen Ort, gesegnet von den Geistern des Windes und des Wassers, das sich an seine Mutter schmiegt.

Er is een lied, een eenvoudige melodie die hoog door de hemelen schalt, en in de ruimte tussen hemel en aarde groei jij, vibrerend en uniek. Van de berg en van de hemel en uit dit alles kom jij, met de kleinste bloem begon jouw bestaan, een jonge zaadplant, levendig en groeiend op de juiste plaats, gezegend door de geesten van de wind en van het water dat zich bij zijn moeder voegt.

There is singing song in motion in simple melody high through the heavens and in the space between sky and earth are you growing and vibrant and unique. Hayashee, Dya-tza-ah all from the mountain and sky and from this you stem from the smallest flower your existence began a little seedling alive and growing in its rightful place blessed by spirits of wind and water.

C'è una canzone, una melodia semplice; echeggia lontano e alta attraverso il cielo, e nello spazio tra cielo e terra cresci tu e sei vibrante ed unica. Dalla montagna e dal cielo e da tutto ciò vieni tu. Con il fiore più piccolo iniziò la tua esistenza, era un minuscolo seme vivo che cresceva nel luogo a lui adatto, benedetto dagli spiriti del vento e dell'acqua e che se ne sta stretto alla madre.

Un chant s'élève, un chant parfait, une mélodie simple traversant le ciel, et dans l'espace entre le Ciel et la Terre, tu grandis, tu vibres, unique. Tes ancêtres sont la montagne et le ciel et tous les éléments, ton existence a commencé par la fleur la plus petite, provenant d'un semis, vivant et grandissant au bon endroit, béni par les esprits du vent et de l'eau qui se blottit contre sa mère.

Есть песнопение, совершенная песнь, простая мелодия; далеко раздаваясь по небу, в пространстве между небом и землей ты растешь, дрожа, неповторимо. Ты родом от горы и от неба, малым цветком началась твоя жизнь — живой росток, благословенный духами ветра и воды, прижимающийся к матери.

Kinder des Weltalls: „Glück vom Haus aus Dämmerung“

Kinderen van het heelal: "Geluk uit het huis van ochtendgloren"

Ein Sirren wie von einer Klapperschlange fiel aus der Nacht durch die Deckenluke des Steinbaus. Jeder Laut erstarb; alle Augen wandten sich nach oben, wo eine Kette aus rotbemalten Körpern und federgeschmückten Masken die Sprossenleiter herabglitt, ohne abzubrechen an den Wänden entlangfloß, dann anhielt. Langsames Trommeln setzte ein, Arme hoben sich und schüttelten Rasseln, die Oberkörper neigten sich vor, und im selben Augenblick, da die Kette mit dem ersten Aufstampfen der Beine nach unten schwang, durchdrang ein Singen, ein tiefer Weltraumklang aus vibrierenden Hartledermasken den Raum. Auf- und abwärts wogte der Kreis; nach bestimmten Intervallen wechselten die „Kachinas“ mit schweren Stampfschritten die Tanzrichtung, ohne daß Bewegung, Gesang und Trommelschläge auseinanderflossen. Heftiger wurden das Stampfen, lauter das Singen. Resonanz baute sich auf zwischen Tänzern und Felsberg, in den die „Kiva“, die „Unterwelt“ eingelassen war. Die „Kachinas“ waren aus der Tiefe des Alls gekommen und jetzt berührten sie die Erde. Für die Hopi-Indianer in der Kiva waren es spirituelle Formen von Naturphänomenen, Botschaften zwischen den Menschen und höheren Mächten wie Sotuqnangu, den Gott des Himmels, der Wolken und des Regens. Die Indianer gaben den Kachinas ihre Bitten um Regen für ihre Maisfelder mit auf den Weg. Die Geistwesen waren vertraute Partner, Erscheinungsformen der Natur, denen sich die Hopis nahe fühlten.
Bei den Nachbarn der Hopis, den Navajos, habe ich miterleben dürfen, welche Rolle solche Rituale, die das Verbundensein des Menschen mit dem Kosmos zelebrieren, auch im täglichen Leben heute noch spielen. Wayne, ein Farmer und Medizinmann, der mich für ein Jahr in seine große Familie aufnahm, betete jeden Morgen zur aufgehenden Sonne, wobei er etwas Maispollen - Symbol für alles Leben - in die Luft streute. Nie werde ich den Tag vergessen, als er mich einmal zu sich rief und sagte, ich solle meine Arme nach Osten ausstrecken und die Luft der Dämmerung zum Mund führen, sie einatmen. Eine Brise lief über das Wüstengras, es roch nach Salbei und Piniennadeln. Ein roter Streifen kündigte die Sonne an, kroch langsam höher, während Wayne eines der alten Gebetslieder „vom Urgroßvater“ übersetzte: „Vom Haus aus frühem Morgen wird Glück zu mir kommen. Von dort, wo das Sonnenlicht als erstes auftrifft, wird Glück zu mir kommen, von gestreiften Felsen, von Echos der Canyonwände und von dort, wo der blaue Wind geht...“ Der frühe Morgen sei die heiligste Zeit, sagte er. „Du begegnest dem 'Heiligen, der spricht'; das Rot am Horizont ist seine äußere Form. Du atmest Gesundheit und Schönheit ein um diese Zeit.“ Schönheit, „Bike hoscho“, ist nach dem Verständnis der Navajos eine zentrale Antriebskraft des Universums. „Saa naghai“ heißt 'langes Leben' und wird als Synonym für „Vater Himmel“ verwendet. Während seiner Heilungszeremonien oder auch an ganz normalen Abenden im Kreis der Familie wurde Wayne nicht müde, sein Credo von den Wechselbeziehungen aller Dinge des Kosmos' zu wiederholen. Der runde Navajohogan sei nicht nur Behausung, sondern auch ein heiliger Ort sowie eine Ermahnung, die gegebenen Verhältnisse nicht zu vergessen: Das Domdach der Hütte stehe für das Universum, der nackte Lehmboden für Mutter Erde, und die Zeremonien erinnerten die Teilnehmer an ihren Platz in der Weltordnung. „Denkt an das Feuer, das Wasser, die Erde“, sagte er oft. „Ihre inneren Formen leiten uns. Dankt Mutter Erde und Vater Himmel, dankt der Sonne und der Luft. Dankt dem Regen, er dringt zu den Wurzeln des Mais', läßt ihn gedeihen. Wir sitzen in einem Hogan aus Holz und Erde. Alles hängt zusammen. Alles Gute wird kommen heute Nacht, vom Osten, Süden, Westen und Norden...“ Gab es Zwist in der Familie, versammelten sich alle, Großeltern, Eltern und Kinderschar, um ein Häufchen glühender Holzkohlen, die Wayne in der Mitte des Hüttenbodens aufschüttete. Der Konflikt wurde diskutiert, aber so, daß jeder in Richtung des Feuers sprach. Der „kosmische Rahmen“ des Hogans, so schien es mir, förderte dabei die Kompromißbereitschaft so mancher gekränkten Seele. Es berührte mich oft tief, wie offen und mit welcher Wärme drei Generationen einer Familie vor dem Feuer ihre Zuneigung zueinander und zu „Mutter Erde und Vater Himmel“ bekundeten, wie sie das gemeinsame Dahinfließen in Zeit und Raum, das Eingebettetsein in eine gemeinsame Ordnung und Identität noch als wirklich erlebten, und wie sie trotz aller Kulturbrüche ein mit dem Kosmos gemeinsames Schicksal zu fühlen schienen. Zumindest für die Großeltern und Eltern entsprach das Weltbild noch in großen Zügen der Realität ihres Lebens. Mit ihren Mais- und Bohnenfeldern, den Schafen und Ziegen konnten sie wie ihre Vorfahren in relativ stabilem Rhythmus durch die Jahreszeiten gehen, sich auf die Wiederkehr der Sonne, des Regens, des Schnees verlassen, von denen ihr Überleben mit abhing. Für sie machte es noch Sinn, nach alter Navajoart beim Reden Erde in die Hand zu nehmen und sie zwischen den Fingern zu Boden rieseln zu lassen.

Door het plafondluik van het stenen bouwwerk viel uit de nacht een zoemen als van een ratelslang. Elk geluid bestierf; alle ogen werden naar boven gericht, waar een keten van rood beschilderde lichamen en met veren getooide maskers over de sporten van de ladder omlaag gleed, onafgebroken langs de wanden vloeide, en dan bleef staan. Langzaam getrommel zette in, armen verhieven zich en klepten met ratels, de bovenlichamen neigden zich naar voren en op hetzelfde ogenblik waarop de keten met het eerste stampen van de benen omlaag zwaaide, vervulde een zingen, met diepe wereldruimteklanken uit vibrerende hardleren maskers de ruimte.
De kring golfde op en neer; na bepaalde intervallen wisselden de "katchinas" met zware stampende passen de dansrichting, zonder dat beweging, gezang en trommelslagen uit elkaar vloeiden. Het stampen werd heviger, het zingen harder. Er werd een resonantie opgebouwd tussen de dansers en de rotsberg waarin de "kiva", de "onderwereld", was aangebracht. De "katchinas" waren van diep uit het heelal gekomen en nu kwamen zij langs de aarde. Voor de Hopi-Indianen in de kiva waren het spirituele vormen van natuurfenomenen, missies tussen de mensen en hogere machten zoals Sotuqnangu, de god van de hemel, de wolken en de regen. De Indianen gaven de katchinas beden om regen voor hun maïsvelden mee op weg. De geestwezens waren vertrouwde partners, verschijningsvormen van de natuur, waarmee de Hopi's zich nauw verbonden voelden.
Bij de buren van de Hopi's, de Navajo's, heb ik mogen meebeleven welke rol zulke rituelen die het verbonden zijn van de mens met de kosmos celebreren, ook vandaag nog in het dagelijkse leven spelen. Wayne, een farmer en medicijnman, die mij voor een jaar in zijn grote familie opnam, bad elke ochtend tot de opgaande zon. Daarbij strooide hij een beetje maïsstuifmeel - symbool voor al het leven - de lucht in. Nooit zal ik de dag vergeten toen hij mij eens tot zich riep en zei dat ik mijn armen naar het oosten moest uitstrekken en de lucht van het ochtendgloren naar mijn mond moest brengen om haar in te ademen. Er ging een zachte bries over het woestijngras, het rook naar salie en pijnboomnaalden. Een rode streep kondigde de zon aan, kroop langzaam hoger, terwijl Wayne een van de oude gebedsliederen "van overgrootvader" vertaalde: "Van het huis van vroege ochtend zal geluk naar mij komen. Van daar, waar het zonlicht het eerst optreedt, zal geluk naar mij komen, van gestreepte rotsen, van echo's der canyonwanden en van daar, waar de blauwe wind waait..."

Children of the universe: "Happiness from the house of early morning"

"De vroege ochtend is de heiligste tijd", zei hij. "Je ontmoet 'de heilige die spreekt'; het rood aan de horizon is zijn uiterlijke vorm. Om deze tijd adem je gezondheid en schoonheid in." Schoonheid, "bike hosjo", is volgens het begrip van de Navajo's een centrale aandrijvingskracht van het universum. "Saa naghai" betekent "lang leven" en wordt gebruikt als synoniem voor "Vader Hemel". Werd Wayne er niet moe van zijn credo van de onderlinge samenhang tussen alle dingen van de kosmos te herhalen: De ronde Navajo-hogan is niet alleen behuizing, maar ook een heilige plaats en ook een vermaning om de gegeven omstandigheden niet te vergeten; het koepeldak van de hut staat voor het universum, de naakte lemen vloer voor Moeder Aarde, en de ceremonies herinneren de deelnemers aan hun plaats in de wereldorde. "Denk aan het vuur, het water, de aarde", zei hij vaak. "De innerlijke vormen ervan leiden ons. Dank Moeder Aarde en Vader Hemel, dank de Zon en de Lucht. Dank de Regen, want hij dringt naar de wortels van de maïs, laat hem groeien. Wij zitten in een hogan van hout en aarde. Alles hangt met elkaar samen. Al het goede zal komen deze nacht, van het oosten, zuiden, westen en noorden..." Als er ruzie in de familie was, kwamen allen bijeen, grootouders, ouders en kinderen, rondom een hoopje gloeiende houtskolen die Wayne in het midden van de vloer van de hut bijvulde. Er werd over het conflict gediscussieerd, maar zo dat iedereen in de richting van het vuur sprak. Het "kosmische kader" van de hogan, zo leek het mij, bevorderde daardoor bij menig gekrenkte ziel de bereidheid tot compromissen. Ik was vaak diep getroffen hoe open en met welke warmte drie generaties van een familie voor het vuur hun genegenheid voor elkaar en voor "Moeder Aarde" en "Vader Hemel" toonden, hoe zij het gemeenschappelijke voortstromen in tijd en ruimte, het ingebed zijn in een gemeenschappelijke orde en identiteit nog als een werkelijkheid beleefden, en hoe zij ondanks alle cultuurbreuken een met de kosmos gemeenschappelijk lot schenen te voelen. Tenminste voor de grootouders en ouders kwam het wereldbeeld in grote lijnen nog overeen met de realiteit van hun leven. Met hun maïs- en bonenvelden, schapen en geiten konden zij zoals hun voorouders in een relatief stabiel ritme door de jaargetijden gaan, op de terugkeer van de zon, regen en sneeuw vertrouwen, waarvan hun overleven mede afhankelijk was. Zij zagen er nog de zin van in om volgens een oude Navajotraditie bij het spreken aarde in de hand te nemen en deze tussen de vingers op de grond te laten glijden.

A whirring like the sound made by a rattlesnake came through the hole that was in the ceiling of the stone building, interrupting the stillness of the night; all eyes looked up to where a chain of red-painted bodies and feathered masks slid down the peg ladder and, without stopping, moved along the wall, and then came to a halt. A slow drumming rhythm started up; arms were raised and rattles shaken, the upper parts of the bodies swayed forward and at the very moment at which the group stamped its feet and bowed downwards, singing filled the room - a deep cosmic tone coming from vibrating hard leather masks. Up and down the circle swayed; and at given intervals the "Kachinas" switched the direction of the dance, still making heavy pounding steps, without losing the rhythm of movement, song and drums. The pounding became heavier, the singing louder. There was an echo between the dancers and the rocky mountain where "Kiva", the "underworld" was to be found. The "Kachinas" had come from the depths of the universe and now they were touching the Earth. For the Hopi indians in the Kiva world, the dancers symbolized spiritual forms of natural phenomena, conveyed messages between Man and higher powers like Sotugrangu, the God of the Sky and the Clouds and the Rain. The indians passed on to the Kachinas their entreaties for rain for their fields of corn. These spiritual beings were familiar partners, outward images of a nature to which the Hopis felt very close. Amongst the Hopi's neighbours, the Navajos, I was myself able to experience what role such rituals celebrating man's links with the cosmos play today in daily life also. Wayne, a farmer and medicine man, who took me into his large family for a year, prayed to the rising sun every morning and in doing so he strewed some corn pollen - the symbol for all life - into the air. Never will I forget the day when he called me to him and said that I should stretch out my arms to the East and waft the air of the dawn to my mouth and breathe it in. A gentle breeze swept through the desert grass and it smelled of sage and pine needles. A red strip announced the coming of the sun and slowly crept up while Wayne translated one of the old prayer songs "from his great-grandfather": "From the house of early morning happiness will come to me. From the place where sunlight can first be perceived, happiness will come to me, from striped rocks, from the echoes of the canyon walls and from where the blue wind goes ..." The early morning, he said, is the holiest time. "You meet the 'Holy One Who Speaks'"; the red flush on the horizon is his outer form. You breathe in health and beauty at this time. "Beauty", "bike hoscho" is, as the Navajos see it, a central driving force of the universe. "Saa maghai" means "long life" and is used as a synonym for "Father Sky". In the course of his healing ceremonies or also on quite normal evenings in the circle of his family, Wayne never tired of repeating his credo of the reciprocity of all things in the cosmos. The round Navajo hogan was not just a dwelling but also a holy place and an ad-monishment not to forget the given circumstances: the vaulted roof of the hut stands for the universe, the bare loam floor for Mother Earth, and the ceremonies remind those participating of their place in the world order. "Think of fire, water and earth," he often said. "Their inner forms guide us. Thank Mother Earth and Father Sky, thank the sun and the air. Thank the rain that finds its way to the roots of the corn and makes it thrive. We are sitting in a hogan made of wood and earth. Everything is linked. Everything good will come tonight from the East, the South, the West and the North..." If there was dissension within the family, then everyone gathered, the grandparents, the parents and all the children, and all sat round a heap of glowing charcoal which Wayne had poured out on the middle of the floor in the hut. The conflict was then discussed, with everyone involved speaking in the direction of the fire. The "cosmic framework" of the hogan, it seemed to me, assisted many an offended soul in his readiness to compromise. It often touched me how openly and how warmly three generations of one family declared their affection for each other and for "Mother Earth and Father Sky", how they experienced as real that flowing away of all things in time and space, that sense of being embedded in a joint order and identity, and how in spite of all the breaks in the development of their culture, they seemed to sense a common fate shared with the cosmos. For the grandparents and the parents at least, this picture of the world corresponded to a large part with the reality of their lives. With their fields of corn and beans, their sheep and their goats, they had been able like their ancestors to move through the seasons in a relatively stable rhythm and to rely on the return of the sun, the rain and the snow, which their survival depended on. For them it still made sense to take earth into their hand while talking, in old Navajo style, and to crumble it between their fingers and let it trickle to the ground.

Le lampade dei preastronauti, nove soli di troppo, con sangue e senza sangue, la colonna distrutta

Tutti gli anni, nella notte tra il 31 luglio ed il 1° di agosto, in prossimità di Kumamoto, sulla costa giapponese, si svolge una solenne cerimonia per celebrare il misterioso fuoco col quale esseri provenienti dalle stelle, approdarono sulle acque illuminando il percorso con un pullulare di lanterne rosse.
È un evidente fatto da ascrivere alla storia preastronautica, che comunque trova riscontro altrove. Il 27 settembre 1098, ad esempio, un anonimo Monaco Cassinese, scrive nei suoi Annali: "E in questo anno durante tutta la notte apparve in cielo una via di fuoco."
Terzo avvenimento stralciato dal nostro nutrito archivio, ci trasferisce in Cina alla corte dell'Imperatore Tch'en della dinastia Han, ove, nel 32 a.C. all'incirca, si verificò una cospicua pioggia di meteoriti contemporanea al crollo del monte Ming. Gli astrologi, ovviamente, ci misero mano deducendo che quella dinastia era arrivata al capolinea, cosa che successe puntualmente.
Sono fatti esteriori della realtà cosmica, marginali e spiegabili; ma che richiamano ai misteri del magico verso il quale bisogna andare cauti apprestando, quando possibile, qualche difesa, qualche deterrente. E che altro se non i riti propiziatori ed i sacrifici per cercare di arginare calamità e tristi inaspettati eventi? Va ricordato che le costellazioni appartengono alla mitologia astrale, dalle cui influenze discendono molteplici, differenziatissime personalizzazioni riguardanti il pubblico ed il privato, la storia dei popoli, numeri e colori con tutti i loro influssi e significati, guerre e carestie, periodi fecondi e serenità, da ingraziarsi o placare, in ogni caso cercare di rendere benigni.
Nel Madagascar lo fanno intagliando i crani degli animali bloccati nei ceppi e questo per celebrare gli dei provenienti dalle stelle. La identificazione delle divinità con gli stessi corpi celesti è, peraltro, frequente come è frequente la comparsa di creature inspiegabili provenienti dal cosmo. L'antica tradizione cinese descrive l'alba con nove soli collocati fra i rami giganteschi dell'albero "po", più un decimo staccato dal gruppo. Ebbene, arrivò il tempo delle catastrofi in quanto il concentrato, insostenibile calore dei dieci astri sulla terra, la Cina nel nostro caso, portavano allo incendio e all'incenerimento imminenti. Per fortuna intervenne l'arciere Yi, il quale, con il suo arco fatato, ne colpì nove, lasciando il decimo a percorrere il cielo sul suo carro trainato da dragoni. Per la terra era più che sufficiente, e la faccenda fu definitivamente risolta senza più bisogno di riti e di sacrifici. Altrimenti questi appaiono più che indispensabili e neppure separati, facendo parte il rituale del sacrificio medesimo. Il quale costituisce una rinunzia eccezionale e volontaria in previsione di un bene futuro proprio o per altri. Può essere senza sangue o con sangue, con una casistica perfino suggestiva nel primo caso ed agghiacciante nel secondo, con riti adeguati al significato del tributo o della immolazione che si svolgevano prevalentemente in ogni locale santuario. Sacrifici che potevano derivare dall'implorazione di grazia per colpe commesse. Per il mancato rispetto o addirittura offesa verso i possessori del cielo. I quali, peraltro, possono diventare improvvisamente tali come racconta Ovidio in una delle sue tante "Metamorfosi": "Ed ecco un giorno avanzare il nipote di Licaone, Arcade, ignaro della sorte toccata alla sua genitrice. Aveva ormai quasi quindici anni. Inseguiva la selvaggina, sceglieva gli anfratti più adatti e stendeva reti flessibili attorno alle selve dell'Erimanto, quando ad un tratto s'imbatté nella madre. Essa trasalì e si arrestò, come lo vide, e parve proprio che lo riconoscesse. Lui arretrò spaventato - perché non sapeva - da quegli occhi che immobili lo fissavano senza fine, e quando essa accennò ad avvicinarglisi, si preparava a trapassarle il petto con un dardo micidiale. L'Onnipotente lo impedì: li bloccò entrambi, e insieme bloccò il delitto, e sollevatili in aria con un vento veloce li collocò nel cielo facendone due costellazioni vicine."
Alcuni re irlandesi, prima della cristianizzazione, si sottoponevano al rito della morte sacrificale, diventandone protagonisti e vittime, bruciandosi nel palazzo o annegandosi in un tino di vino per raggiungere la purificazione ed essere degni del cielo. Un cielo dal quale si è continuamente separati ed uniti come dice il "Shu Ching", il classico documento dell'antica Cina, che racconta di una colonna centrale che collegava il cielo con la terra, molto utile in verità per l'intercambio fra le due situazioni che creava sì una certa confusione, ma che toglieva ai sacrifici quel ruolo gerarchico successivo del basso verso l'alto. La colonna, infatti, non esiste più.

Giuseppe Brunamontini

Enfants de l'Univers: "Bonheur de la maison venue du crépuscule"

Un bourdonnement nocturne ressemblant au bruit d'un serpent à sonnettes entra par la lucarne du toit de la maison de pierre. Le silence se fit; tous les yeux se tournèrent vers le haut: une chaîne faite de corps peints en rouge et de masques ornés de plumes glissa le long de l'échelle à barreaux, longea le mur sans se disloquer pour finalement s'arrêter. Le bruit de tambours au rythme lent s'éleva; des bras se levèrent et agitèrent des crécelles, les torses se penchèrent en avant et au moment où les jambes commencèrent à frapper le sol, la chaîne se balança; un chant semblant venir de l'espace cosmique, un son grave émis par les vibrations des masques en cuir dur, remplit la pièce. Le cercle ondulait, montait puis retombait; au bout de certains intervalles, les "Kachinas" changeaient le sens de la dance d'un pas lourd sans que le mouvement, le chant et le bruit des tambours se désunissent. Les pieds frappèrent plus fortement le sol, le chant s'amplifia, la résonance s'établit entre les danseurs et le roc où se trouvait la "Kiva", les "Enfers". Les "Kachinas" étaient venus des profondeurs de l'espace et touchaient maintenant la Terre. Pour les Indiens Hopis dans la Kiva, il s'agissait de formes spirituelles des phénomènes naturels, de messages entre les hommes et les forces supérieures comme Sotugnangu, le Dieu du ciel, des nuages et de la pluie. Les Indiens confiaient aux Kachinas leurs prières appelant la pluie pour leurs champs de maïs, afin qu'ils les acheminent jusqu'au Dieu. Les esprits étaient des partenaires familiers, des phénomènes de la nature dont les Hopis se sentaient proches. Chez les voisins des Hopis, les Navajos, j'eus le privilège de voir quel rôle ces rituels qui célèbrent l'union de l'homme et du cosmos joue aujourd'hui encore dans la vie quotidienne. Wayne, fermier et médecin-sorcier qui m'accueillit dans sa grande famille pour un an, faisait chaque matin une prière au soleil couchant en répandant un peu de pollen de maïs, symbole de toute existence. Jamais je n'oublierai le jour où il me fit venir et me dit de tendre les bras vers l'Est et d'amener à ma bouche l'air du crépuscule, de l'inspirer. Une brise caressait l'herbe du désert; une odeur de sauge et d'aiguilles de pin planait dans l'air. Une traînée rouge annonçait le soleil, grimpait lentement pendant que Wayne traduisait l'une des vieilles chansons de prières "de l'arrière grand-père": "De la maison du grand matin, le bonheur viendra vers moi. De là où tombe le premier rayon du soleil, le bonheur viendra vers moi, des rochers striés, de l'écho des parois du canyon et de là où souffle le vent bleu..."
Le grand matin est le moment sacré, me dit-il.

Дети Вселенной: «счастье от дома из раннего утра»

"Tu rencontres le "Sacré qui parle"; le rouge à l'horizon est sa forme extérieure. Tu inspires la santé et la beauté à ce moment de la journée". La beauté "Bike hoscho", est pour les Navajos une force motrice essentielle de l'Univers. "Saa naghai" signifie "longue vie" et est employé comme synonyme de "Père Ciel". Au cours de ses cérémonies de guérison et de soirées tout à fait normales au sein de sa famille, Wayne ne se lassait pas de répéter son credo des corrélations de tous les éléments du cosmos. Le hogan rond des Navajos n'est pas seulement un habitat, c'est également un lieu sacré et l'exhortation à ne pas oublier la situation donnée: le toit en coupole de la hutte représente l'Univers, le sol d'argile nu symbolise la Terre Mère et les cérémonies rappellent aux participants leur place dans l'ordre du monde. "Pensez au feu, à l'eau, à la terre", disait-il souvent. "Leurs formes intrinsèques nous conduisent. Remerciez la Terre Mère et le Père Ciel, remerciez le soleil et l'air. Remerciez la pluie qui pénètre jusqu'aux racines de notre maïs et le fait pousser. Nous sommes assis dans un hogan de bois et de terre. Tout est relié. Tout le bien viendra cette nuit, de l'Est, du Sud, de l'Ouest et du Nord..." En cas de dispute dans la famille, tous se rassemblaient, grands-parents, parents et marmaille, autour d'un petit tas de charbon de bois incandescent que Wayne amoncelait au milieu du sol de la hutte. Le conflit était discuté, chacun parlant en direction du feu. Le "cadre cosmique" du hogan, me semblait-il, favorisait l'acceptation d'un compromis par l'âme offensée. Cela me touchait souvent profondément de voir avec quelle franchise et avec quelle chaleur les trois générations d'une famille témoignaient devant le feu leur amour réciproque et leur amour de la "Terre Mère" et du "Père Ciel", comme elles ressentaient la réalité de l'écoulement mutuel du temps et de l'espace, du fait d'être enveloppés dans une identité et un ordre mutuels et comme, malgré toutes les cassures dans leur civilisation, elles semblaient ressentir un destin commun avec le cosmos. Du moins pour les grands-parents et les parents, l'image qu'ils avaient du monde correspondait encore en gros à la réalité de leur existence. Avec leurs champs de maïs et de haricots, leurs moutons et leurs chèvres, ils pouvaient traverser les saisons comme leurs ancêtres à un rythme relativement stable, compter sur le retour du soleil, de la pluie et de la neige dont dépend leur survie. Pour eux, prendre selon une vieille coutume des Navajos de la terre dans sa main en parlant et la laisser ruisseler entre ses doigts avait encore une signification.

Звук, будто от змеиной погремушки, донесся из ночи через лаз вверху каменной норы. Все глаза обратились вверх, где цепочка раскрашенных тел с пернатыми масками на головах скользила по приставной лестнице, текла вдоль стен, потом замирала. Раздались мерные удары в барабан, поднялись руки с трещотками, торсы склонились, и, когда вся цепочка, топнув в такт ногами, присела, вдруг из-под кожаных масок раздалось пение — низкое, вибрирующее, неземное. Круг то сужался, то раздавался; тяжело топая ногами, «кахина» меняли направление движения в танце, не сбиваясь с ритма и не нарушая пения. Топание и пение становились все громче, скала, в которой была заключена «кива», т. е. «подземный мир», резонировала. «Кахина» явились из глубины космоса, теперь они были на земле. Для индейцев гопи, собравшихся в кива, это были духовные формы природных явлений, посредники между людьми и высшими силами, например богом неба, облаков и дождя Сатугнангу. Индейцы напутствовали кахина просьбами о дожде для кукурузных полей. Духи были для них добрыми помощниками, формой проявления природы, которая была им близка и понятна. У соседей гопи, индейцев навахо, мне довелось увидеть, какую роль играют в повседневной жизни древние ритуалы, посвященные связи человека с космосом. Фермер и лекарь Уэйн, принявший меня на год в свою большую семью, каждое утро молился восходящему солнцу и сыпал кукурузную пыльцу — символ всего живого. Никогда не забуду день, когда он позвал меня к себе и сказал, что я должен протянуть руки к востоку и поднести ко рту рассветный воздух, вдохнуть его. По траве пустыни бежал бриз, пахло шалфеем и хвоей пиний. Красная полоска на небе, медленно поднимаясь, возвещала солнце, и Уэйн переводил для меня древнюю молитвенную песнь о «прадеде»: «От дома из раннего утра придет ко мне счастье. Оттуда, где солнечный свет появляется первым, придет ко мне счастье, от полосатых скал, от эха каньона и оттуда, где ходит голубой ветер...»

Он сказал, что раннее утро — самое святое время. «Ты встречаешь 'Святого, который говорит'; красный свет на горизонте — это его внешняя форма. В эту пору ты вдыхаешь здоровье и красоту.» Красота, «бике хошо» — это, по представлениям навахо, главная движущая сила вселенной. «Саа нагхай» означает 'долгая жизнь' и является синонимом «Отца Небо». Во время обрядов исцеления или просто вечером в кругу семьи Уэйн не уставал повторять свое кредо о взаимосвязи всех вещей в космосе. Круглый хоган навахо — это не только жилище, но и святое место, и символ-напоминание о связи вещей: крыша хижины символизирует вселенную, глиняный пол — Мать Землю, а обряды напоминают участникам об их месте в мироздании. «Думайте об огне, земле и воде», — часто повторял Уэйн. «Их внутренние формы направляют нас. Благодарите Мать Землю, благодарите Отца Небо, благодарите солнце и воздух. Благодарите дождь: он проникает к корням кукурузы и питает их. Мы сидим в хогане, который сделан из дерева и земли. Все друг с другом связано. Все доброе придет нынче ночью, с востока, юга, запада и севера...» Когда бывали семейные неурядицы, все — старики, родители, дети — собирались вокруг раскаленных угольев, которые Уэйн насыпал на земляной пол посредине хогана. Они обсуждали конфликт, но каждый говорил, обращаясь к огню. Космическое обрамление жилища, как мне казалось, способствовало готовности пойти на компромисс, даже для обиженной души. Меня глубоко трогало, с какой откровенностью и с каким теплом все три поколения одной семьи высказывали перед огнем свои симпатии друг к другу и свою любовь к «Матери Земле и Отцу Небу», как естественно и реально переживали они свою сплоченность в потоке времени и пространстве, свою принадлежность к общему порядку и роду и как они, вопреки всем культурным потрясениям, чувствовали общность своей судьбы с космосом. Хотя бы для старшего и среднего поколения реальная жизнь в общем и целом еще отвечала их мировоззрению. Со своей кукурузой и бобами, с овцами и козами они, как их предки, могли в устойчивом ритме идти сквозь череду времен года, твердо полагаясь на солнце, на дождь и на снег, с которыми тесно связана их жизнь. Для них пока еще не был лишен смысла обычай племени навахо — во время разговора взять пригоршню земли и тихонько сыпать ее между пальцами.

PROFANE

COSMOS

Harrah's

EXCALIBUR

In bizarre Ersatzwelten gerät der Besucher im „Neuen Las Vegas“. Die Casino-Biosphären sind fast völlig autark, zumindest was Konsum- und Unterhaltungswünsche betrifft. Die „Virtual Reality Shows“ des „Luxor“, „MGM“ und „Mirage“ lassen die Grenzen zwischen echt und artifiziell verschwimmen.

In bizarre imitatiewerelden komt de bezoeker in het "Nieuwe Las Vegas" terecht. De casinobiosferen zijn bijna geheel autarkisch, tenminste wat consumptie- en amusementswensen betreft. De "Virtual Reality Shows" van "Luxor", "MGM, en "Mirage" laten de grenzen tussen echt en namaak vervagen.

The visitor to the "New Las Vegas" enters bizarre substitute worlds. The casino biospheres are also completely self-sufficient, at least as regards consumption and entertainment wishes. The "Virtual Reality Shows" of the "Luxor", the "MGM" and the "Mirage" blur the boundaries between what is real and what is artificial.

Chi visita la "Nuova Las Vegas" si ritrova a vagare in bizzarri mondi virtuali. Le biosfere dei casinò sono quasi completamente autarchiche, per quanto concerne la soddisfazione di desideri di consumo e divertimento. I "Virtual Reality Shows" dei"Luxor", "MGM" e "Mirage" fanno confondere i confini tra reale e artificiale.

Le visiteur du "Nouveau Las Vegas" se retrouve dans des mondes bizarres. Les biosphères des casinos sont quasiment autarciques, du moins en ce qui concerne consommation et amusement. Les "Virtual Reality Shows" du "Luxor", "MGM" et "Mirage" font s'évanouir les limites entre le réel et l'artificiel.

В «Новом Лас-Вегасе» посетитель попадает в странный эрзац-мир. «Биосфера» казино почти автономна, можно развлекаться и делать покупки. «Виртуальные шоу» — «Луксор», «МГМ» и «Мираж» стирают грань между реальным и искусственным миром.

TRI OCEANIC COR

CASH
Most Cards Accepted
Simply Try Yours!
STAR
PLUS
Cirrus
VISA
HONOR
Express Cash

ROBOT

Noch nie zuvor hat man sich auf dieser Erde der Aufgabe, das Unmögliche möglich zu machen, so energisch zugewandt wie heute; der Aufgabe, unsere kurzen Jahre hier mit allen damit verbundenen Freuden bis in die Ewigkeit zu verlängern. Und trotzdem lauerte hinter dem Unternehmungsgeist das Gefühl, daß der ganze Glanz nur Schein war, daß all die massiven Mittel nur einen traumähnlichen Zustand herbeiführen und die Tatsachen nicht im geringsten ändern würden.

Nog nooit tevoren heeft men zich op deze aarde met zoveel energie gewijd aan de taak het onmogelijke mogelijk te maken; de taak onze korte jaren hier met alle ermee verbonden vreugden tot in alle eeuwigheid te verlengen. En desalniettemin loert achter de ondernemingsgeest het gevoel dat de hele schitterende pracht alleen maar schijn was; dat al die intensieve middelen slechts een droomachtige toestand teweegbrengen en de feiten niet in het minst zullen veranderen.

Never on this earth was the task of turning the impossible into the possible addressed with so much energy; the task of extending the brief span of man's years together with all of its delights, into eternity. Nevertheless, behind the enterprise there lurked the feeling that all of the splendor was only illusory; that all the massive means would produce only a haunting dreamlike state of existence and not really change the facts the least bit. The body, in spite of the magic, still would not be alive.

Mai prima d'ora ci si era dedicati con energia a rendere possibile l'impossibile; al compito di prolungare il tempo concesso all'uomo su questa terra, con tutte le gioie che la vita può comportare. E tuttavia, dietro allo spirito d'intraprendenza si nasconde la sensazione che questo splendore non sia altro che apparenza; che questi potenti mezzi non siano volti a produrre niente altro che uno stato della coscienza simile al sogno e che non riescano a cambiare neanche minimamente la realtà.

Jamais encore sur cette terre on ne s'est fixé avec tant d'énergie le but de rendre possible l'impossible, le but de prolonger jusqu'à l'éternité nos courtes années grâce à tous les plaisirs qu'elles procurent. Et pourtant, derrière cette entreprise, un sentiment nous guête qui nous souffle que toute cette splendeur n'est qu'illusoire, que tous les moyens employés ne produisent qu'un état chimérique et ne changeront en rien les faits. Malgré toute la magie deployée, le corps ne pourra pas rester en vie.

Никогда прежде на свете не брались за задачу сделать невозможное возможным с такой энергией: мол, нашу краткую жизнь со всеми ее радостями надо продлить до бесконечности. И все же дух предприимчивости не заглушил чувство, что весь этот блеск только кажущийся, что все могучие средства способны только дать иллюзию, но ничуть не могут изменить факты.

STUPAK

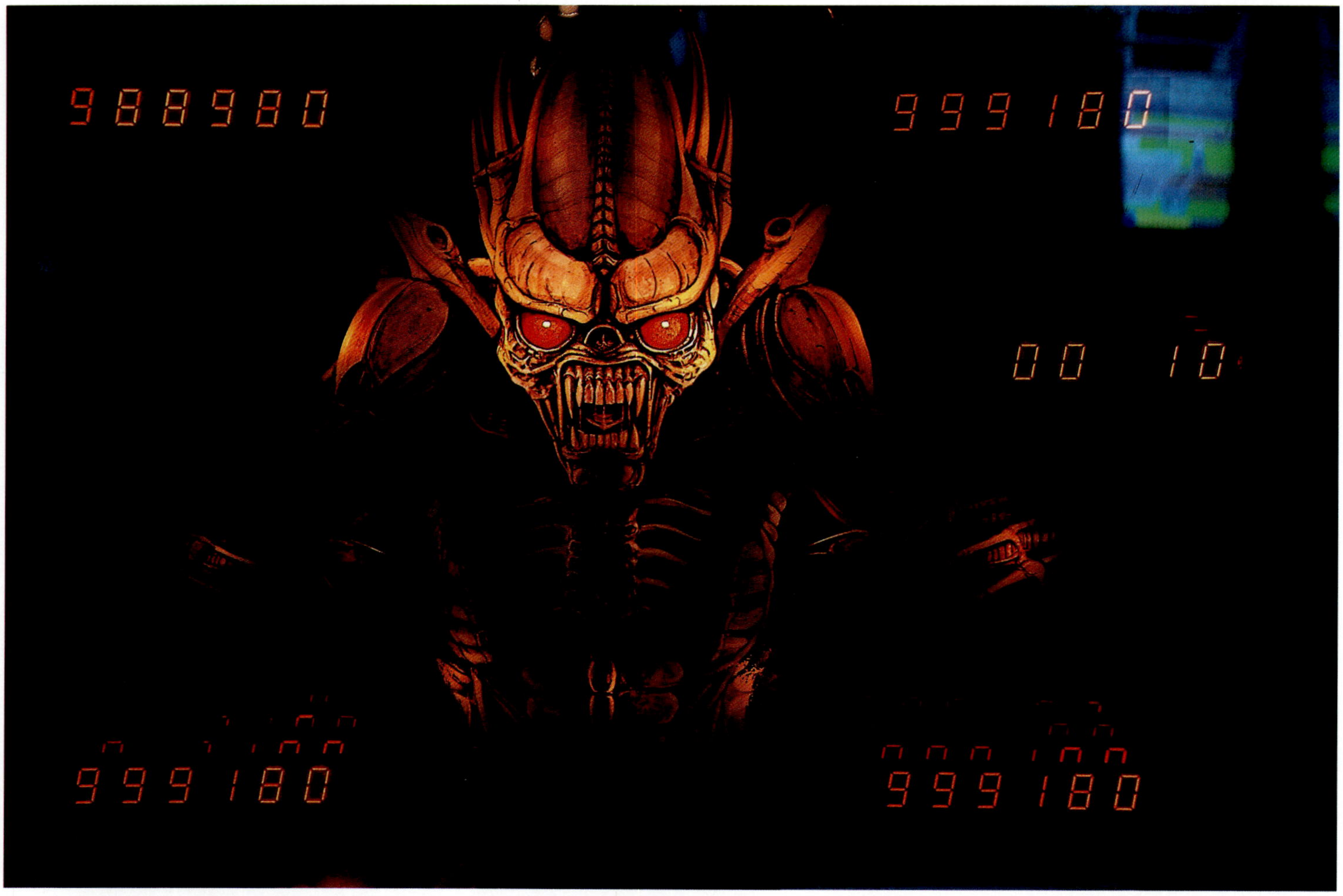
988980
999180
00 10
999180
999180

TERMINATOR 2
JUDGMENT DAY
Pull Trigger
To Shoot Ball

Jede Stadt ist eine ewige Stadt: das zivilisierte Geld hält für immer; die Zeit und die Stadt gewinnen hinzu. Bloß ausharren heißt den Tod zu besiegen. Die Zivilisation ist ein Versuch, den Tod zu besiegen. Der Ehrgeiz des zivilisierten Menschen wird in der Stadt offenbart - hier ruht seine Hoffnung auf die Ewigkeit... Diese Menschen hielten sich für Gott. Das heißt, sie waren verrückt und noch dazu wurden sie in diesem Glauben unterstützt, geschmeichelt und ermutigt durch ihre Geistlichen, ihre Frauen und von allen, die sich auch für Gott hielten.

Elke stad is een eeuwige stad: het geciviliseerde geld houdt eeuwig; de tijd en de stad krijgen er iets bij. Alleen geduldig wachten betekent de dood overwinnen.
De civilisatie is een poging de dood te overwinnen. De eerzucht van de geciviliseerde mensen wordt in de stad geopenbaard - hier rust zijn hoop op eeuwigheid: hier wordt zijn rente op rente opgestapeld... Deze mensen dachten dat zij God waren. Zij waren dwaas en werden bovendien in hun geloof gesteund door hun geestelijken, hun vrouwen en door allen die zich ook God waanden.

Every city is an eternal city: civilized money lasts forever. It endures; time and the city accumulate. But to endure is to conquer death. Civilization is an attempt to overcome death. The ambition of civilized man is revealed in the city where reposes both the hope of immortality and the fruit of compound interest... These people supposed that they were god. That is to say, they were madmen. Moreover, they were supported in this belief, taught, flattered, and encouraged, by their clergy, parents, wives, and all who also thought they were god.

Daniel Brown

Ogni città è una città eterna: il denaro della civiltà dura per sempre. Tener duro significa avere la meglio sulla morte. La civiltà è un tentativo di superare la morte. L'ambizione dell'uomo civilizzato si rivela nella città, qui egli ripone le sue speranze di immortaità, qui si accumulano gli interessi degli interessi... Questi individui credevano di essere dèi. Erano cioè pazzi e come se non bastasse questa loro illusione era nutrita, adulata ed incoraggiata dai loro sacerdoti e dalle loro donne e da tutti coloro che credevano anch'essi di essere degli dèi.

Chaque ville est une ville éternelle: l'argent civilisé est impérissable; le temps et la ville amoncèlent. Persévérer signifie vaincre la mort. La civilisation essaye de surmonter la mort. L'ambition de l'homme civilisé se révèle dans la ville où repose à la fois son espoir d'immortalité et le fruit de ses intérêts composés... Ces gens se prenaient pour Dieu. Autrement dit, ils étaient fous. De plus, leur clergé, leurs parents, leurs épouses et tous ceux qui se prenaient aussi pour Dieu appuyaient leur foi, les instruisaient, les flattaient et les encourageaient.

Каждый город — вечный: цивилизованные деньги вечные. Выстоять — значит победить смерть. Цивлизация — попытка преодолеть ее. Тщеславие цивилизованного человека проявляется в городе: здесь его надежда на вечность, здесь растут его проценты на проценты... Эти люди считали себя богами. Они были сумасшедшими и к тому же в этой вере их укрепляли, поучали и ободряли священники, жены и те, кто тоже считали себя богами. Все общество было сумасшедшим.

RARA

DISE

Virtuelle Welten: Seltsame Tage in Las Vegas

Mein Gott, was war es schön, nach drei Tagen in Las Vegas wieder zurück zu sein in der Wüste! Die Stadt lag nur eine halbe Autostunde entfernt hinter einer Hügelkette, aber hier in dem abgelegenen Tal der Mojave Desert hätte man glauben können, auf einem unberührten Planeten zu sein. Ich setzte mich auf von meinem Nachtlager in einem trockenen Bachbett, fuhr mit der Hand durch den Sand - und fühlte mich auf eine seltsame Weise wohl dabei. Wie belebend hier alles war, selbst das Zusammenspiel der Dünenrippen und Eidechsenspuren, der Kräuterdüfte und des Gräserflüsterns. Es gab Klarheit und unsichtbare Ordnung, die einen hineinzogen in die Wirklichkeit jenseits der menschgemachten Welt - eine Wirklichkeit, in der man sonderbar leicht dahinging, um dann von woanders als aus dem Kopf heraus zu sagen, daß es ganz wunderbare Tage gibt.

Noch freilich spukten in meinem Gehirn andere Bilder, völlig andere Gedanken. Ich sah die Menschen vor mir, wie sie Reihe um Reihe an den Spielautomaten saßen, Robotern in Fabrikhallen gleich, Zehntausende von ihnen, den Blick starr auf die rotierenden Scheiben gerichtet, in immer gleichen Bewegungen Münzen einwerfend, scheinbar losgelöst von der Welt, wie gefangen in der eigenen Realitätskapsel, und dies Stunde um Stunde rund um die Uhr. Hektisches Rauchen und fahrige Griffe zum Cocktailglas mochten etwas von dem verraten, was wirklich vorging in manchen Spielern. Blieb man lange an einem Platz stehen und saugte die Emotionen ein und betrachtete die Gesichter, dann war es manchmal als würde man sie spüren können, die Gier nach schnellem Glück, die das innerste Getriebe der Stadt in Gang hielt. Gerade war ein „Neues Las Vegas" eröffnet worden - mit viel High-Tech aufgepeppte Ersatzkosmen, autarke Kunstwelten, die man zu keiner Zeit mehr verlassen mußte, um sich einen Konsumwunsch zu erfüllen. Der wirkliche Kosmos - Wüste, Sonne, Mond und Sterne - kam mit den neuen Milliarden-Dollar-Bauten nicht mehr in Berührung: Klimaanlagen, verspiegelte Wände und sanft getöntes Licht sorgten dafür, daß im synthetischen Kosmos Tag und Nacht angenehmste Bedingungen herrschten. Selbst für Leute, die es hin und wieder nach etwas Natur und unterhaltsamer Bildung verlangte, hatte man mit einem „Themen"-Sammelsurium aus dem Bilderbuch der Romantik vorgesorgt. Sozusagen in einer verglasten „Grotte" zogen weiße Tiger zwischen herabströmenden Wassern und überfließenden Teichen ihre neurotische Kreisbahn, anderswo sprangen Delphine durch Ringe oder es amüsierten lebensgroße Stofftiere mit Motorantrieb. Holographisch in die Wasserschleier eines Springbrunnens gezaubert, schwebte geisterhaft Tut-Ench-Amuns Goldmaske vor der Cheops-Pyramide und der Sphinx. In einer „Schatzinsel"-Lagune wiederum schlugen alle 90 Minuten 22 Matrosen und Piraten eine Seeschlacht mit echten Feuern und Explosionen, aufwühlender Musik und einem sinkenden Schiff. Und gleich nebenan inszenierte ein Computer spektakuläre „Vulkanausbrüche": Von verborgenen Gasleitungen gespeiste Flammen loderten, knatterten und röhrten, Wasserfontänen schossen himmelwärts, Wogen erhoben sich brodelnd, weißer Dampf zischte, magisches Licht explodierte in Gelb, Rot und Violett. Ständig wurden irgendwo der Langeweile vorbeugende Kitschhäppchen angeboten in den hypereklektischen Ersatzwelten mit Namen wie „Mirage" und „Mystère", „Sahara" und „Dunes", „Excalibur" und „Tropicana". Als Cäsar und Kleopatra kostümierte Schauspieler schlenderten da im „Caesars Palace" würdevoll grüßend zwischen den Spielenden dahin. Im „MGM" verfolgten Kinder professionell gespielte Märchenhandlungen in einem künstlichen Wald, im „Luxor" schwebte dem Besucher ein beachtlich echt wirkendes Geistwesen vor der Nase herum, wenn er für die „Virtual Reality Show"mit dem Titel „Der verlorene Obelisk" eine 3-D-Brille aufgesetzt hatte.

Ließ man sich auf diese Weise treiben in den bizarren Casino-Kosmen, beschlich einen das Gefühl, hier werde der künftige Way of Life der gesamten Welt erprobt: ein Lebensstil, der sich noch weitergehend als bisher an den Bedürfnissen der Menschen orientiert, den natürlichen Kosmos aber kaum mehr berücksichtigt. In Las Vegas, der derzeit am schnellsten wachsenden Stadt der USA, hat etwa extrem verschwenderischer Wasserverbrauch bereits dazu geführt, daß jetzt Grundwasser der Mojave-Wüste angezapft werden muß, um Golfplätze, Rasen und künstliche Seen weiter versorgen zu können. 20 Millionen Menschen aus aller Welt besuchen alljährlich die Stadt - und es wird alles getan, um ihnen das Gefühl zu geben, in einem Schlaraffenland gelandet zu sein, nicht in einem fragilen Wüstenbiotop. Dessen Tiere und Pflanzen haben die Jahrtausende nur überlebt, indem sie sich durch genügsame Lebensweise an die natürlichen Verhältnisse angepaßt haben. Daß in Boom-Zeiten niemand eine solche Botschaft hören möchte, ist verständlich. Viel populärer ist die Illusion, daß wir Menschen sicher seien in unseren menschgemachten Welten.

Virtuele werelden: Merkwaardige dagen in Las Vegas

Hemeltje lief, wat was het heerlijk na drie dagen in Las Vegas weer terug te zijn in de woestijn! De stad lag op een afstand van maar een half uurtje autorijden achter een heuvelrug. Maar hier in het afgelegen dal van de Mojave Desert had men kunnen geloven op een onbewoonde planeet te zijn. Ik stond op van mijn bivak in een droge bedding van een beek, streek met mijn hand door het zand - en ondervond daarbij een merkwaardig prettig gevoel. Wat voor een stimulerende werking alles hier had. Zelfs het samenspel van de duinribbelingen en de sporen van een hagedis, de kruidengeuren en het fluisteren van het gras. Er was duidelijkheid en onzichtbare orde waardoor men in de werkelijkheid aan gene zijde van de door de mens geschapen wereld werd getrokken - een werkelijkheid waarin men merkwaardig gemakkelijk liep om dan met een gevoel dat ergens heel anders vandaan komt, te zeggen dat er zeer wonderbaarlijke dagen bestaan.

Natuurlijk spookten er in mijn hersenen nog andere beelden, geheel andere gedachten. Ik zag de mensen voor mij hoe zij rij na rij achter de speelautomaten zaten, net als robots in fabriekshallen, tienduizenden ervan, de blik strak gericht op de roterende schijven, er met altijd dezelfde bewegingen munten in gooiend, schijnbaar afgescheiden van de wereld, als gevangenen in hun eigen realiteitscapsule, en dat urenlang, dag en nacht. Koortsachtig trekjes aan de sigaret en ongecontroleerde bewegingen naar het cocktailglas brachten iets aan het licht van hetgeen zich bij sommige spelers werkelijk afspeelde. Als men lang op één plaats bleef staan, de emoties inademde en de gezichten bekeek, was het soms net alsof de zucht naar het snelle geluk die de interne machinerie van de stad in gang hield, te bespeuren was. Zojuist was een "Nieuw Las Vegas" geopend - met veel hightech opgepepte imitatiekosmossen, autarkische kunstwerelden die je nooit meer hoefde te verlaten om een consumptiewens te vervullen. De werkelijke kosmos - woestijn, zon, maan en sterren, dakloze bedelaars - kwam met de nieuwe miljoenen dollar kostende gebouwen niet meer in aanraking: airconditioning-installaties, verspiegelde wanden en zacht getint licht zorgden ervoor dat er in de synthetische kosmos dag en nacht alleraangenaamste omstandigheden heersten. Zelfs voor mensen die zo af en toe het verlangen naar een beetje natuur en cultureel amusement hadden, was voorzorg genomen door een mengelmoes van onderwerpen uit het prentenboek der romantiek. Zo liepen witte tijgers in een "grot" van glas neurotisch tussen overstromende vijvers en naar beneden

Virtual worlds: Strange days in Las Vegas

stortende wateren te ijsberen, ergens anders sprongen dolfijnen door ringen of zorgden levensgrote pluche beesten met motoraandrijving voor amusement. Holografisch in de watersluiers van een fontein getoverd zweefde het gouden masker van Toetanchamon voor de piramide van Cheops en de sfinx. In een lagune van "Schatteneiland" sloegen om het anderhalf uur 22 matrozen en piraten weer een zeeslag met echte beschietingen en explosies, met aangrijpende muziek en een zinkend schip. En direct ernaast ensceneerde een computer spectaculaire "vulkaanuitbarstingen": Vlammen, via verborgen gasleidingen gevoed, laaiden, knetterden en loeiden, waterfonteinen schoten hemelwaarts, golven borrelden gistend omhoog, witte damp siste, magisch licht explodeerde in geel, rood en violet. Voortdurend werden ergens om verveling tegen te gaan, kitsch-hapjes aangeboden in de hypereclectische imitatiewerelden met namen als "Mirage" en "Mystère", "Sahara" en "Dunes", "Excalibur" en "Tropicana". Als Caesar en Cleopatra verklede acteurs liepen daar in "Caesar's Palace" plechtig groetend tussen de spelers door. In "MGM" volgden kinderen professioneel gespeelde sprookjeshandelingen in een kunstbos, in "Luxor" zweefde de bezoeker een opmerkelijk echt ogend geestwezen voor de neus rond, als hij voor de "Virtual Reality Show" met de titel "De verloren obelisk" een 3-D-bril had opgezet. Wanneer men zich op deze wijze in de bizarre casino-kosmossen liet drijven, kreeg men het gevoel dat hier de toekomstige "way of life" van de gehele wereld werd uitgetest: een levensstijl die nog meer dan tot nu toe gericht is op de behoeften van de mens, maar die met de natuurlijke kosmos nauwelijks meer rekening houdt. In Las Vegas, de momenteel snelst groeiende stad van de USA, heeft bijvoorbeeld het extreem kwistige waterverbruik er nu al toe geleid dat grondwater van de Mojavewoestijn moet worden afgetapt om golfterreinen, gazons en kunstmatige meren verder van water te kunnen voorzien. Ieder jaar brengen 20 miljoen mensen uit de gehele wereld een bezoek aan de stad. Er wordt alles voor gedaan om hun het gevoel te geven dat ze in een luilekkerland, en niet in een fragiele woestijnbiotoop zijn beland. De planten en dieren van deze woestijn hebben de duizenden jaren alleen maar overleefd, doordat ze zich door een genoegzame levenswijze aan de natuurlijke omstandigheden hebben aangepast. Het is begrijpelijk dat niemand graag zo'n boodschap wil horen wanneer het zakenleven bloeit. De illusie, dat wij mensen in onze door mensen gemaakte werelden veilig zijn, is veel populairder.

My god how nice it was to be back in the desert again after three days in Las Vegas! The city was only a half an hour away by car behind the chain of hills but here in this remote valley in the Mojave Desert you could easily believe yourself to be on a virgin planet. My camp for the night was in the dry bed of a brook. I sat up straight and ran my fingers through the sand - and felt strangely at ease as I did so. How invigorating everything was here, even the interplay of the ribbed patterns on the dunes and the tracks of the lizards, the fragrance of the herbs and the whispering of the grass. There was clarity and an invisible order which drew one in into a reality beyond the man-made world - a reality in which one moved with such a strange lightness, to then say from somewhere other than the head, that life holds absolutely wonderful days.
But in my mind I of course still saw other pictures, and I had thoughts of a completely different kind. I saw the people before my eyes, line upon line of them, as they sat in front of the gaming automats, like robots in factory halls, tens of thousands of them, their eyes fixed on the rotating disks, throwing money in the slots with the same movement all the time, seemingly removed from this world, imprisoned in their own reality-capsule, and this hour after hour, round-the-clock. Heavy smoking and hasty grasps in the direction of their cocktail glasses betray perhaps something of what was really going on in the minds and hearts of some of the gamblers. If one remained for a long time at one spot and sucked up the emotions and observed the faces then it was sometimes as if one could feel that greed for quickly secured happiness which kept the whole inner mechanism of the city going. A "new Las Vegas" had just been opened - with a lot of high-tech-enriched substitute cosmoses, self-supporting artificial worlds which one would not have to leave for a second if it were only a matter to fulfilling a consumer wish of some kind. The real cosmos - desert, sun, sand and stars, beggars without a roof over their head - did not have any contact whatsoever with those new billion-dollar constructions: air conditioning, mirrors on the wall, and glare-free lighting guarantee the most pleasant of conditions day and night in this synthetic cosmos. Even the needs of people who now and again feel the urge to experience a bit of nature and to improve on their general knowledge in entertaining fashion, had been catered for - with a mixture of topics taken from the picture book of Romanticism. White tigers paced their neurotic circles in a glass-covered grotto between waterfalls and overflowing ponds; dolphins sprang through rings; and mechanically-driven oversize cuddly animals served or were meant to serve the purposes of entertainment. Holographically created in the veil of water produced by a fountain, there floated ghostlike Tut Ench Amun's golden mask with the Cheops Pyramid and the Sphinx in the background. In a "Treasure Island" lagoon, 22 seamen and pirates fought out every 90 minutes a sea battle with real fires and explosions, stirring music and a sinking ship. And, right alongside, a computer simulates spectacular "vulcanic eruptions": flames fed by hidden gas pipelines flared up, crackling and roaring, jets of water shot towards the heavens, waves seethed, white steam hissed, magic light exploded in yellow, red and violet colours. Morsels of kitsch were constantly on offer everywhere to prevent boredom arising in the hyper-eclectic substitute worlds bearing names like "Mirage" and "Mystère", "Sahara" and "Dunes", "Excalibur" and "Tropicana". In "Caesar's Palace" actors dressed up as Caesar and Cleopatra strolled about among the customers greeting them in dignified fashion. In "MGM" children followed professionally-acted fairy tale plots in an artificial forest; in the "Luxor" a pretty genuine-looking ghostly being floated in front of the visitor's nose when he put on 3-D glasses for the Virtual Reality show going under the name "The Lost Obelisk". If one drifts around the bizarre casino cosmoses then one gets the feeling that the future Way of Life of the entire world is being tested out: a style of life which, more than has been the case to date, orientates itself according to the needs of men but does not take the natural cosmos into account to any great degree. In Las Vegas, the fastest-growing city in the USA at present, the extremely wasteful consumption of water has already led to the tapping of the ground water of the Mojave Desert, so that golf courses, lawns and artificial lakes can continue to be watered. Twenty million people from all over the world visit the city ever year - and everything is undertaken to ensure that they have the feeling that they have arrived in a Land of Milk and Honey and not in a fragile desert biotope where the plants and animals have only survived over the millennia by adapting to the prevailing natural conditions through being content with very little. It is understandable that in boom-times no-one wants to hear such a message. What is much more popular is the illusion that we humans are secure in our man-made worlds.

Prudenza con i fiammiferi, il riposo illuminato, riempie di miele i cervelli

„Non bisogna lasciar giocare gli intellettuali con i fiammiferi ..." scriveva Jacques Prévert in una delle sue sorprendenti poesie; per lo meno non tutti, aggiungiamo noi, pensando in che modo gli esoterici si distinguevano per essere tanto selezionati, ovvero pochi e preparatissimi. Esoterico, infatti, vuol dire essere dentro, in modo riservato, segreto. Poi Aristotele allargò cautamente il campo all'esterno, raccogliendo discepoli anche poco acculturati, affidabili forse, per quanto bisognasse esser prudenti con i fiammiferi. Infatti, più di un intellettuale „Lavora arbitrariamente / Erigendo a se stesso", come dice Prévert, "Un auto-monumento".
In ogni caso è sempre il grande Jacques che parla, "Il mondo mentale / Mente / Monumentalmente." Forse è esagerato generalizzare; nell'antichità, tuttavia, nella civiltà della Grecia, le scuole erano a numero chiuso e l'attenzione di Aristotele ritornò ai soli iniziati. Soprattutto quando si parlava di scienza, di riti esoterici che introducevano nei misteri più remoti, la cui conoscenza non poteva essere diffusa fra i profani. La pratica del cielo richiedeva questi parametri, con insite le difficoltà del capire, del dedurre e dello studiare, soprattutto di spiegare gli arcani. Il principio oggi vale più che mai, non per selezione imposta dall'alto, ma per generazionale pigrizia, per diseducazione permanente. Un problema inverso, insomma.
Così che i fenomeni astrali sono assoggettati a sommarie interpretazioni di parte, come avvenne per il mistero dei dischi volanti, considerati macchine spia di uno dei blocchi contrapposti, prima della caduta del muro di Berlino. Per fortuna c'erano Marte ed i marziani di riserva chiamati in ballo con qualche ragionevolezza ad osservare, forse a giudicare la Terra.
Nel volume "Mythologies", pubblicato a Parigi nel 1957, Roland Barthes, infatti scriveva: "Il dato più significativo è che in tal modo Marte viene implicitamente dotato di un determinismo storico ricalcato su quello della Terra. Se i dischi sono i veicoli di geografi marziani venuti ad osservare la configurazione della Terra, come ha detto chiaro e tondo non so quale scienziato americano, e come molti con certezza pensano tra sé, si deve al fatto che la storia di Marte è maturata con lo stesso ritmo di quella del nostro mondo, e produce geografi nello stesso secolo in cui abbiamo scoperto la geografia e la fotografia aerea. Il solo vantaggio è quello del veicolo in sé, Marte venendo così a configurarsi come una Terra sognata, provvista di ali perfette come in tutti i sogni di idealizzazione. È probabile che se a nostra volta sbarcassimo su Marte quale lo abbiamo costruito non vi troveremmo che la terra stessa ..."
Esiste una setta religiosa indiana, tra pagana e buddista, vale a dire il trantismo, che collega la mezzanotte con il riposo illuminato, splendendo soltanto a quell'ora il sole spirituale contrapposto a quello fisico del mezzodì. Pertanto, l'avviarsi alla conoscenza dei misteri antichi era riservata a quel misterioso sole notturno. Vale a dire, al persuadersi di quel paradosso interiore che si affida alla notte luminosa ed al mezzogiorno oscuro.
Teorie del tutto contrarie alla stupenda poetica di Charles Baudelaire, l'autore de "I fiori del male", che a proposito del Sole, scrive: "Quel padre che nutre ogni cosa, nemico della clorosi, suscita nei campi i versi come le rose; fa evaporare su in cielo gli affanni e riempie di miele i cervelli e le arnie ... e comanda alle messi di crescere e di maturare nel cuore immortale che vuole sempre fiorire!" Secondo la Bibbia, infatti, vedere Dio faccia a faccia significa vederlo nella gloria di mezzogiorno, momento senza ombra, ossia nella luce piena del sole che, peraltro, illumina ogni fatto dell'Antico Testamento.
Persino le impressionanti visioni della gloria divina di Ezechiele, che la scienza esoterica ha sempre cercato di interpretare, oggi più tangibilmente emulata dallo studio dei fenomeni extraterrestri in archeologia, astronomia e mitologia.
In fondo, potrebbe essere considerata una delle diverse piste per trovare la chiave dello ignoto che l'esoterismo ha cercato di raggiungere pure attraverso i tarocchi in relazione con lo Zodiaco e con i pianeti. Infatti i Tarocchi, un gioco di carte fra i più remoti, si basa sull'azione di un intero mondo di simboli ed il cosmo ne è una delle corsie preferenziali. Tarocchi astrologici, dunque, nei cui raggruppamenti zodiacali si caratterizzano l'inizio dell'evoluzione, ossia la nascita, il culmine, il regresso ovvero la caduta.
I segni del Fuoco, della Terra, dell'Aria e della Acqua collegati con l'Ariete, il Toro, i Gemelli, il Cancro, esprimono la prima fase; il culmine è nel Leone, nella Vergine, nella Bilancia e nello Scorpione; mentre alla caduta ci si avvia nel Sagittario, Capricorno, Acquario, Pesci. Una scienza perciò che ha nel cosmo una miniera incantevole di piste.

Giuseppe Brunamontini

Mondes virtuels: étranges journées à Las Vegas

Mon Dieu, quel plaisir de retrouver le désert après trois jours à Las Vegas! La ville n'était qu'à une demi-heure de voiture derrière une chaîne de collines, mais ici, dans la vallée isolée du désert de Mohave, on aurait pu se croire sur une planète vierge. Je m'assis sur ma couche que j'avais installée pour la nuit dans un lit sec de rivière, passai la main dans le sable et ressentis un étrange bien-être. Comme tout était vivifiant, même l'ensemble formé par les dunes cannelées et les traces des lézards, les plantes aromatiques et le chuchotement des herbes. La clarté et un ordre invisible m'attiraient dans une réalité au-delà du monde que l'homme avait fait, une réalité où la vie était étrangement facile et qui faisait dire, autrement qu'avec l'esprit, qu'il y avait des jours absolument merveilleux.
Mon cerveau, bien sûr, était encore hanté par d'autres images, par des pensées totalement différentes. Je voyais des rangées de personnes assises devant les machines à sous, comme des robots dans des ateliers de fabrication, des milliers de robots, le regard fixé sur les disques qui tournent, introduisant leurs pièces d'un mouvement stéréotype, apparemment détachés de ce monde, comme prisonniers dans une capsule contenant leur propre réalité, et cela 24 heures sur 24. Les cigarettes fumées fébrilement et le geste nerveux pour saisir le verre à cocktail trahissaient parfois vaguement l'état psychique de certains joueurs. Si l'on restait longtemps à la même place, s'imprégnant des émotions et observant les visages, on avait parfois l'impression de pouvoir le sentir, ce brûlant désir de faire vite fortune qui graisse les rouages de la ville. On venait juste d'ouvrir un "nouveau Las Vegas" avec des cosmos compensatoires élaborés par la technique de pointe, des mondes artificiels autarciques que l'on n'a plus besoin de jamais quitter pour satisfaire son appétit de consommation. Le véritable cosmos, désert, soleil, lune et étoiles, n'a plus aucun contact avec les nouveaux bâtiments de plusieurs miliards de dollars: la climatisation, les murs recouverts de miroirs et la lumière légèrement colorée assurent jour et nuit des conditions quasiment idéales dans ce cosmos synthétique. Des dispositions avaient même été prises pour les gens ayant un besoin sporadique de nature et de culture distrayante, leur offrant un ramassis de sujet tirés du livre d'images de l'époque romantique. Ainsi, dans une "grotte" vitrifiée, des tigres blancs suivaient inlassablement leurs sentiers neurotiques entre les cascades et les pièces d'eau communicantes; en d'autres endroits, des dauphins sautaient à travers des anneaux ou bien on s'amusait à regarder

Виртуальные миры: странные дни в Лас-Вегасе

des animaux en peluche grandeur nature actionnés par un moteur. La magie de l'holographie faisait apparaître dans le voile aquatique d'une fontaine le masque d'or de Toutankhamon qui flottait, spectral, devant la pyramide de Chéops et le Sphinx. Dans la lagune d'une "île au trésor", toutes les 90 minutes, 22 matelots et pirates se livraient une bataille navale avec incendies et explosions véritables, musique bouleversante et un navire qui sombrait. Et tout à côté, un ordinateur mettait en scène de spectaculaires "éruptions volcaniques": des flammes alimentées par des conduites de gaz invisibles flamboyaient, crépitaient et rugissaient, des fontaines d'eau s'élançaient vers le ciel, des vagues se gonflaient en bouillonnant, de la vapeur blanche sifflait et une lumière magique explosait en jaune, rouge et violet. Constamment et en tous lieux, on offrait des amusements de quatre sous pour prévenir l'ennui dans les mondes factices hyperéclectiques nommés "Mirage" et "Mystère", "Sahara" et "Dunes", "Excalibur" et "Tropicana". Des acteurs déguisés en César et en Cléopâtre déambulaient dans le "Caesars Palace", saluant les joueurs avec dignité. Dans les studios de "MGM", les enfants suivaient des contes joués de manière professionnelle dans une forêt artificielle; au "Luxor" un fantôme qu'on aurait pu croire vrai dansait devant le nez du visiteur qui avait mis des lunettes 3D pour le "Virtual Reality Show" intitulé "L'obélisque perdu".

Si l'on se laissait emporter ainsi dans ce cosmos bizarre de casino, on avait l'impression d'expérimenter ici le futur "Way of Life" du monde entier, un style de vie s'orientant encore plus sur les besoins humains,mais ne tenant pratiquement plus compte du cosmos naturel. A Las Vegas, la ville enregistrant actuellement l'expansion la plus grande aux Etats-Unis, la consommation abusive d'eau a déjà contraint à en soutirer de la nappe souterraine du désert de Mohave pour pouvoir continuer à alimenter les terrains de golf, les espaces verts et les lacs artificiels. Chaque année, 20 millions de personnes venues du monde entier visitent cette ville où tout est mis en oeuvre pour leur donner l'impression d'avoir atterri au pays de cocagne et non dans le fragile biotope d'une région désertique dont la flore et la faune n'ont survécu pendant des millénaires qu'en s'adaptant aux conditions naturelles par un mode de vie frugal. On comprendra certes que personne ne veuille entendre un tel message en période faste. Les humains lui préfèrent l'illusion d'être en sécurité dans les mondes qu'ils ont eux-mêmes créés.

Боже мой, до чего же хорошо было снова очутиться в пустыне после трех дней в Лас-Вегасе! Город лежал всего в получасе езды, за грядой холмов, но здесь, в уединенной долине пустыни Мохаве, можно было вообразить себя на необитаемой планете. Я уселся на своем спальнике, который расстелил в ложе высохшего ручья, провел рукой по песку и почувствовал странное удовлетворение. Какое здесь все было настоящее — гребни дюн и следы ящериц, запах трав и шелест ветра. Во всем были ясность и невидимый порядок, которые увлекали в другую действительность, не связанную с рукотворным миром, и эта действительность воспринималась уже не умом, а только чувствами как чудное и счастливое время.

В моей голове еще проносились совсем другие картины, иные мысли. Я видел перед собой людей, сидящих рядами перед игральными автоматами, десятки тысяч, как роботы в цеху, устремив неподвижный взгляд на вращающийся круг, деревянным движением опуская монету в автомат, отгородившись от мира коконом собственной реальности, — и так час за часом, круглые сутки. Торопливые затяжки сигаретой и рука, нервно схватывающая бокал с коктейлем, выдавали внутреннее состояние некоторых игроков. Если долго стоять на одном месте, впитывая эмоции и разглядывая лица, то, казалось, можно ее ощутить — жажду быстрых денег, которая и приводит в действие потайную пружину этого города. Только что открылся «Новый Лас-Вегас», напичканный электронными «космосами», автономными искусственными мирами, которые уже не надо покидать, если захочется что-то приобрести. Настоящий космос — пустыня, солнце, луна и звезды, бездомные нищие — никак не соприкоснулся с новым, стоившим миллиарды долларов: кондиционеры, зеркальные стены и мягкий свет создают в синтетическом космосе днем и ночью самую приятную атмосферу. Подумали даже о тех, кого порой влечет на природу, кому хочется слегка пополнить свои знания; для тех приготовлен целый винегрет романтических сюжетов из книжки с картинками: в застекленном гроте нервно ходят по кругу среди низвергающихся водопадов белые тигры; в другом месте дельфины прыгают сквозь кольца и развлекают публику механические звери в натуральную величину; голография золотой маски Тутанхамона в оправе из фонтанных струй висит на фоне пирамиды Хеопса и сфинкса; в лагуне «остова сокровищ» каждые 90 минут 22 матроса и пирата инсценируют под бодрящую музыку морское сражение с настоящим огнем, взрывами и тонущим кораблем. И тут же рядом компьютер устраивает впечатляющее «извержение вулкана»: ревущим факелом горит поступающий по трубам газ, взлетают в небо фонтаны воды, вздымаются бурлящие волны, шипит белый пар, феерически меняется освещение: желтый, красный, фиолетовый свет. Где-нибудь всегда имеется в качестве средства от скуки очередной кич из эрзац-миров с зазывными названиями «Мираж», «Тайна», «Сахара», «Дюны», «Тропикана». Актеры в костюмах Цезаря и Клеопатры прогуливаются во «дворце цезаря», церемонно раскланиваясь с игроками. В «Метро Голдвин Мейер» дети следят за сюжетом профессионально разыгрываемых сказок в искусственном лесу; в «Луксоре» перед носом посетителя, надевшего обязательные стереоочки, появляется мастерски сделанное жуткое привидение, участвующее в «виртуальном шоу» под названием «Потерянный обелиск».

Если погрузиться в эти балаганные космосы, ощутишь, что здесь опробируют будущий образ жизни всего мира, стиль жизни, ориентированный на потребности людей, но совершенно не учитывающий реального космоса. В Лас-Вегасе, наиболее быстро растущем городе США, расточительное использование воды привело, например, к тому, что теперь для полива полей для гольфа, газонов и питания искусственных озер пришлось брать грунтовые воды пустыни Мохаве. Каждый год город посещают 20 млн. человек со всего мира, и здесь делается все, чтобы они почувствовали себя в сказочной стране, а не в хрупком биотопе пустыни. Животные и растения тысячелетиями жили в пустыне за счет неприхотливости и приспособляемости к естественным условиям. Понятно, во времена бума никто не хочет слышать об этом. Но еще шире распространено заблуждение, будто мы, люди, можем чувствовать себя в наших рукотворных мирах в безопасности.

Space pop

SATIRICAL

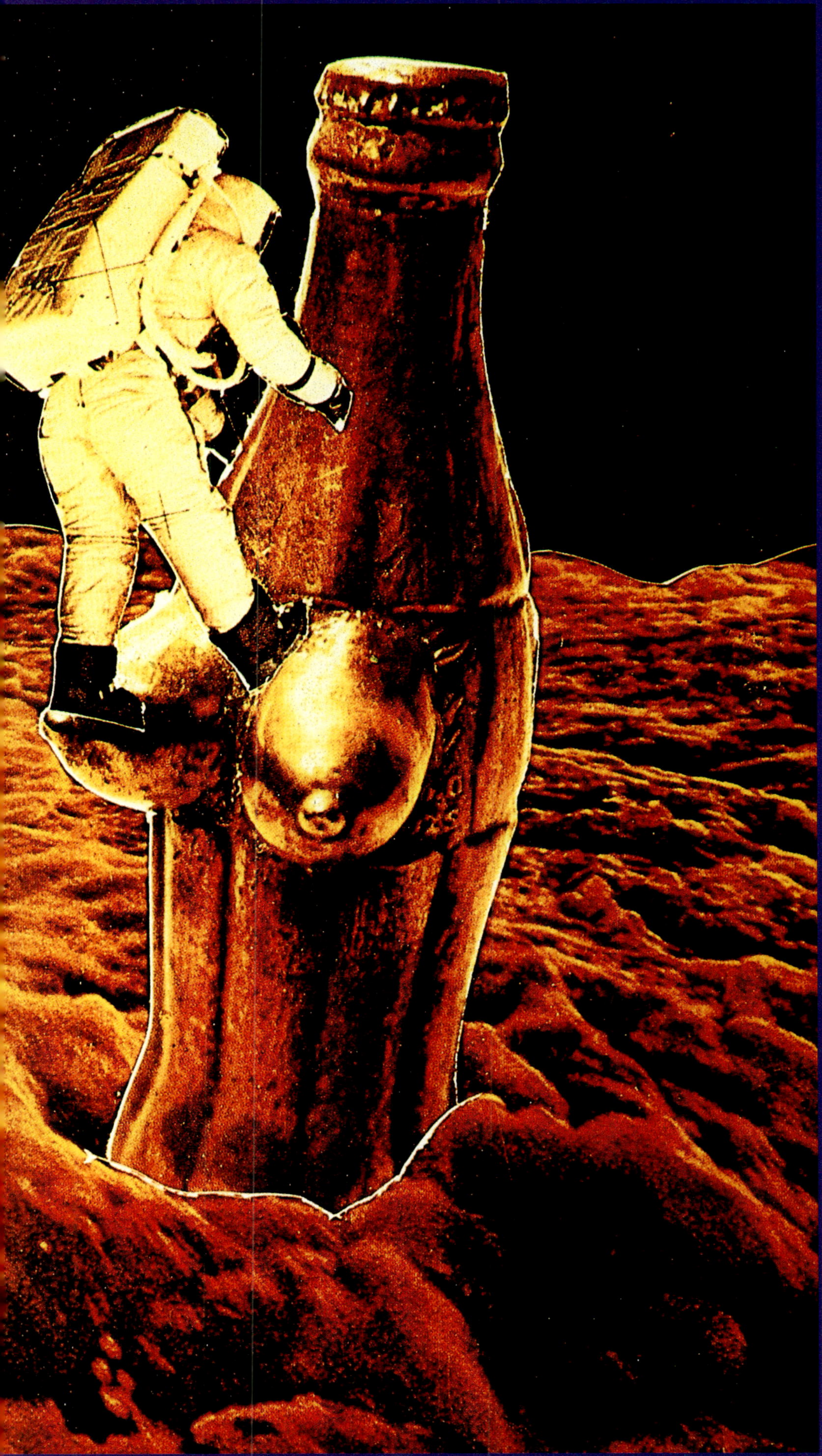

COSMOS

Venus on venus

Sternschnuppen schießen Quanten
in den dunklen Bauch des Universums,
wo die Mond-Eier warten,
um Berge, Affen und Göttinnen zu gebären.

Vallende sterren schieten quanta
in de donkere buik van het universum,
waar maaneieren liggen te wachten
om bergen, apen en godinnen te baren.

Falling stars shoot quanta into
the dark belly of the universe
where moon ova wait to bring forth
mountains, monkeys, goddesses.

Le stelle cadenti sparano i quanti
nel ventre buio dell'universo,
dove lune-uova li attendono.
Nasceranno montagne, scimmie e dee.

Des étoiles filantes envoient des quanta
dans le sombre ventre de l'Univers où atten-
dent les ovules de la Lune pour accoucher
de monts, de singes et de déesses.

Метеориты несут кванты в темное
чрево Вселенной, где лунные яйца
ждут, чтобы родить горы,
мартышек и богинь.

Lillian Morrison

Gates of heaven

The dream

13

Luncheon on the moon

Andys installation

Steven Steinman, Bildhauer, Kunsterzieher und Science Fiction-Anhänger mit zahlreichen Ausstellungen und Preisen ihm zu Ehren, zelebriert in seiner Kunst ironische intergalaktische Transaktionen... Manets „Frühstück im Freien" wird zum Picknick auf dem Mond und Botticellis Venus erfährt eine Wiedergeburt auf Venus...

Steven Steinman, beeldhouwer, leraar plastische kunsten en liefhebber van science fiction, met talrijke tentoonstellingen en prijzen, celebreert in zijn kunst ironische intergalactische omzettingen... Manet's "Déjeuner sur l'herbe" wordt een picknick op de maan en Botticelli's Venus wordt op de planeet Venus weer geboren.

Steven Steinman, sculptor, educator in the art field and lifelong science-fiction-devotee with numerous exhibitions and awards to his credit celebrates in his art ironic intergalactic transpositions... Manet's "Luncheon on the Grass" is converted into a picnic on the moon and Botticelli's "Birth of Venus" is reborn on the planet Venus...

Space installation

Steve Steinman, scultore, insegnante d'arte nonché cultore di "science fiction", numerose mostre alle spalle e detentore di molti premi, celebra nelle sue opere degli ironici trasferimenti intergalattici... In un suo quadro, la "Colazione sull'erba" di Manet si trasforma in un picnic sulla luna, mentre la Venere di Botticelli rinasce su Venere...

Steven Steinman, sculpteur, éducateur d'art, passionné de science fiction, honoré par des expositions et prix multiples, célèbre pour son art réalisant d'ironiques transpositions intergalactiques... "Le déjeuner sur l'herbe" de Manet est transformé en pique-nique sur la lune et la Vénus de Botticelli renaît sur la planète Vénus...

С. Стейнмен, скульптор, педагог и любитель фантастики, создает иронические «трансгалактические произведения»... «Завтрак на траве» Мане превращается у него в «Пикник на Луне», а «Венера» Ботичелли возрождается на Венере...

The rendezvous

Signum der Kreativität: Astro-Expeditionen und tellurische Spaziergänge

Kosmogonie, Kosmologie, Kosmographie und sonstige kosmosophische Betrachtungen haben den Menschen seit jeher zur praktisch-philosophischen Erforschung des Metaphysischen, des Seins als Ganzes animiert - durch Abstraktion von dem rein physikalischen All-Terminismus und in wesensimmanenter Assoziation von unendlichen weltraumfassenden Kräften.

Von der spirituellen Erkundung einer unveränderten, gleichbleibenden Unendlichkeit des Raums (für lange Zeit die unumstößliche Lehre des Giordano Bruno) über Albert Einsteins Modell-Aufstellung eines endlichen,statischen Kosmos', die von E. P. Hubble konstatierte Expansion des Weltalls bis hin zur P. A. M. Dirac'schen Hypothese einer veränderlichen Kosmos-Entwicklung, bietet L. Boltzmann mit seiner darüber hinausgehenden jüngeren Verfechtung der ständigen Zunahme von „Unordnung" und „Unsymmetrie" des Entropie-Wachstums eine willkommene Einladung an alle Ethiker, Naturphilosophen, Künstler, i. e., alle Freidenker, zu weltraumumfassenden Explorationen.

Wie natürlich und selbstverständlich, wie notwendig wäre es doch, alle spirituellen und ethischen Werte als kosmisch feste Bestandteile, als hochenergetische Ladungsträger der Ultrastrahlungs-Kaskaden im Makro-Kosmos begreifen und unterlegen zu können!

Die nichtkonstante Unendlichkeit des Weltraumgebäudes mit ihren „schwimmenden Kontinentalschollen", ihren rätselhaften Explosions-Vorgängen bei der Entstehung von Sternen, das Beieinander von kosmischen Gaswolken und durch Urknallstrahlung evozierte Spiralnebel öffnet den unbegrenzten All-Freiraum für die menschliche Fähigkeit, in schöpferischer Phantasie und spielerischer Mah-Jongg-Manier das Weltall, den Kosmos, neu zu entwerfen. Gegebene Zustände lassen sich, frei von normativem Denken, zurechtrücken. Dieses in Bewegung fließende, atmende Universum schließt in seiner multidimensionalen Möglichkeits-Vielfalt nicht nur jeden Begriff von „Unmöglichkeit" aus, sondern macht die „königliche Kunst" des Freimaurerstrebens nach humanitärer Ethik umsetzbar - nach aristotelischem Prinzip, durch Anstreben des entgegengesetzten Extrems, zur Mitte zu gelangen.

In Otl Aichers Philosophie läßt sich das All durch Neuentwurf erfahren, und schöpferische Erkenntnis geht einher mit dem „Entwurf als Erzeugen von Welt durch Machen". Die sinnlich-geistige Erfahrung von Wahrheit wird durch individuelles „Machen", z. B. durch Hinstellen, Wegnehmen, Greifen, Konstruieren, Bauen in Masonnerie-Façon und durch andere praktische Modell-Aktivitäten erkennbar und real, über jede Vision hinaus. Karl Poppers wissenschaftliche Wahrheitsfindung ist jederzeit durch das Auffinden einer neuen Wahrheit ersetzbar. Bei Wilhelm von Ockham beruht die Erfahrung von Wahrheit auf „evidenter, sinnlicher Erkenntnis", die abstraktiv nicht möglich ist, sondern nur intuitiv.

Der Macher und Philosoph Otl Aicher geht darüber hinaus: für ihn ist das entscheidende Kriterium der Wahrheit die individuelle Subjektivität, und: „Jeder Entwurf ist wahr, der selbst verantwortet wird". Selbstverantwortung jedoch erhält erst durch Mitverantwortung eine tragfähige ethische Dimension. Aicher belegt entlang seiner philosophischen Vorbilder Wilhelm von Ockham, Immanuel Kant und Ludwig Wittgenstein die Notwendigkeit des spielerischen freien Umrückens „gegebener Zustände" im All. Er fordert die mehrdimensionale Konzeption eines Modell-Entwurfs als „komplexestes Gebilde geistiger Tätigkeit und schöpferischer Entfaltung" als „Erzeugen von Welt", die neue Wirklichkeit und neue Einsichten freigibt in die subtile enge Relation zwischen Denken, Körper und Universum.

Diesen vielfacettigen Domänen verleiht der Astro-Künstler Steven Steinman leichtfüssig-diabolische Weihen. Die subjektiven „Wahrheiten" seiner Entwürfe vermitteln mehrdeutige Einblicke in seine Modelle des Weltganzen. In seinen Collagen entfaltet Steinman seinen Künstlertraum, „Kunst" auf andere Planeten zu transferieren und die Galaxis und ihre unendlichen Umfelder als „Grenzen-lose" Ausstellungs-Galerie zu nutzen. Mit diesen „Art and Other Planets"-Bildern konzipiert und projiziert Steinman vorgegebene terrestrische objets d'art in eigens kreierte visionäre Weltraum-Landschaften.*

Seine inszenierten Collagen zeigen unterhaltsame methodische Disziplin im verbrämten aristotelischen Zuschnitt von Kasten-Kugel, versteinerter Welle, fließendem Granit, seidenglatter Bodenkruste, tonnenschwerer Feder.

Die Mitteilungsfunktion der Bilder erfährt durch das Spiele-Spektrum von „Comédie larmoyante", „Dramolett", bis zu „Klamauk" symbolische Bandbreite. Der Künstler evoziert einen mit Sarkasmus und Humor gesäumten ironischen Weltenentwurf, vielfacettige Bühnenbilder, die zu phantasievollen Regie-Debuts einladen - in einem Kosmos mit multidimensionalen Bild- und Gedankenelementen wie „reflektierender Urteilskraft". „Come quick, danger"!

Ingrid T. Schick

Signum van Creativiteit: Astro-expedities en tellurische wandelingen

Kosmogonie, kosmologie, kosmografie en andere kosmosofische benaderingen hebben de mens van oudsher aangemoedigd tot praktisch-filosofisch onderzoek van het metafysische, van het zijn als geheel - door het abstraheren van het zuiver natuurkundige heelal-terminisme en in wezensimmanente associatie van oneindige wereldruimomvattende krachten.

Van de spirituele verkenning van een onveranderde, gelijkblijvende oneindigheid van de ruimte (gedurende lange tijd de onomstotelijke leer van Giordano Bruno) via Albert Einsteins opstelling van het model van een eindige, statische kosmos, en de door E.P. Hubble geconstateerde expansie van het heelal tot en met de door P.A.M. Dirac opgestelde hypothese van een veranderlijke ontwikkeling van de kosmos, biedt L. Boltzmann recentelijk met zijn nog verder gaande voorstaan van de voortdurende toename van "wanorde" en "onsymmetrie" der entropievermeerdering, een welkome uitnodiging aan alle ethici, natuurfilosofen, kunstenaars, dat wil zeggen aan alle vrijdenkers, tot wereldruimomvattende verkenningen.

Hoe natuurlijk en vanzelfsprekend, hoe noodzakelijk zou het toch zijn om alle spirituele en ethische waarden als kosmisch vaste bestanddelen, als hoogenergetische ladingdragers van de ultrastraling-cascades in de macrokosmos, op te vatten en interpreteren te kunnen!

De niet-constante oneindigheid van het wereldruimtegebouw met haar "drijvende continentale schollen", haar raadselachtige explosieprocessen bij het ontstaan van sterren, het samenzijn van kosmische gaswolken en door oerknalstraling opgeroepen spiraalnevels, opent de onbegrensde vrije ruimte van het heelal voor het menselijke talent om in scheppende fantasie op speelse mahjong-manier het heelal, de kosmos opnieuw te ontwerpen. Gegeven toestanden kunnen, vrij van normatief denken, worden rechtgezet. Dit vloeiend bewegende, ademende universum sluit met zijn multidimensionale mogelijkheidsscala niet alleen ieder begrip van "onmogelijkheid" uit, maar maakt de "koninklijke kunst" van het streven der vrijmetselaars naar humanitaire ethiek omzetbaar - volgens aristotelisch principe, door naar het tegenovergestelde extreem te streven naar het midden te komen.

In de filosofie van Otl Aicher kan het heelal door een nieuw ontwerp worden ervaren, en creatief inzicht gaat gepaard met het "ontwerp als voortbrengen van wereld door doen". De zinnelijk-geestelijke ervaring van waarheid wordt door individueel "doen", bijvoorbeeld door neerzetten, wegnemen, grijpen, con-

Signum of creativity: Astro-expeditions and tellurian strolls

strueren, bouwen à la maçonnerie en door andere praktische modelactiviteiten, zichtbaar en reëel, meer dan ieder visioen. De wetenschappelijke vaststelling van de waarheid is bij Karl Popper te allen tijde te vervangen door het vinden van een nieuwe waarheid. Bij Wilhelm von Ockham berust het ervaren van waarheid op "evident, zinnelijk inzicht", dat abstractief niet mogelijk is, maar alleen intuïtief.
De doener en filosoof Otl Aicher gaat nog verder: voor hem is het doorslaggevende criterium van de waarheid de individuele subjectiviteit, en: "Elk ontwerp dat zelf wordt verantwoord, is waar." Zelfverantwoording verkrijgt echter pas door medeverantwoording een draagkrachtige, ethische dimensie.
Aicher toont langs zijn filosofische voorbeelden - Wilhelm von Ockham, Immanuel Kant en Ludwig Wittgenstein - de noodzaak van het speelse, vrije verzetten van "gegeven toestanden" in het heelal aan. Hij verlangt de meerdimensionale conceptie van een modelontwerp als "het meest complexe gewrocht van geestelijke activiteit en creatieve ontplooiing" als "voortbrengen van wereld", die nieuwe werkelijkheid en nieuwe inzichten vrijgeeft in de subtiele nauwe relatie tussen denken, lichaam en universum.
De astrokunstenaar Steven Steinman geeft aan deze veelfacettige domeinen een lichtvoetig-diabolisch cachet. De subjectieve "waarheden" van zijn ontwerpen verschaffen een veelduidige kijk op zijn modellen van het wereldgeheel. In zijn collages ontvouwt Steinman zijn kunstenaarsdroom om "kunst" op andere planeten te transfereren en het galactische stelsel en zijn oneindige omgeving als "grenzeloze" tentoonstellingsgalerie te gebruiken. Met deze "Art and Other Planets" beelden concipieert en projecteert Steinman bepaalde terrestrische objets d'art in speciaal gecreëerde wereldruim-landschappen.
Zijn geënsceneerde collages tonen vermakelijke methodische discipline in de verbloemde aristotelische formatie van blok-bal, versteende golf, vloeiend graniet, zijdegladde aardkorst, loodzware veer.
De mededelingsfunctie van de beelden ondergaat door het spelspectrum van "Comédie larmoyante", "sketch" tot "slapstick" een symbolische bandbreedte. De kunstenaar evoceert een met sarcasme en humor omzoomd ironisch wereldontwerp, veelfacettige decors, die tot fantasierijke regiedebuten uitnodigen - in een kosmos met multidimensionale beeld- en gedachtenelementen zoals "reflecterend oordeelsvermogen".
"Come quick, danger!"

Cosmogony, cosmology, cosmography and other cosmosophic contemplations have at all times animated Man to a practical-philosophical investigation of things metaphysical, of Being as a Whole - by abstracting from the purely physical space terminology and through intrinsic associations of infinite universal forces.
From the spiritual exploration of an unchanged, constant endlessness of space (for a long time the irrefutable teaching of Giordano Bruno) to Albert Einstein's model formulation of a finite, static cosmos; from E. P. Hubble's confirmation of the expanding universe to P. A. M. Dirac's hypothesis of a variable development of the cosmos, L. Boltzmann, with his latest assertion of the constant increase in "disorder" and "dissymmetry" of entropy-growth, goes a step further and in doing so extends a welcome invitation to all moral philosophers, natural philosophers and artists, that is to all free-thinkers to indulge in explorations encompassing the entire universe. How natural and matter-of-course, how necessary it would be to be able to comprehend and to underlay all spiritual and ethical values as firm constituent parts of the cosmos, as high-energy charge carriers of the ultra-radiation cascades in the macrocosm.
The non-constant infinitude of the structure of space with its "floating continental plates", its mysterious explosive processes as new stars come into being, and the co-existence of cosmic clouds of gas and spiral nebulas called forth by Big Bang radiation, opens unlimited, universal manoeuvrability for the human capability of drafting anew the universe, the cosmos, in creative fantasy, in playful Mah-Jongg manner. Set circumstances can be "adjusted", free from any kind of normative thinking. This flowing, breathing universe in its multi-dimensional manifoldness of possibilities does not just exclude every definition of "impossibility", instead it makes the "royal art" of Masonic striving for humanitarian ethics realizable - according to the Aristotelion principle of reaching the middle ground by aspiring to the opposite extreme. In Otl Aicher's philosophy, the universe can be experienced by designing it anew, and creative knowledge goes hand in hand with "creating the world by doing". The sensual and spiritual experience of truth becomes recognizable and real, above and beyond any vision, through individual "doings", e.g. by placing upright, by taking away, by grasping, by constructing, by building in masonry fashion and by other practical model activities. Karl Popper's scientific location of truth is at all times replaceable by the discovery of a new truth. Where William of Ockham is concerned, the experience of truth is based on "evident, sensual knowledge", something which is not abstractly possible, only intuitively.
The "doer" and philosopher Otl Aicher goes a step further: in his eyes, the decisive criterium of truth is individual subjectivity: "Every draft which is the product of personal responsibility is true." Self-responsibility, however, only acquires an acceptable ethical dimension through joint responsibility. Reflecting his philosophical paragons William of Ockham, Immanuel Kant and Ludwig Wittgenstein, Aicher provides evidence for the necessity of the playful, free juggling of "given circumstances" in the universe. He calls for the multi-dimensional conception of a model-draft as "the most complex structure based on spiritual activity and creative unfolding" as a means of "producing world", enabling new reality and new insights into the subtly close relationship between thinking, the body and the universe.
These multi-facetted domains are lent light-footed, diabolical solemnity by the astro-artist Steven Steinman. The subjective "truths" of his imaginative works convey ambiguous insights into his models of the Whole. In his collages Steinman unfolds his artist's dream of transferring art to other planets and exploiting the galaxies and their endless environs as a boundless exhibition gallery. With these "Art and Other Planets" pictures Steinman conceives of and projects given terrestrial objets d'art into visionary space landscapes created especially for this purpose.
His carefully-staged collages reveal an entertaining and methodic discipline in the embellished Aristotelian style of a square circle, a petrified wave, flowing granite, the earth's crust as smooth as silk, a feather weighing tons.
The communicative function of the pictures covers a wide symbolic range reflecting the spectrum from "comédie lamoyante", through the "sketch" to "slapstick". With his ironic view of the world lined with sarcasm and humor the artist evokes multi-facetted settings which prove inviting for stage debuts full of fantasy - in a cosmos with multi-dimensional pictorial and contemplative elements like "reflecting power of judgement".
"Come quick, danger"!

Prosciutto e pane di cielo, le donne infedeli, astronave a rischio, gli specchi ustori dell'umorismo

Se a qualcuno venisse in mente di organizzare o di allargare nel cosmo le catene del fastfood, ovvero del pasto veloce, sul tipo di quello offerto dai McDonald's, sappia che lassù è stato già provveduto. Infatti. "Dio, dopo avé ccreato in pochi ggiorni / Cuello che cc'è de bello e cc'è de bbrutto, / In paradiso o in de li su' contorni / Creò un rampino e ciattaccò un prosciutto."
Una bella iniziativa, dunque, atta a scoraggiare qualsiasi concorrenza, soprattutto perché i maiali del cielo dovevano essere prelibati. Se non che, ne vincolò il consumo ad una clausola piuttosto preoccupante per le donne che avessero mancato di fedeltà verso i loro uomini. "E ddisse: Cuella femmina che in tutto / Er tempo che ccampò nun messe corni, / N'abbi una fetta, acciò nun magni asciutto / Er pandescèlo de li nostri forni". Cioè non mangi pane solo, anche se si tratta di "Panem de caelo", dei forni di lassù. Altro elemento dissuasivo per l'intraprendenza umana. Questa poesia è del grande Giuseppe Gioacchino Belli e la scrisse il 26 gennaio 1832, quando tentavano di alzarsi verso il cielo i primi palloni aerostatici. La coincidenza, dunque, è piuttosto singolare, tenendo conto che la satira del tempo su tali straordinari avvenimenti ne dileggiava i protagonisti raffigurandoli come una manica di gaudenti. In ogni caso il prosciutto, il primo di un fornitissimo deposito, è sempre lì, perché le donne, da Eva in poi, via via che morivano non avevano la prerogativa richiesta. Il sonetto, intitolato "Er companatico der Paradiso", delizioso nella sua arguzia, così conclude. "E ttutte cuante còr cortello in mano, / Cuanno furno a ttajjà, fesceno scecca (cilecca) / So' sseimil'anni, e cquer presciutto è ssano."!
Al mondo non può mai accadere alcunché di buono senza che, sulle prime, qualcuno "si prenda il gusto di riderne"!, scrisse Charles Dickens. E allora, come avrebbe reagito la satira durissima del Belli, se fosse stato testimone della partenza per lo spazio della astronave delle tentazioni, avvenuta di recente? Quali, queste tentazioni? Il pane ed il sesso! Nella cinquantesima missione Shuttle, infatti, fra i sette componenti dell'equipaggio figurava la dottoressa Mae Jemison, la prima astronauta donna di colore di 35 anni, scatenata per il ballo e per il jazz. Poteva, volendo, ben badare a se stessa e non era certo a rischio come Jan Davis e Mark Lee, due bei ragazzi, regolarmente sposati, anch'essi facenti parte dell'equipaggio. "Siamo in missione scientifica, non in viaggio di nozze", hanno risposto seccamente a chi avanzava qualche facile ironia. Per il pane, invece, caldo e profumato, il divieto di mangiarlo rasentava la crudeltà verso l'intero equipaggio. Ma consumandolo, non si sarebbe potuto verificare in Terra, se in assenza di peso lievita e cuoce compiutamente. Orbene, su quanto finora esposto, è difficile che qualcuno non si sia preso il gusto di riderne. Quanto meno al primo impatto, come probabilmente intendeva sottolineare Dickens. Altrimenti, la Satira e soprattutto l'Umorismo, di cui essa fa parte, sono sinonimo di intelligenza critica, serenità, equilibrio interiore. Non a caso, nella lingua inglese, il verbo "to humour" si usa per indicare azioni che richiedono adattamento, tatto, tolleranza.
Niente a che vedere, perciò, quando il dato comico viene appesantito e degradato da scurrilità, oscenità, turpiloquio; ovvero, nel caso in cui un intento satirico finisce per trasformarsi in volontà di vilipendere. Ecco quello che non deve fare l'uomo se vuole proiettare nello spazio i propri valori umani. "Dovremmo batterci ogni tanto - scrive Orazio in una delle sue famose "Satire" - con le nocche della dita, come fossimo vasi, per sentire le nostre incrinature ed i nostri difetti. Simili ai campi incolti, siamo pieni di erbacce da bruciare." Occhio, dunque. Perché l'umorismo è pure uno specchio - alla stregua dell'amore, della sessualità, dell'amicizia - nel quale le donne e gli uomini possono osservare le proprie immagini da tutti i punti focali. Ma attenzione, esso è pure un archimediano specchio-ustorio, sebbene il "fuoco" che accende sia distruttore soltanto nei confronti della parte peggiore della natura umana. Per contro può generare quel fuoco purificatore, sprigionato dall'atto comico che squaglia ogni scoria di disumanità, possibilmente con una risata omerica. Perché la satira e l'umorismo costituiscono quel linguaggio capace di decongestionare le esasperazioni della vita umana ormai proiettata nel cosmo. "Lassù, lassù è la patria dei nostri sogni...", scriveva Goethe.

Giuseppe Brunamontini

Signe de créativité: expéditions astronautiques et promenades telluriques

Cosmogonie, cosmologie, cosmographie et autres considérations cosmosophiques ont toujours animé l'homme à explorer, par la philosophie pratique, la métaphysique et l'existence dans son intégralité, en faisant abstraction du terminisme universel purement physique et en association immanente à toute chose de forces infinies influant sur l'ensemble de l'espace cosmique.
De l'identification spirituelle d'une infinité, immuable de l'espace (pendant longtemps la théorie irréfutable de Giordano Bruno), en passant par le modèle mis en place par Albert Einstein d'un cosmos infini et statique, tout comme l'expansion de l'Univers constatée par E.P. Hubble, jusqu'à l'hypothèse de P. A. M. Dirac d'un développement variable du cosmos, L. Boltzmann, qui défend sa récente théorie, allant plus loin encore, de l'augmentation constante du "désordre" et de l'"asymétrie" de la croissance de l'entropie, invite tous les moralistes, philosophes de la nature, artistes et libres penseurs, qui lui en sont d'ailleurs reconnaissants, à entreprendre des explorations englobant l'Univers.
Comme il serait naturel et légitime, comme il serait nécessaire de pouvoir comprendre et attribuer à toutes les valeurs spirituelles et morales le caractère d'éléments fixes du cosmos, de porteurs de charges ultraénergétiques des cascades d'ultraradiation dans le macrocosme!
L'infinité non constante de la structure spatiale avec ses "blocs continentaux flottants", ses mystérieuses explosions à la naissance des étoiles, le voisinage de nuées cosmiques gazeuses et de nébuleuses spirales générées par la radiation de l'explosion originelle ouvre l'espace libre illimité de l'Univers à la faculté humaine d'élaborer, grâce à son imagination créatrice et aux règles du jeu de mah-jong, une nouvelle conception de l'Univers, du Cosmos. Libérées de la pensée normative, les circonstances données peuvent être arrangées d'une autre manière. Cet Univers qui bouge, glisse, respire, n'exclut pas seulement, de par sa multiplicité multidimensionnelle de possibilités, toute notion d'"impossibilité", il rend également réalisable l'"art royal" de la morale humanitaire auquel aspire la Franc-Maçonnerie et qui consiste, selon le principe d'Aristote, à atteindre le milieu en essayant de parvenir à l'extrême opposé.
Dans la philosophie de Otl Aicher, on peut faire l'expérience de l'Univers en le concevant sous une forme nouvelle et la prise de conscience créative va de paire avec le "concept de créer le Monde en le faisant".
L'expérience de la vérité, rendue possible par un mélange matériel-immatériel, devient

Символ творчества: астро-экспедиции и теллурические прогулки

perceptible et réelle, au-delà de toute vision, par une "organisation' individuelle: placer, enlever, saisir, construire, maçonner ou autres activités pratiques sur maquette. La vérité scientifique de Karl Popper peut être à tout moment renversée par la découverte d'une nouvelle vérité. Pour von Ockham, l'expérience de la vérité repose sur une "prise de conscience évidente, sensible", possible par l'intuition et non par l'abstraction.
Le "constructeur" et philosophe Otl Aicher va plus loin: pour lui, le critère décisif de la vérité est la subjectivité individuelle et: "Tout concept justifiable aux yeux de son créateur est vrai". Cette justification personnelle n'atteint cependant une dimension éthique admissible qu'en union avec une corresponsabilité. S'appuyant sur ses idoles Wilhelm von Ockham, Immanuel Kant et Ludwig Wittgenstein, Aicher prouve la nécessité d'une réorganisation libre et ludique des "conditions données" dans l'espace. Il demande la conception multidimensionnelle d'un modèle représentant la "structure la plus complexe de l'activité intellectuelle et du développement créatif", la "production d'un monde", un concept qui libère la vraie réalité et de nouvelles perspectives dans l'étroite et subtile relation entre la pensée, le corps et l'Univers.
Steven Steinman, artiste "astral", prête à ces domaines aux innombrables facettes un caractère légèrement diabolique. Les "vérités" subjectives de ses concepts donnent des aperçus polysémiques de ses modèles représentant le monde. Dans ses collages, Steinman déploie son rêve de transférer l'"art" sur d'autres planètes et d'utiliser la galaxie et ses champs périphériques comme galerie d'exposition "illimitée". Avec ces images d'"Art and Other Planets", Steinman conçoit et projette des objets d'art terrestres existant déjà dans des paysages spatiaux visionnaires spécialement créés par lui pour les recevoir.
Le scénario de ses collages atteste une discipline méthodique mais divertissante dans la découpe aristotélique chamarrée de boules carrées, de vagues pétrifiées, de granite ruisselant, de croûte terrestre satinée et de plumes pesant des tonnes.
Le spectre ludique allant de la "Comédie larmoyante", du "Dramolett" au "Burlesque" confère à la fonction communicative de ces images une ampleur symbolique. L'artiste génère une conception du monde ironique, bordée de sarcasme et d'humour, des décors de théâtre multiformes qui enjoignent à s'essayer comme régisseur dans un cosmos aux éléments multidimensionnels tirés d'images et d'idées tels que la "faculté réflectrice de jugement". "Come quick, danger"!

Космогония, космология, космография и прочие космософские построения с древних времен вели людей к практико-филососфскому освоению метафизики, бытия в целом, отвлекаясь от чисто физического детерминизма в постоянной ассоциации с бесконечными мировыми силами.
Идя от умозрительного проникновения в неизменную бесконечность пространства (неопровержимое в течение долгого времени учение Джордано Бруно), через эйнштейнову модель конечного, статического космоса, через установленное Хабблом расширение Вселенной и вплоть до гипотезы Дирака о переменчиво развивающемся космосе, Больцман своей позднейшей идеей постоянного роста неупорядоченности, асимметрии, энтропии дал возможность всем моралистам, натурфилософам, художникам, всем вольнодумцам заняться изысканиями в космическом масштабе.
Как было бы естественно, как необходимо осознать и истолковать все духовные и этические ценности в качестве постоянных космических элементов, неких высокоэнергетических носителей заряда в каскадах космического излучения макрокосмоса!
Непостоянная бесконечность здания космоса с его «блинами», загадочными взрывами при рождении звезд, сосуществованием газовых облаков и спиральными туманностями открывает для человека, обладающего творческой фантазией и склонностью к игре, неограниченные возможности для конструирования все новых моделей космоса. Существующий порядок вещей можно, не обременяясь нормативным мышлением, установить по-новому. Эта текучая, дышащая Вселенная не только исключает в своем многомерном разнообразии понятие невозможного, но и делает осуществимым аристотелевский принцип: стремясь к противоположной крайности, достичь середины. В философии Отла Айхера Вселенная познается через проект, и творческое познание сопутствует «проекту как строительству миров простым деланием». Духовно-чувственное постижение истины становится возможным и реальным через индивидуальное «делание», например добавление, изъятие, конструирование, строительство, и превосходит любое представление. Поиски научного знания К. Поппера в любой момент заменимы какой-то новой истиной. У Оккама опыт истины опирается на «очевидное, чувственное знание», возможное только на уровне интуиции, но никогда — абстракции.
Философ Отл Айхер идет еще дальше: для него решающим критерием истины является индивидуальная субъективность: «истинен каждый объект, который можно оправдать». Однако личная ответственность обретает приемлемый этический масштаб только в коллективной ответственности. Опираясь на философские идеи Оккама, Канта и Витгенштейна, О. Айхер приводит необходимость свободной игры «существующим порядком вещей» в космосе. Он отстаивает многомерную концепцию проекта как «наиболее комплексного построения на основании духовной деятельности и творческого развития» для «создания миров», что открывает новую действительность и новый взгляд на сложную и тесную связь между мышлением, телом и Вселенной.
Астро-художник С. Стейнмен придает всей этой многомерности дьявольскую грациозность. Субъективные «истины» его проектов дают многозначное представление о его моделях мира. В коллажах Стейнмена раскрывается мечта художника перенести искусство на другие планеты и использовать галактики в качестве безграничной «выставочной галереи». В картинах, называемых им «Art and Other Planets», Стейнмен проецирует заданные земные «предметы искусства» на созданные его воображением космические ландшафты.
Его манерные коллажи отличаются занимательной методической дисциплиной в обрамлении аристотелевского принципа: «ящик-шар», «окаменевшая волна», «текучий гранит», «шелковистая земная кора», «многотонное перо».
Коммуникативная функция этих картин обретает символическое значение в диапазоне игры — от «комедии слез» и «маленькой драмы» до «потехи». Художник создает иронический «проект мира», обрамленный иронией и сарказмом, многоплановые декорации, будящие режиссерскую фантазию, в космосе из многомерных изобразительных и мыслительных элементов как «рефлектирующий рассудок».
«SOS!»

IMAGINARY

COSMOS

Chesley Bonestell, gerühmter „Dean of Space Art“ (1888-1986), veröffentlichte 1944 seine ersten Weltraum-Bilder. Er offenbarte einer ganzen Generation Traum und Verheissung des Universums.

Chesley Bonestell, geprezen als "Dean of Space Art" publiceerde zijn eerste schilderijen van de ruimte in 1944. Hij openbaarde een hele generatie het wonder van het geloof in de ontsluiting ervan.

Chesley Bonestell praised as "Dean of Space Art" (1888-1986), published his first space paintings in 1944. He conveyed to an entire generation the wonder and promise of the universe.

Chesley Bonestell, "Decano dell'arte spaziale", espose i suoi primi dipinti spaziali nel 1944. Egli ha rivelato ad una intera generazione il sogno e la promessa dell'universo quale entità tangibile.

Chesley Bonestell, prôné comme le "Dean of Space Art", publia en 1944 ses premières peintures de l'espace. Il ouvrit à toute une génération le miracle et la promesse de la tangibilité de l'Univers.

Бонестелл (1888-1986), «мэтр космоискусства», выставил в 1944 первые космические картины. Он дал целому поколению мечту о космосе и вселил веру в его освоение.

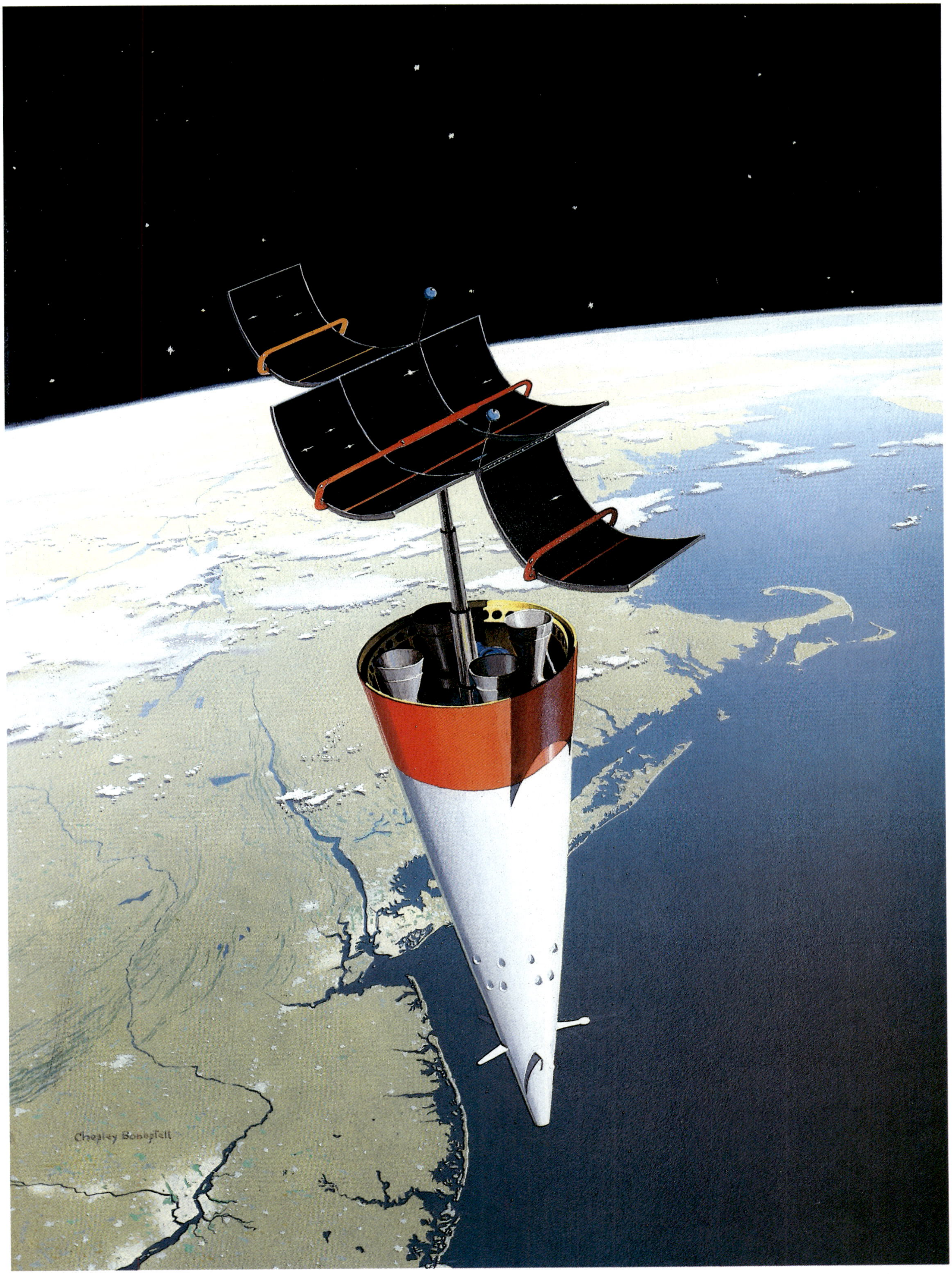
Chesley Bonestell

Chesley Bonestell ©

Die Oberfläche des Abendsterns Venus ist möglicherweise nichts weiter als ein öder, staubgefüllter Krater. Vor dem dunstigen, von dunklen Wolken verhangenen Himmel erheben sich phantastische, hochaufragende Steinformationen. Mächtige, vom Wind hochgewirbelte Staubfahnen lassen die schroffen Felsen in bizarren, ständig wechselnden Schattengestalten erscheinen.

Het is mogelijk dat het oppervlak van de avondster Venus niets anders dan een kale met stof gevulde krater is. Voor de nevelige, met donkere wolken bedekte hemel rijzen fantastische, hoog uitstekende steenformaties op. Geweldige, door de wind omhoog gedwarrelde stofvlagen laten de grillige rotsen als bizarre, voortdurend wisselende schaduwfiguren verschijnen.

It is indeed very possible that the surface of the bright evening star Venus is in fact nothing more than a massive, barren, dust-filled crater. Fantastic, protruding stone formations rise up against a hazy sky draped with dark clouds. Mighty, wind-blown walls of dust cause the precipitous rocks to appear in bizarre, constantly changing shadowy shapes.

La superficie di Venere, la stella della sera, non è niente altro che un desolato cratere pieno di polvere. Davanti allo scenario di un cielo caliginoso, coperto da nubi plumbee, si levano delle svettanti formazioni rocciose dalle sembianze fantastiche. Possenti folate di polvere sollevate dal vento fanno apparire sulle rupi delle ombre bizzarre che cambiano continuamente forma.

La surface de Vénus, l'astre du soir, n'est peut-être rien d'autre qu'un cratère désert, rempli de poussière. Des formations rocheuses, hautes, élancées, fantastiques, se découpent devant le ciel vaporeux, couvert de nuages sombres. D'imposantes masses de sable soulevées par le vent laissent apparaître les falaises abruptes aux silhouettes bizarres, constamment changeantes.

Поверхность Венеры, по-видимому, представляет собой пыльную пустыню. На фоне мглистого, завешенного темными тучами неба поднимаются фантастические каменные колонны. Огромные тучи пыли, гонимые ветром, придают отвесным скалам призрачный вид.

X/11

Chesley Bonestell

Chesley Bonestell

Vierzig Kilometer über dem Gipfel eines Mondbergs: Ein atemberaubender Blick auf den mächtigen Mount Pico von schwindelnder Höhe (30 000 Meter). Ihm fast zur Seite stellt sich in einiger Mond-Entfernung das andere Extrem dar: Plato, ein riesiger Krater, ebenso erhaben und unergründlich. Noch weiter in der Ferne ragt das einsame Massiv der Tenerife Mountains empor.

Veertig kilometer boven de top van een maanberg: een adembenemend uitzicht op de imposante Mount Pico met een duizelingwekkende hoogte (30.000 meter). Bijna naast hem is het andere extreem zichtbaar: Plato, een reusachtige krater, net zo majestueus en ondoorgrondelijk. Nog meer verderop rijst het eenzame bergmassief van de Tenerife Mountains op.

Forty kilometers up over the peak of an impressive moon mountain: a breathtaking view of mighty Mount Pico from a dizzy height. Seemingly almost alongside it but in reality some distance away, the other extreme: Plato, a gigantic crater, sublime and unfathomable. Still further away in the distance the lonely massif of the Tenerife Mountains rises up.

Quaranta chilometri sopra alla vetta di una montagna lunare: uno spettacolo mozzafiato sul possente Mount Pico che può vantare un'altezza da capogiro (30 000m). Quasi al suo fianco si pone, ad una certa distanza dalla luna, l'opposto: Platone, un enorme cratere, grandioso ed imperscrutabile. Ancora più lontano si leva il massiccio solitario delle Tenerife Mountains.

Quarante kilomètres au-dessus du sommet d'un mont lunaire: une vue fabuleuse sur le Mont Pico d'une hauteur étourdissante (30 000 mètres). A quelque distance, le côtoyant presque, l'autre extrême: Platon, un cratère géant, tout aussi majestueux et insondable que le philosophe dont il porte le nom. Encore plus loin se dresse le massif solitaire des montagnes de Ténérife.

Сорок километров над вершиной лунной горы: захватывающий вид могучей Пико (высота 30 000 м). На некотором расстоянии от нее — другая достопримечательность: громадный бездонный кратер Платон. Вдали высится одинокий горный массив Тенерифа.

Fast spielerisch versucht der Fotograf Anselm Spring mit Momentaufnahmen von Chaos- und Ordnungsprozessen Perspektiven moderner Kosmologien zu vermitteln. Im Makrokosmos des Alls spielen sich dieselben Sequenzen von Instabilität und Ordnung ab, wie im Mikrokosmos von Filmmaterial, das den kreativen Kräften kosmischer Gesetze ausgesetzt wurde.

Bijna speels probeert de fotograaf Anselm Spring met momentopnames van chaos- en ordeningsprocessen perspectieven van modernekosmologieën over te brengen. In de macrokosmos van het heelal spelen zich dezelfde sequenties van instabiliteit en orde af als in de microkosmos van filmmateriaal dat aan de creatieve krachten van kosmische wetten werd blootgesteld.

With instantaneous shots of chaos and order processes, the photographer Spring attempts in playful fashion to convey perspectives of modern cosmologies. In the macrocosm of the universe the same sequences of instability and subsequent order take place as in the microcosm of film material which has been exposed in imaginative fashion to the creative forces of cosmic laws.

Con istantanee tra arte e gioco, raffiguranti i processi del caos e dell'ordine, A. Spring ci comunica qualcosa dell'universo così come si presenta dal punto di vista della cosmologia moderna. Come nel macrocosmo dello spazio, anche nel microcosmo di un film si avvicendano caos e regolarità, l'instabilità e l'ordine che le succede quando il calore "disturba" una diapositiva.

Avec des photos instantanées des processus de chaos et d'ordre, le photographe Anselm Spring essaie de présenter l'Univers de la cosmologie moderne. Dans le macrocosme de l'espace on retrouve les mêmes séquences d'instabilité et d'ordre ultérieur que dans le microcosme des pellicules exposées avec beaucoup de fantaisie aux forces créatrices des lois cosmiques.

Почти играя, фотограф А. Спринг в моментальных снимках хаотических и упорядоченных процессов пытается передать перспективы современной космологии. В макрокосмосе Вселенной неустойчивость сменяется порядком, в чем проявляется действие космических законов.

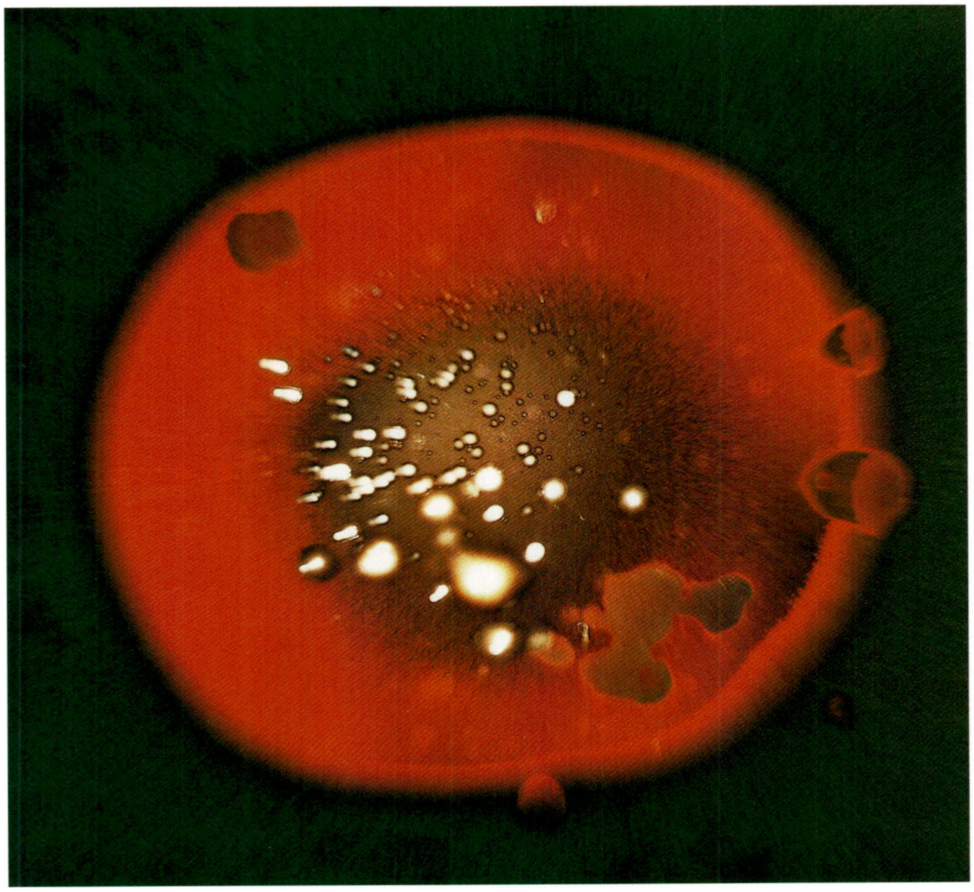

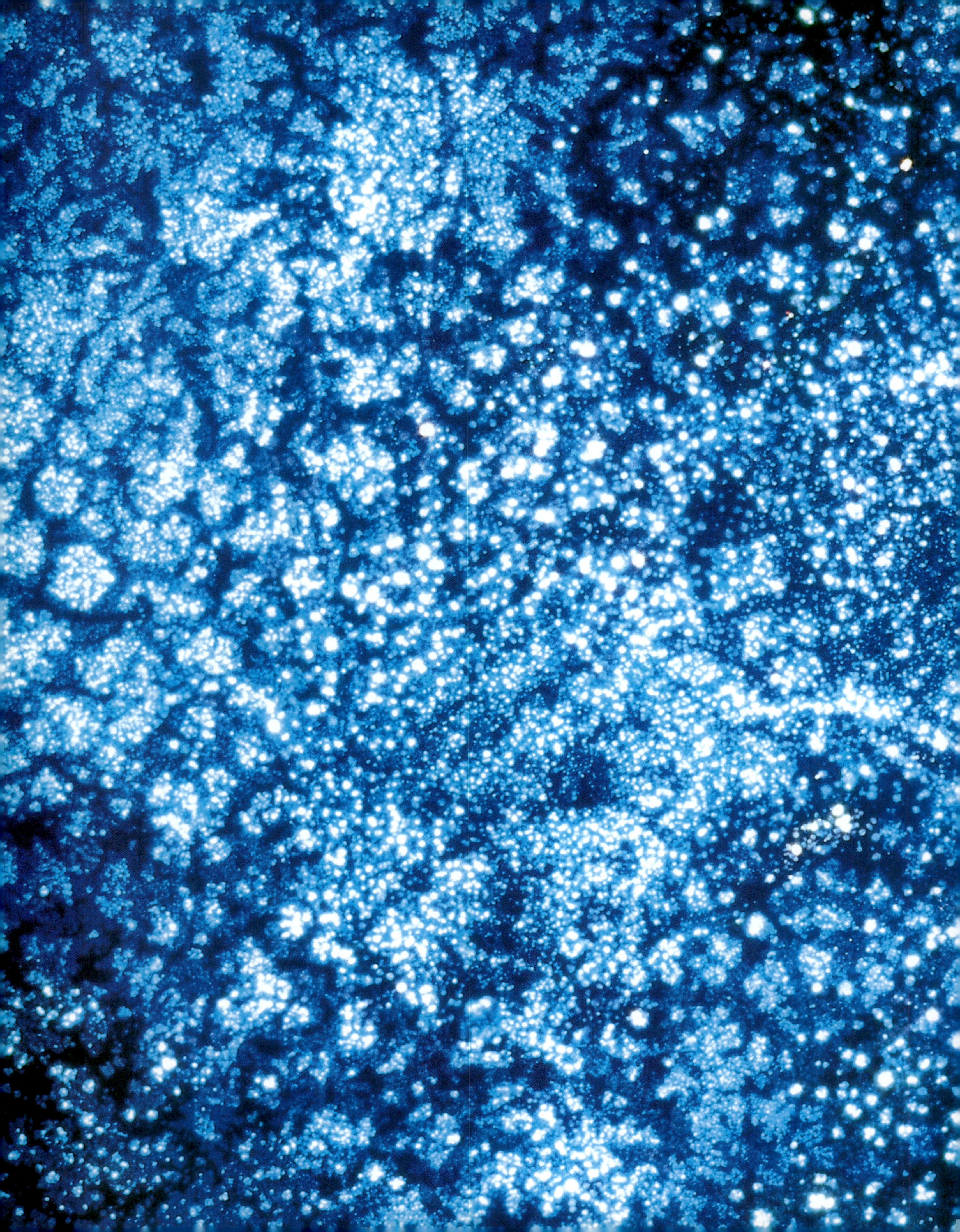

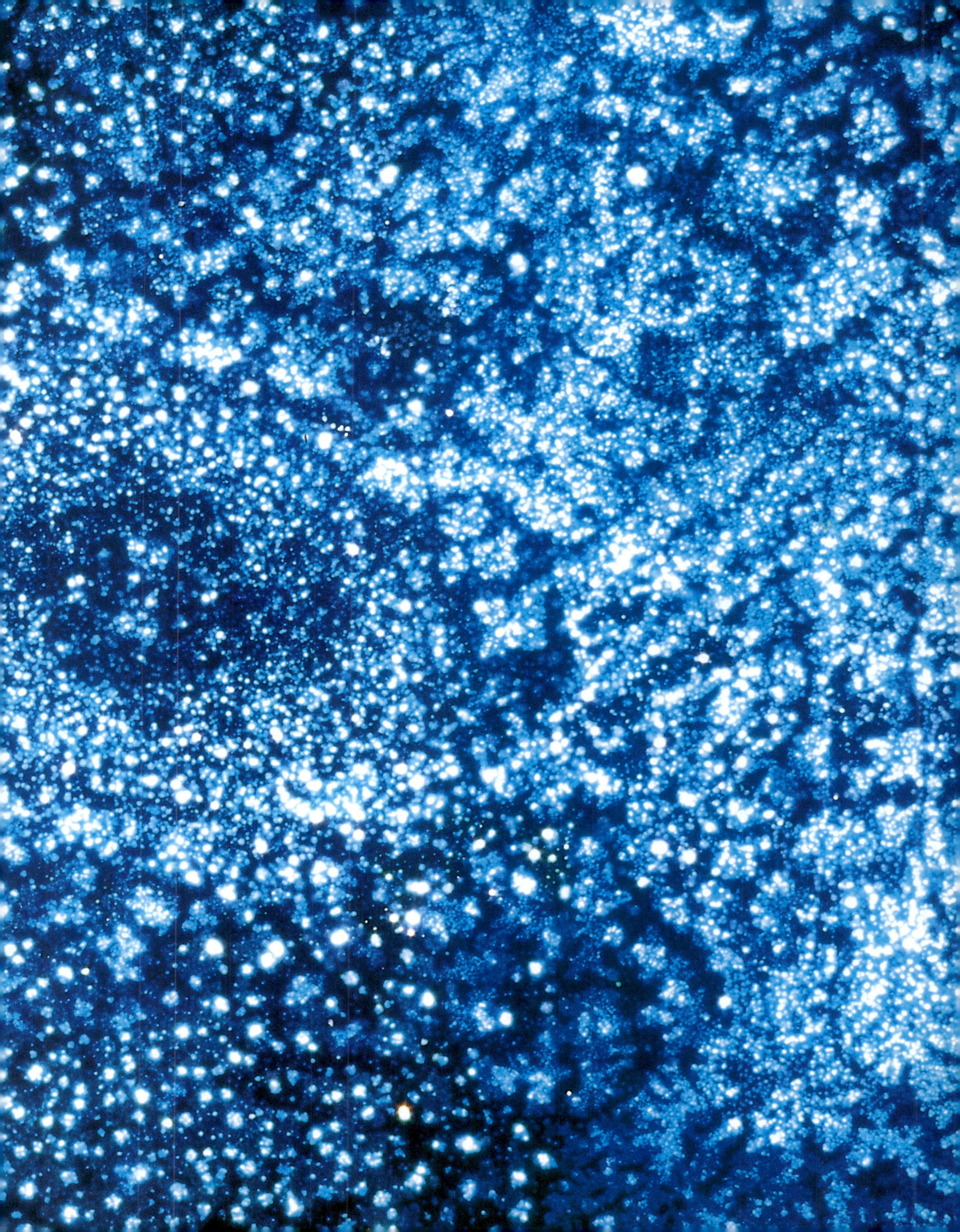

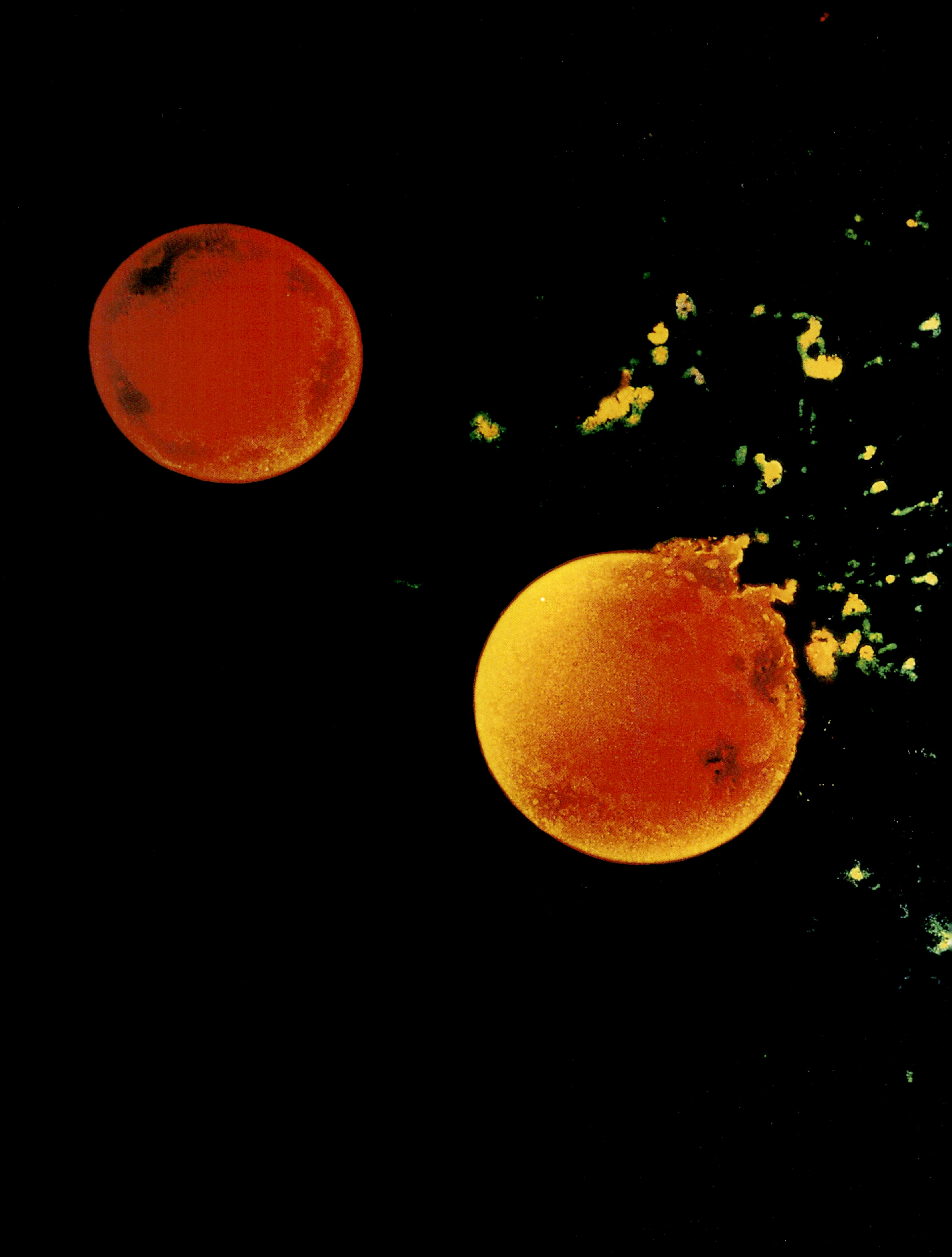

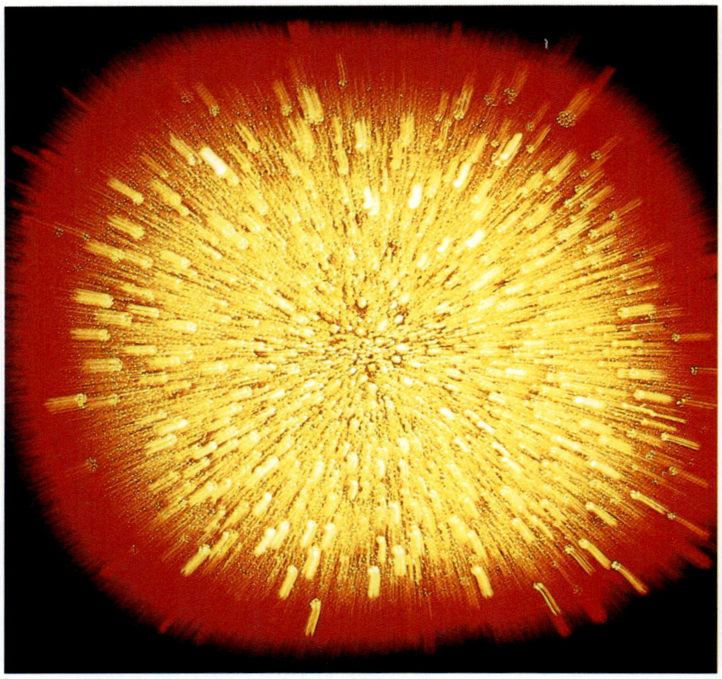

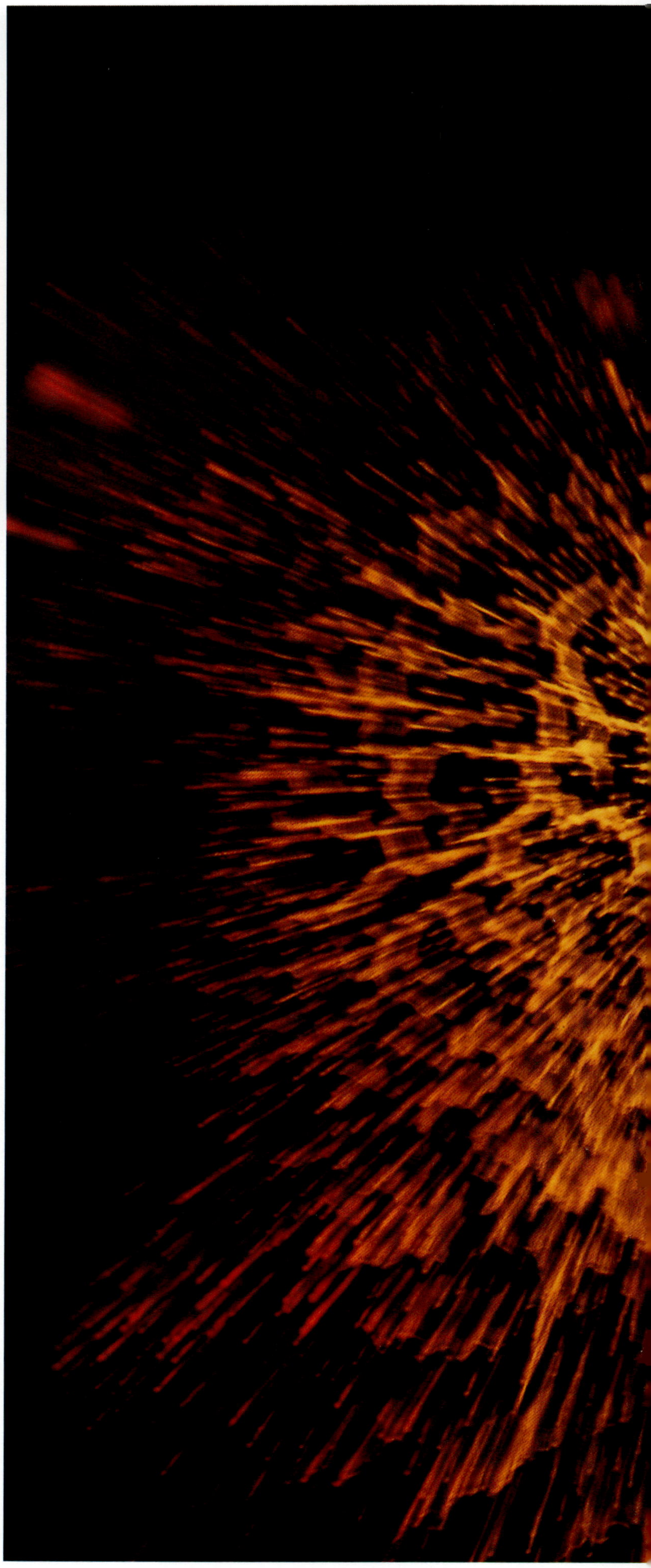

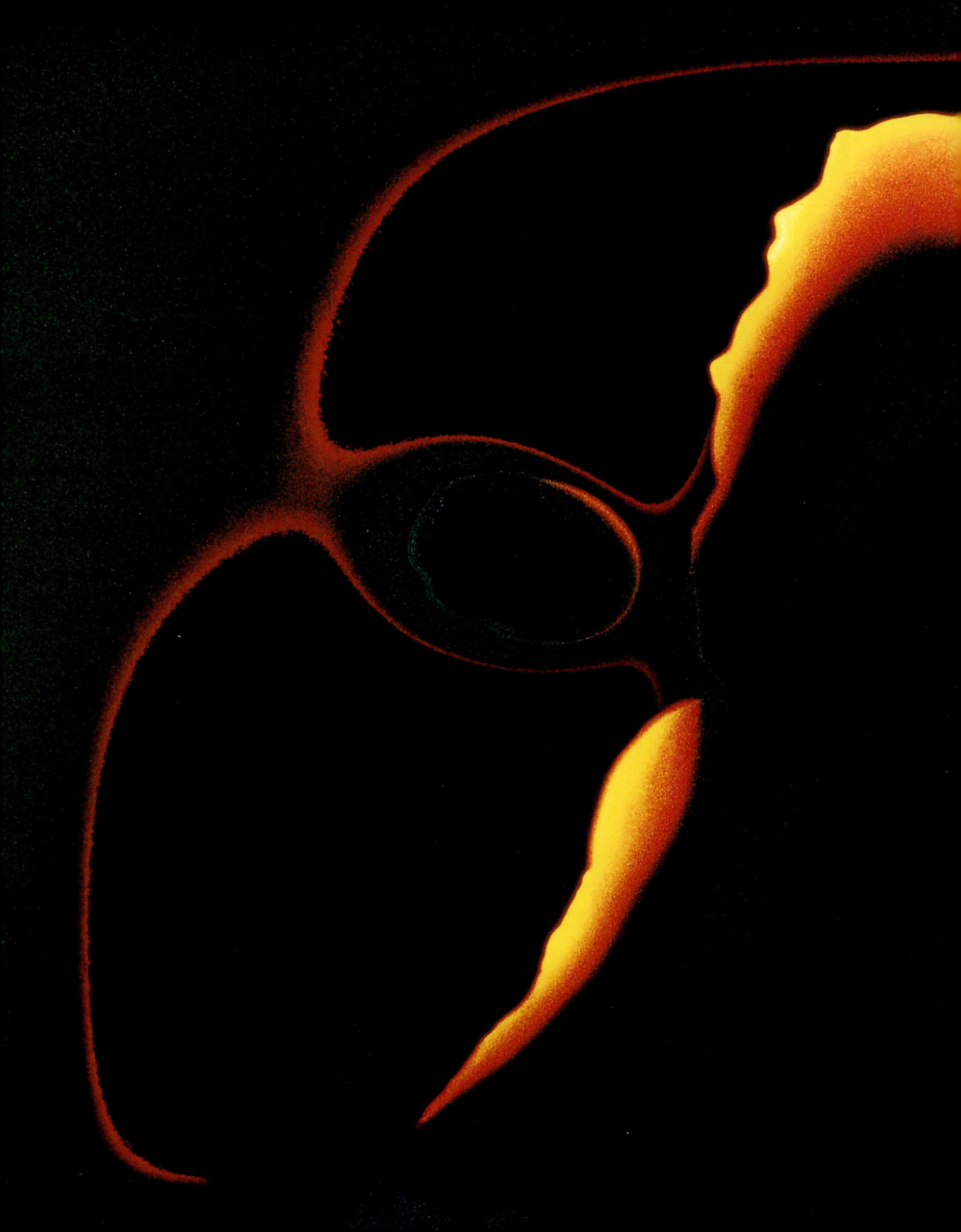

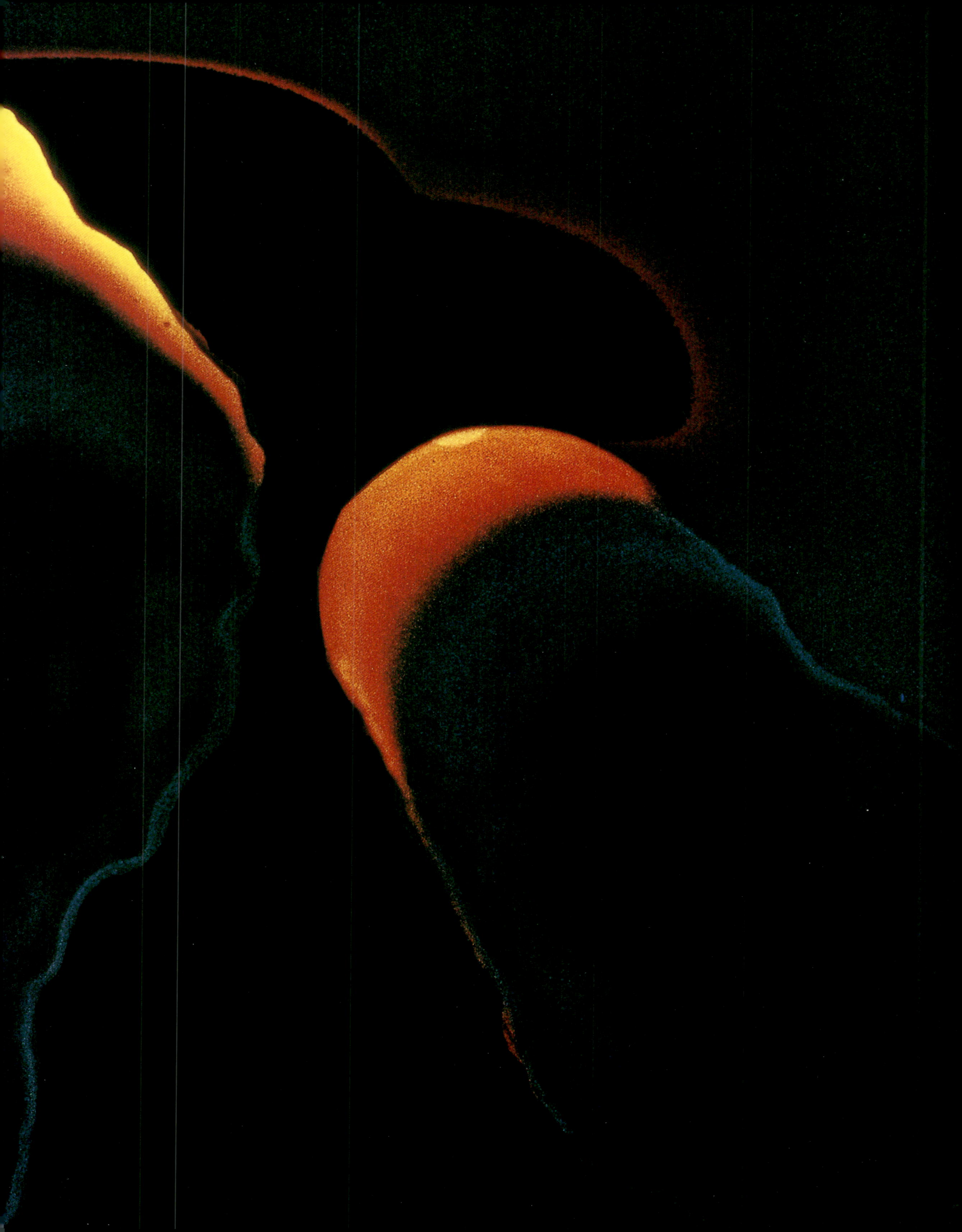

Kosmotopia: Flucht vor der Apokalypse

Man mag die Sorge darüber für verfrüht halten, aber an der Tatsache selbst gibt es wenig zu rütteln: Die Erde ist zum Untergang verdammt. Mit jedem Tag nimmt die Leuchtkraft der Sonne zu, und in etwa sieben Milliarden Jahren wird sich ihre äußere Atmosphäre so weit ausgedehnt haben, daß sie unseren Planeten verschlingt. Schon lange vorher freilich, in 900 Millionen bis 1,5 Milliarden Jahren, wird die Sonne die Erdoberfläche bereits so stark aufheizen, daß alles Leben zugrunde geht. Will die Spezies Mensch diesem Schicksal entgehen, muß sie vorher die Erde verlassen und das Weltall kolonisieren.
Was sonst eher Stoff von Science-Fiction-Romanen ist - die Flucht ins All vor drohendem Untergang - findet in diesen Tagen auch in Büchern renommierter Wissenschaftler ihren Niederschlag. Allen voran liegt die „Physik der Unsterblichkeit" des Kosmologen Frank Tipler, ein konsequent durchdachtes, mathematisch untermauertes Thesenpapier zur langfristigen Zukunft der Menschheit. Das Universum, sagt Tipler, wird mit ziemlicher Sicherheit weitere 100 Milliarden Jahre existieren - ausreichend Zeit für uns, die Quantenunbestimmtheit in den physikalischen Gesetzen zu nutzen, „um die Entwicklung des Universums auf eine sehr begrenzte Zahl möglicher „Zukünfte" hin zu zwingen. Unser schieres Überleben gebietet es, dem Kosmos eine Ordnung aufzuerlegen! Die praktische Umsetzung des stolzen Programms soll so aussehen, daß eine technologisch „ewig weltweit fortschreitende Menschheit" im Laufe der kommenden Jahrmilliarden den gesamten Kosmos erobert, und zwar mit Hilfe intelligenter Robotersonden, die sich selbst programmieren, reparieren und beliebig oft reproduzieren können: „Beispielsweise werden wir vermutlich eine Sonde zum Proxima-Centauri-Sonnensystem schicken, die dort auf Planeten landet und mit Hilfe geeigneter Materialien Kopien von sich herstellt. Die Kopien könnten dann Kurs auf Alpha Centauri, Sirius, Epsilon Eridan und Prokyon nehmen." Sobald die neuen Sonden diese Sterne erreicht haben, würde das Ganze wiederholt und nochmals wiederholt werden, bis die Sonden - und später menschliche Pioniere - alle Sternensysteme der Galaxie erobert hätten. Mit der Kolonisierung unserer Milchstraßen-Galaxie „könnte man Mitte des nächsten Jahrhunderts beginnen." Die nächstgelegene große Galaxie, der Andromeda-Nebel, könnte, meint Tipler, binnen 70 Millionen Jahren vereinnahmt werden. Kritisch wird die Lage für die das gesamte All beherrschende Raumfahrerspezies der Zukunft, wenn das expandierende Universum seine maximale Größe erreicht haben wird und beginnt, sich zusammenzuziehen. Aber auch für diese Krise weiß Tipler ein Gegenmittel: Um das endgültige Kollabieren des Raumes hinauszuzögern, wird der Kosmos „gezwungen werden, in zwei Richtungen sehr schnell zu kollabieren, während er in der dritten Richtung die gleiche Größe beibehält. Um das Universum zu solchem Verhalten zu zwingen, muß Leben es jedoch völlig erobern. Die Möglichkeit, in einem begrenzten Bereich zu bleiben, steht ihm nicht offen, für die Herrscher des Universums ist es schließlich - wenn der „Big Crash" ins Schwarze Loch naht - auch kein Problem mehr, sich selbst, den verstorbenen Vorfahren und allen denkbaren genetischen Variationen davon zu ewigem Leben zu verhelfen, komplett mit einem wiederauferstandenen Universum. In einer virtuellen Wirklichkeit existierende Computer von unbegrenzter Kapazität werden, so der stets positiv denkende Astrophysiker, perfekte Computersimulationen aller möglichen Menschen, anderer Lebensformen und sogar paralleler Welten herstellen. „Es wird unvermeidlich die Zeit kommen, in der genügend Computerkapazität vorhanden sein wird, um unsere heutige Welt einfach durch schiere Kraft zu simulieren, nämlich durch eine exakte Simulation aller logisch möglichen Varianten unserer Welt... Somit könnten alle möglichen Menschen allein aufgrund des in der DNA kodierten Wissens auferweckt werden. Ein simulierter Mensch wird aus simulierten menschlichen Zellen bestehen, gebildet aus simulierten Elektronen, Quarks und Gluonen. Keine experimentelle Apparatur wird zwischen Original und Simulation unterscheiden können." Die auf diese Weise zu ewigem Leben erweckten Personen „würden sich als genauso real und ihren Körper als genauso materiell erleben, wie wir das heute tun.
Beim Lesen dieser fast 600 Seiten langen Vision einer schönen neuen Welt kommt man natürlich immer wieder ins Kopfschütteln über so viel naiven Fortschrittsglauben. Man muß dem Mathematiker und Kosmologen aber zugute halten, daß er sich als erster mehr als nur beiläufige Gedanken über eine Fluchtmöglichkeit vor der Big-Crash-Apokalypse gemacht hat.
Ein Weiser auf dem Weg des Buddha oder des Tao würde freilich sehr, sehr milde lächeln über so viel Eroberungswillen und Angst vor dem Verlust von Kontrolle, Welt und Ego. Wie sagte doch Laotse so schön: „Die, welche die Welt regieren wollen, können es nicht fassen, das sehe ich; denn die Welt ist ein geistiges Gefäß und kann nicht gezwungen werden. Wer es zwingt, verdirbt es. Wer es festhält, verliert es."

Kosmotopia: Op de vlucht voor de Apocalypse

De zorgen hierover kan men wel prematuur vinden, maar aan het feit op zich valt weinig te tornen: de Aarde is tot ondergang gedoemd. Met elke dag neemt de lichtkracht van de zon toe, en over ongeveer zeven miljard jaren zal haar buitenste atmosfeer zo ver zijn uitgedijd dat deze onze planeten opslokt. Uiteraard al veel eerder, over 900 miljoen tot 1,5 miljard jaren, zal de zon het aardoppervlak reeds zo hevig verwarmen, dat al het leven te gronde zal gaan. Als de species mens aan dit lot wil ontsnappen, moet hij de Aarde van te voren verlaten en in het heelal kolonies stichten.
Wat anders eerder stof voor toekomstromannetjes is - het heelal in vluchten voor een dreigende ondergang - vindt dezer dagen ook in boeken van gerenommeerde wetenschappers zijn neerslag. Helemaal vooraan staat de "Physik der Unsterblichkeit" van de kosmoloog Frank Tipler, een consequent doordachte, wiskundig gefundeerde, op papier gebrachte stelling over de toekomst van de mens op lange termijn. Tipler zegt dat het universum met grote waarschijnlijkheid nog 100 miljard jaar verder zal bestaan - genoeg tijd voor ons om de quantumonbepaaldheid in de natuurkundige wetten te gebruiken "om de ontwikkeling van het universum tot een zeer begrensd aantal mogelijke 'toekomsten' te dwingen". Voor het pure overleven van ons is het noodzakelijk dat de kosmos orde wordt opgelegd! De praktische omzetting van het indrukwekkende programma zal er zo uitzien dat een technologisch "eeuwig wereldwijd vooruitgaande mensheid" in de loop van de komende miljarden jaren de gehele kosmos verovert, en wel met behulp van intelligente robotsondes die zichzelf programmeren, repareren en onbeperkt reproduceren kunnen: "Bijvoorbeeld zullen wij vermoedelijk een sonde naar het Proxima Centauri zonnestelsel sturen, die daar op planeten landt en met behulp van geschikt materiaal kopieën van zichzelf produceert. Deze kopieen zouden dan koers op Alfa Centauri, Sirius, Epsilon Eridanus en Procyon kunnen nemen." Zodra de nieuwe sondes deze sterren hebben bereikt, zal het geheel herhaald en nogmaals herhaald worden, totdat de sondes - en later menselijke pioniers - alle galactische sterrenstelsels veroverd hebben. Met het koloniseren van ons Melkwegstelsel "zou men in het midden van de volgende eeuw kunnen beginnen". Het dichtstbij gelegen grote galactische stelsel, de Andromedanevel, zou volgens Tipler binnen 70 miljoen jaar veroverd kunnen worden. Kritiek wordt de situatie voor de toekomstige ruimtevaartspecies, die dan het gehele heelal beheerst, als het uitdijende

Cosmotopia: Flight in the face of the apocalyse

universum zijn maximale grootte bereikt zal hebben en begint te schrompelen. Maar ook voor deze crisis heeft Tipler een tegenmiddel. Om het definitieve collaberen van de ruimte te vertragen zal de kosmos "worden gedwongen in twee richtingen zeer snel te collaberen, terwijl hij in de derde richting dezelfde grootte behoudt. Om het universum tot zo'n gedrag te dwingen, moet het evenwel volledig door leven zijn veroverd. Het heeft niet de mogelijkheid om in een begrensd gebied te blijven. Voor de heersers over het universum zal het tenslotte - als de "Big Crash" in het Zwarte Gat nadert - ook geen probleem meer zijn om zichzelf, de overleden voorouders en alle denkbare genetische variaties ervan, het eeuwige leven te bezorgen, compleet met een herrezen universum. Computers van onbegrensde capaciteit, die in een virtuele werkelijkheid bestaan, zullen - aldus de immer positief denkende astrofysicus - perfecte computersimulaties van alle mogelijke mensen, andere levensvormen en zelfs van evenwijdige werelden vervaardigen. "Er zal onvermijdelijk de tijd komen waarin genoeg computercapaciteit aanwezig zal zijn om onze huidige wereld eenvoudig door pure kracht te simuleren, namelijk door een exacte simulatie van alle logisch mogelijke varianten van onze wereld. Zodoende kunnen alle mogelijke mensen alleen op grond van de kennis die in het DNA is gecodeerd, tot leven worden gewekt. Een gesimuleerde mens zal uit gesimuleerde menselijke cellen bestaan, gevormd uit gesimuleerde elektronen, quarks en gluonen. Geen enkele experimentele apparatuur zal tussen origineel en simulatie kunnen onderscheiden." De op deze wijze tot eeuwig leven gewekte personen "zouden zichzelf als even reëel en hun lichaam als even materieel ervaren als wij dat heden doen".
Bij het lezen van dit bijna 600 pagina's tellende visioen van een mooie nieuwe wereld wordt natuurlijk altijd weer hoofdschudden teweeggebracht over zoveel naief vertrouwen in vooruitgang. Maar je mag het een wiskundige of kosmoloog niet kwalijk nemen dat hij als eerste min of meer serieuze gedachten laat gaan over mogelijkheden om aan de Big Crash Apocalypse te ontkomen.
Een wijze op de weg van Boeddha of Tao zou uiteraard alleen maar glimlachen over zoveel veroveringsdrift en vrees voor het verliezen van controle, wereld en ego. Hoe zei Lao Tse toch zo mooi: "Degenen die de wereld willen regeren, kunnen het niet vatten, dat zie ik; want de wereld is een geestelijk vat en kan niet worden gedwongen. Wie het dwingt, bederft het. Wie het vasthoudt, verliest het."

One can consider it too early to worry about but it is an unalterable fact that the Earth is doomed to destruction. With each passing day, the luminosity of the sun increases and in about seven thousand million years the outer atmosphere will have expanded to such a degree that it will devour our planet. Long before this, admittedly, in 900 million to 1.5 thousand million years, the sun will have already heated up the surface of the Earth so strongly that all life will be destroyed. If mankind wishes to escape this fate, it will have to leave the planet beforehand and colonize space.
What is normally more the subject-matter of science-fiction novels - the flight into space in the face of threatening destruction - is also something which is now finding expression in the books of reputed scientists. Top of the list is "Physics of Immortality" by the cosmologist Frank Tipler, a consistently thought out, mathematically corroborated series of theories on the long-term future of mankind. The universe, Tipler states, will with considerable certainty exist for another 100 thousand million years - enough time for us to use the quantum indeterminacy in the physical laws "to restrict the development of the universe to a very limited number of possible 'futures'. Our sheer survival is dependent on our imposing an order on the cosmos!" Putting this proud program into practice involves the conquest of the entire cosmos in the course of the coming thousands of millions of years by a technologically "eternally progressing humanity", with the aid of intelligent robot probes which program themselves, repair themselves and which can be reproduced as often as required: "We will for example send a probe to the Proxima-Centauri Solar System which will land on planets there and with the help of suitable materials will make copies of itself. These copies could then take course for Alpha Centauri, Sirius, Epsilon Eridan and Prokyon." As soon as the new probes have reached these stars then the whole thing will be repeated and repeated again until the probes - and later human pioneers - will have conquered all the systems of stars in the galaxy. "One could begin in the middle of the next century" with the colonization of our Milky Way Galaxy. The next large galaxy, the Andromeda Nebula, could, Tipler feels, be reached within 70 million years. The whole thing will become critical for the astronaut species of the future and their attempt to explore the whole of space, when the expanding universe attains to its maximum dimensions and then starts to shrink. But Tipler also knows an answer for this crisis, too: in order to delay the collapse of space the cosmos will "be forced to collapse very quickly in two directions while retaining the same size in the third direction. In order to compel the universe to behave in such a manner, life must have fully conquered it. There is no means of remaining in a limited area; for the rulers of the universe it is after all, as the Big Crash into the Black Hole nears, no problem to help themselves, and deceased ancestors, and all conceivable genetic variations thereof, to Eternal Life, including a resurrected universe. Computers of unlimited capacity existing in a virtual reality, the always positively thinking astro-physicist believes, will produce perfect computer simulations of everyone, of other forms of life and even of parallel worlds. "A time will inevitably come in which sufficient computer capacity will exist to simulate our entire world simply by sheer energy, by an exact simulation of all the logically possible variations of our world... Thus everyone can be resurrected on the sole basis of the knowledge codified in his DNA. A simulated man will consist of simulated human cells, formed from simulated electrons, quarks and gluons. No experimental apparatus will be able to distinguish between the original and the simulation." The persons revived in this way to eternal life "would feel themselves to be just as real and would experience their bodies in the same material way as we do today".
When one reads this almost 600-page-long version of a beautiful, new world then one of course repeatedly shakes one's head about so much naive faith in progress. One must however say in the defence of the mathematician and cosmologist that he was the first to do more than just casually muse about a way-out in the face of the threat of the Big Crash apocalyse.
A wise man on the path of Buddha or of Taoism would, admittedly, just smile laconically at so much will to conquer and so much fear of loss of control, the world and the ego. How did Laotse put it?: "Those who rule the world, cannot get hold of it, I see that; for the world is a spiritual vessel and cannot be constrained. He who constrains it, ruins it. He who tries to hold it fast, loses it."

La verità della immaginazione, la fantascienza di Ezechiele, Verne e l'Apollo 11, dal vescovo Wilkins a Cyrano di Bergerac

Nei secoli d'oro della scienza Greca, irripetuti nella storia del mondo, esistevano scienziati enciclopedici, che avevano il buon senso di riconoscere i diritti della fantasia. Ossia di concedere possibilità di cittadinanza ad opere poetiche che avessero relazione - ed è il nostro caso - con il cielo e le costellazioni. Eratostene, nato a Cirene nel 276, ne curò per l'appunto una raccolta, intitolandola "Catasterismi", che riuniva storie mitiche di poeti in rapporto alle costellazioni. Ma la fantascienza, perché di essa si tratta, va ben più addietro nel tempo, in quanto intesse l'intera storia del genere umano a cominciare da quella delle religioni e perciò dalla mitologia. Un modo di esporre, diceva Platone, dovuto alla opinione anziché alla certezza scientifica. Alla percezione fantasiosa che non attesta realtà esatte, bensì verità dell'immaginazione. Ma fino a che punto le due cose si escludono? Esistono argomenti che la scienza odierna tenta di sovrapporre con risultati, occorre dirlo, affidabili. Tratti dalla Bibbia, che al di là del suo supremo significato religioso, è uno straordinario libro di fantascienza. Ezechiele, nel descrivere l'oggetto di fuoco venuto dal cielo, precisa: "L'aspetto delle ruote e la loro forma eran come l'aspetto del crisolito; tutte e quattro si somigliavano; il lor aspetto e la loro forma eran quelli d'una ruota. Quando si muovevano andavano tutte quattro dal proprio lato, e, andando, non si voltavano. Quanto ai loro cerchi, essi erano alti e formidabili; e i cerchi di tutte e quattro eran pieni d'occhi d'ogn'intorno. Quando gli esseri viventi camminavano, le ruote si movevano allato a loro; e quando gli esseri viventi s'alzavan su da terra, s'alzavano anche le ruote. Dovunque lo spirito voleva andare, andavano anch'essi; e le ruote s'alzavano allato a quelli, perché lo spirito degli esseri viventi era nelle ruote."
Letteratura fantascientifica, dunque. Ma fino a che punto? Gli studiosi di pre-astronautica non vedono altro nella cronaca di Ezechiele che l'atterraggio di una nave spaziale con i suoi astronauti. E non per semplici ipotesi, bensì con il sostegno di prove concrete che dimostrarono matematicamente e tecnicamente l'affidabilità di simili opinioni. A proposito di verifiche è significativa la sovrapposizione tra il viaggio descritto nel romanzo di Jules Verne, "Dalla Terra alla Luna", e la impresa dell'Apollo 11 con a bordo i tre astronauti americani: perché pure tre erano i passeggeri del romanzo, analogo il posto di lancio e del successivo ammaraggio nello Oceano Pacifico, simile il razzo a più stadi e prevista l'assenza di peso nel corso del volo spaziale.
Il dottor Robert H. Goddard, il padre della tecnologia spaziale americana, l'aveva letto da ragazzo e gli era rimasta dentro una irresistibile aspirazione di rendere concreta la storia del grande romanziere francese. Chissà se ad Isaac Asimov, il grande scrittore contemporaneo di fantascienza, toccherà la stessa sorte.
Prima di Verne erano comparse assurde diligenze aeree ricolme di gaudenti viaggiatori, frutto della mentalità settecentesca, derivate dalle pretese scoperte lunari, attribuite allo astronomo Herschel; oppure seggiolini trainati da gigantesche pariglie di oche e cigni come scriveva nel 1638 il romanziere inglese Francesco Godwin. Oltremanica, frattanto, il vescovo presbiteriano di Chester, Giovanni Wilkins (1614-1642), era persuaso che la luna fosse abitabile ed abitata e che con essa si potessero instaurare buoni commerci tramite veicoli volanti. Sostenne questo in un volume intitolato "Scoperta di un nuovo mondo", con forti intuizioni, soprattutto quando esprime la certezza della "possibilità d'una navigazione aerea, fondata sul medesimo principio della navigazione ordinaria, ossia sulla proprietà delle navi di sostenersi sulla superficie di un fluido, purché siano riempite di una sostanza più leggera".
Dal versante europeo dello stretto, provvederà quel diavolo di Saviniano Cyrano di Bergerac (1620-1655) a movimentare l'ambiente scrivendo due opere giudicate molto positivamente dalla critica letteraria successiva, dai titoli: "Il viaggio nella luna" e "La storia comica degli stati e degli imperi del sole".
C'è poi quel millantatore celebre, vale a dire Girolamo Carlo Federico, Barone di Munchausen, che viaggia verso la luna per andarsi a riprendere un'accetta lanciata verso l'alto con impeto eccessivo; per arrivare alfine in Italia ove, per non parlare di Ludovico Ariosto, il bolognese Pier Jacopo Martelli (1665-1727) scrive il poemetto "Gli occhi di Gesù" con il preciso scopo di collocare il Paradiso sulla Luna.
Un catalogo appena accennato, ma sufficiente a verificare quale posto la fantascienza occupi nella letteratura.

Giuseppe Brunamontini

Cosmotopia: la fuite devant l'Apocalypse

On peut considérer que ces soucis sont prématurés, mais on ne peut rien y changer: la terre est condamnée à périr. Chaque jour, l'intensité lumineuse du soleil augmente et dans environ sept milliards d'années, son atmosphère extérieure se sera tant élargie qu' elle engloutira notre planète. Bien longtemps avant, certes, dans 900 millions à 1,5 milliards d'années, le soleil chauffera déjà tellement la surface de la terre que toute vie sera anéantie. Si le genre humain veut échapper à ce destin, il lui faut quitter auparavant la terre et coloniser l'espace.
Ce qui généralement se joue plutôt dans les romans de science-fiction, la fuite dans l'espace pour échapper à une destruction imminente, apparaît maintenant dans les livres de savants renommés. On citera tout d'abord la "Physique de l'immortalité" du cosmologue Frank Tipler, une thèse logique et bien pensée, reposant sur des fondements mathématiques et traitant l'avenir à long terme de l'humanité. L'univers, dit Tipler, existera très probablement encore 100 milliards d'années, une durée qui nous laisse suffisamment de temps pour tirer parti de l'indétermination des quanta dans les lois de la physique "en vue d'obliger l'univers à se développer selon un nombre très limité d'"avenirs" possibles. Notre survie impose au cosmos l'établissement d'un ordre précis! La réalisation pratique de cet orgueilleux programme prévoit qu'une "humanité en progrès technique éternel dans le monde entier" conquerra l'ensemble du cosmos au cours des milliards d'années à venir à l'aide de sondes intelligentes qui se programment et se réparent elles-mêmes et peuvent se reproduire à loisir: "Nous enverrons probablement une sonde dans le système solaire Proxima Centauri qui atterrira là-bas sur des planètes et fabriquera des copies d'elle-même avec les matériaux appropriés". Les copies pourront alors mettre le cap sur Alpha Centauri, Sirius, Epsilon Eridani et Procyon." Dès que les nouvelles sondes auraient atteint ces étoiles, le processus se répéterait une fois, puis deux, jusqu'à ce que les sondes (et plus tard les pioniers humains) aient conquis tous les systèmes stellaires de la galaxie. Selon lui, on pourrait commencer la colonisation de notre galaxie, la Voie lactée, "au début du prochain siècle". La grande galaxie la plus proche, le brouillard d'Andromède, pourrait être conquis en l'espace de 70 millions d'années. La situation deviendra critique pour les astronautes de l'avenir qui domineront tout l'espace quand l'univers en expansion aura atteint son ampleur maximale et commencera à rétrécir. Mais là aussi, Tipler connaît un remède: pour ajourner l'effondre-

Космоутопия: бегство от апокалипсиса

ment définitif de l'espace, on obligera le cosmos à s'effondrer très vite dans deux directions tout en gardant les mêmes dimensions dans la troisième. Pour obliger l'univers à se comporter ainsi, les êtres vivants devront cependant le conquérir totalement. Ils n'auront pas la possibilité de rester dans un espace restreint; pour les maîtres de l'Univers, il sera finalement facile (quand le "Big Crash" dans le trou noir approchera) d'assurer, dans un Univers ressuscité, une vie éternelle à soi-même, aux ancêtres disparus et à toutes les variations génétiques imaginables. Dans une réalité virtuelle, les ordinateurs disposant d' une capacité illimitée fabriqueront alors (c'est ainsi du moins que le voit l'astrophysicien à la pensée toujours positive) des simulations parfaites sur ordinateur de tous les humains, d'autres formes vivantes et même de mondes parallèles. "Le temps viendra inévitablement où la capacité des ordinateurs sera suffisante pour simuler uniquement par la force notre monde actuel, ceci par une simulation exacte de toutes les variantes logiquement possibles de notre monde... Ainsi, tous les hommes pourront être réanimés par les simples informations codées dans l'ADN. Un homme simulé sera composé de cellules humaines simulées, formées d'électrons, de quarks et de gluons simulés. Aucun appareillage expérimental ne pourra faire la différence entre l'original et la simulation". Pour les personnes ainsi appelées à une vie éternelle, "l'existence sera aussi réelle et leur corps aussi matériel que nous le ressentons aujourd'hui".
En lisant cette vision du Meilleur des mondes, longue de presque 600 pages, on secoue souvent la tête face à une foi si naïve dans le progrès. Il faut cependant savoir gré à ce mathématicien et cosmologue d'avoir été le premier à réfléchir intensément à une possibilité de fuite avant l'apocalypse du "Big Crash". Un sage sur le chemin de Bouddha ou du Tao sourirait certes avec condescendance face à ce désir intense de conquête et à la peur de perdre le contrôle, le monde et son ego. Comme Lao-Tseu l'a si bien dit: "Ceux qui veulent gouverner le monde ne peuvent pas le comprendre, je le vois; car le monde est un récipient spirituel qu'on ne peut contraindre. Qui le contraint le détériore. Qui le retient le perd."

Может быть, беспокоиться об этом еще рано, но факт остается фактом: Земля обречена на гибель. С каждым днем светимость Солнца увеличивается, и примерно через семь миллиардов лет его внешняя атмосфера, расширяясь, поглотит нашу планету. Задолго до этого, через 900 млн. или 1,5 млрд. лет, Солнце так нагреет поверхность Земли, что всякая жизнь погибнет. Если человек как вид хочет избегнуть этой судьбы, он должен заблаговременно покинуть Землю и колонизовать Вселенную.
То, что прежде было темой научно-фантастических романов — бегство от грозящей гибели во Вселенную —, теперь встречается в книгах авторитетных ученых. На первом месте стоит «Физика бессмертия» космолога Ф. Типлера, представляющая собой последовательно продуманные и математически обоснованные тезисы о долговременном будущем человечества. По Типлеру, Вселенная с большой вероятностью просуществует еще 100 млрд. лет — достаточно для нас, чтобы воспользоваться квантовой неопределенностью физических законов и «пустить развитие Вселенной по ограниченному количеству возможных моделей будущего». Задача выживания заставляет нас навязать космосу свой порядок. Практическое осуществление грандиозной программы должно выглядеть так: технологически «вечно прогрессирующее в глобальном масштабе человечество» завоевывает в ближайшие миллиарды лет весь космос с помощью умных роботов-зондов, способных самопрограммироваться, саморемонтироваться и неограниченно самовоспроизводиться: «например, мы посылаем такой зонд в солнечную систему Проксима Центавра; там он садится на планету и, с помощью соответствующих материалов, изготовляет собственные копии. Копии берут курс на Альфа Центавра, Сириус, Ипсилон Эридан и Прокион». Когда новые зонды достигнут этих звезд, все повторится еще и еще раз, пока зонды — а позднее и первые колонисты — не завоюют все звездные системы в галактике. Колонизацию нашей галактики можно начать «в середине следующего века». Ближайшая к нам крупная галактика — туманность Андромеды — может быть завоевана, как считает Типлер, за 70 млн. лет. Критическим станет положение завоевавшего Вселенную человечества, когда расширяющаяся Вселенная достигнет максимальных размеров и начнет сжиматься. Но и против этого у Типлера есть средство: чтобы оттянуть коллапс пространства, «космос надо заставить очень быстро сжиматься в двух направлениях, а в третьем оставаться неизменным. Но для этого надо сперва его полностью завоевать. Возможности оставаться в ограниченном пространстве у людей нет, но даже угроза падения в «черную дыру» не испугает властелинов Вселенной: будет нетрудно обеспечить вечную жизнь себе, умершим предкам и любым генетическим вариациям в полностью возрожденной Вселенной. Неизменно оптимистически мыслящий астрофизик считает, что существующие в виртуальной действительности компьютеры неограниченной мощности создадут совершенные модели всех людей, других форм жизни и даже параллельных миров. «Неизбежно придет время, когда мощности компьютеров хватит на то, чтобы смоделировать нынешний мир во всех его логически мыслимых вариантах... Тогда можно будет воскресить всех людей на основании кода ДНК. Смоделированный человек будет состоять из смоделированных человеческих клеток, образованных смоделированными электронами, кварками и глюонами. Никакая аппаратура не сможет отличить оригинал от копии.» Воскресшие таким образом в вечную жизнь люди «будут так же реально и материально ощущать себя и свое тело, как мы».
Читая эти 600 страниц, описывающих прекрасное новое время, все время качаешь головой: какая наивная вера в прогресс! Но заслуга этого математика и космолога в том, что он первый не просто упомянул, а подробно описал возможность бегства от апокалипсиса. Мудрец, следующий путем Будды или дао, конечно, очень мягко улыбнется по поводу такого экспансионизма и страха потерять контроль, потерять мир и свое «Я». Как прекрасно сказал Лао-цзы: «Те, кто хотят править миром, не могут понять, я это вижу; ибо мир есть духовный сосуд и не может быть принуждаем. Кто его принуждает — тот его портит. Кто его удерживает — тот его теряет.»

ETHICA HUMANA

Zuverlässigkeit
Betrouwbaarheid
Reliability
Affidabilità
Fiabilité
Надежность

Wege zur Zuverlässigkeit in eine Ordnung des Ganzen

Vorwort und Begleitung der Texte

Die allgewaltigen Kräfte der Natur, Sonnenaufgang und -untergang, Säen und Ernten, Wachsen und Vergehen, Ebbe und Flut, sind für uns Menschen nach wie vor Symbole für die Zuverlässigkeit und deren Synonyme Verläßlichkeit, Unbestechlichkeit, Gewissenhaftigkeit, Genauigkeit oder Beständigkeit am Beispiel der Ordnung der Schöpfung: „Wenn sich die untergehende Sonne dem Horizont nähert, verschmelzen die vier Elemente Wasser, Luft, Feuer und Erde zu einem. Rückkehr zum ersten Schöpfungstag..."
In der „unerschütterlichen Beständigkeit" im Leben seines Vaters sah der römische Kaiser Marc Aurel eine zuverlässige Leitschnur für seinen eigenen Lebensweg, Voraussetzung für klärende Selbstkritik in seinen berühmten „Selbstbetrachtungen".
Naturwissenschaften und Geisteswissenschaften gehörten für die großen Gelehrten an der Schwelle des Mittelalters zur Neuzeit noch untrennbar zusammen. Kritisch wurde die Rolle des Menschen im Unendlichen des Universums von ihnen hinterfragt. „Was ist nun der Mensch in der Welt?", fragte der Physiker, Mathematiker und Religionsphilosoph Blaise Pascal, den 1654, nach langjährigem weltmännischen Leben in Paris und Rouen ein mystisches Gotteserlebnis vom Freigeist zum Verteidiger des Christentums wandelte: „Alle Dinge sind hervorgegangen aus dem Nichts und reichen bis ins Unendliche. Wer wird diesen erstaunlichen Schritten folgen?..."
Die alte und vor allem die neue Welt bereiste und erforschte zwei Jahrhunderte später Alexander von Humboldt, der in seinem Hauptwerk „Kosmos" einen umfassenden „Entwurf einer physischen Weltbeschreibung" gab und dabei die Welt als ein durch innere Kräfte bewegtes und belebtes Ganzes darstellt, auf das die Menschen sich verlassen konnten und „den Geist der Natur" ruhig „ergreifen" sollten.
Zweifel an dem „im Glauben oder Unglauben befestigten Menschen" in der Welt von heute werden dann bei den Philosophen und Wissenschaftlern unseres 20. Jahrhunderts laut. Der „Gott, der einst die wohletablierte Ordnung der Werte... bestätigte und der als Gewölbe dieser Ordnung galt", wurde für nicht mehr existent betrachtet. Das stellte der polnische Philosoph Leszek Kolakowski angesichts der Erfahrung nach zwei apokalyptischen Kriegen und eines Weltchaos' fest, das jeder zuverlässigen Ordnung spottet.
Die heutigen Wissenschaftler und Denker vor der kommenden Jahrtausendwende beschwören erneut wieder eine „Ordnung des Ganzen": „Nur wer das Ganze sieht... ist in der Lage, sich in seinem Teilbereich zu begrenzen, um das Ganze zu erhalten." Es wird das neue Gottesbild eines kosmischen Christus aufgerichtet und als Religion der Zukunft postuliert.
Und zur „Frage nach der Zukunft" stellte Carl Friedrich von Weizsäcker in seinem opus magnum „Zeit und Wissen" kürzlich fest:„Die Bilder vom Diesseits und Jenseits, welche die überlieferten Religionen boten, waren nicht bloße Naivität... Der Weg zu meditativen Wahrnehmungen ist offen. Neue Erfahrungen, neue gedankliche Bilder warten auf uns... Der Weg zu den Erfahrungen wird durch tiefes Leid gehen. Und durch tiefes Glück."

Hans Dollinger

Wegen naar betrouwbaarheid in een geordend geheel

Voorwoord en begeleiding van de teksten

De oppermachtige krachten van de natuur, zonsop- en -ondergang, zaaien en oogsten, groeien en vergaan, eb en vloed, zijn voor ons mensen nog steeds zinnebeelden van trouw en hun synoniemen zijn betrouwbaarheid, onschendbaarheid, nauwgezetheid, nauwkeurigheid of bestendigheid naar het voorbeeld van de orde van de schepping: "als de ondergaande zon de horizont nadert, smelten de vier elementen water, lucht, vuur en aarde samen. Terugkeer naar de eerste scheppingsdag..."
In de "onwrikbare bestendigheid" in het leven van zijn vader zag de Romeinse keizer Marcus Aurelius een betrouwbare leidraad voor zijn eigen levensweg, beding voor opklarende zelfkritiek in zijn beroemde "zelfbeschouwingen".
Natuurwetenschappen en geesteswetenschappen hoorden voor de grote geleerden aan de drempel van de middeleeuwen naar de moderne tijd nog onafscheidelijk samen. Kritisch vroegen zij naar de rol van de mens in de oneindigheid van het universum. "Wat is nu de mens in de wereld?", vroeg de natuurkundige, wiskundige en godsdienst-filosoof Blaise Pascal die in1654 na een lang werelds leven in Parijs en Rouen door een mystische godsbelevenis van vrijdenker in een verdediger van het christendom werd veranderd: "Alles is uit het niets ontstaan en reikt tot in de oneindigheid. Wie zal deze verwonderlijke stappen volgen?..."
De oude en voornamelijk de nieuwe wereld bereisde en doorvorste twee eeuwen later Alexander von Humboldt die in zijn hoofdwerk "Kosmos" een omvangrijk "ontwerp van een fysieke wereldbeschrijving" gaf en daarbij de wereld als een door innerlijke krachten bewogen en gestimuleerd geheel

beschrijft waar de mensen op vertrouwen konden en "de geest van de natuur rustig grijpen" mochten.
Twijfel omtrent de "in het geloof of ongeloof versterkte mens" in de wereld van vandaag rijst dan bij de filosofen en wetenschappers van onze twintigste eeuw. De "God die ooit de zo gevestigde orde van de waarden ... bevestigde en die als gewelf van deze orde gold", werd als niet meer existent beschouwd. Dat concludeerde de Poolse filosoof Leszek Kolakowski denkende aan de ervaring na twee apocalyptische oorlogen en een chaotische wereld die elke betrouwbare orde tart.
De huidige wetenschapbeoefenaars en denkers voor de aanstaande millennium-wisseling manen opnieuw tot een "geordend geheel" aan "Alleen wie het geheel ziet ... is in staat zich tot zijn deelgebied te beperken, om het geheel te behouden." Het nieuwe godsbeeld van een kosmische Christus wordt opgericht en als godsdienst van de toekomst gepostuleerd.
En ten aanzien van de vraag naar de toekomst concludeerde Carl Friedrich von Weizsäcker onlangs in zijn opus magnum Tijd en Wetenschap: "De beelden van het aardse leven en het hiernamaals die de traditionele godsdiensten boden, waren niet pure naïviteit... De weg naar meditatieve waarnemingen is geëffend. Nieuwe ervaringen, nieuwe ideële beelden wachten op ons... . De weg naar de ervaringen leidt door diep leed. En door diep geluk".

Hans Dollinger

Ways to reliability by giving order to the whole

Foreword and accompaniment to the texts

The almighty forces of nature, sunrise and sunset, sowing and harvesting, growing and decaying, ebb and flow, are for us humans, as they have always been, symbols of reliability and its synonyms in relation to the order of creation: dependability, unerringness, accuracy, exactitude or constancy: "When the setting sun nears the horizon, the four elements, water, air, fire and earth, merge into one. A return to the first day of creation..."
In the "unshakeable steadfastness" in the life of his father, the Roman Emperor Marcus Aurelius perceived a reliable guide for his own life, the precondition for the clarifying self-criticism of his famous "The Meditations".
In the eyes of the great scholars on the threshold of the Middle Ages and Modern Times, the natural sciences and the humanities are inextricably bound up. The role of Man in the infinity of the universe was critically assessed. "What is Man in this world?" asked the physicist, mathematician and religious philosopher Blaise Pascal who in 1654, after living an urbane life in Paris and Rouen for many years, had a mystical experience which converted him from being a free-thinker into a defender of Christianity: "Everything has emerged from nothing and stretches to infinity. Who can fathom these astounding steps?..."
Two centuries later, Alexander von Humboldt travelled the globe, especially the New World, and in his greatest scientific work "Kosmos" presented a comprehensive "draft of a physical description of the world" depicting it as a whole influenced by inner forces on which Man could rely.
Doubt about Man "anchored in faith or unbelief" in today's world was then expressed in the work of the philosophers and scientists of our 20th century. The "God who once confirmed the well-established order of values... and who was looked upon as the canopy of this order" was now considered to be non-existent. This is the conclusion drawn by the Polish philosopher Leszek Kolakowski in the light of the experience of two apocalyptic wars and a world chaos which ridicules every form of reliable order.
Today's scientists and thinkers on the threshold of the turn of the millennium evoke once again an "order of the Whole": "Only he who can see the Whole...is in a position, to limit himself to his own area and thus preserve the Whole." A new picture of God, that of a Cosmic Christ, is being established and postulated as the religion of the future.
In relation to the question of what the future will bring, Carl Friedrich von Weizsäcker states in his opus magnum "Zeit und Wissen": "The pictures of this world and of the world beyond which were put forward by the traditional religions were not just naivety... The path of meditative perception is open. New experiences, new ideational pictures are awaiting us... The way to these experiences will involve great suffering but it will also bring great happiness.".

Hans Dollinger

L'affidabilità in un ordine globale

Prefazione ed accompagnamento dei testi

Le forze onnipresenti della natura, alba e tramonto, semina e raccolto, crescita e consunzione, flusso e riflusso, sono per gli uomini tuttora simboli dell'affidabilità e sinonimi di fidatezza, incorruttibilità, coscenziosità, precisione o costanza ed esemplificano l'ordine della creazione: "Quando al tramonto il sole si avvicina all'orizzonte, i quattro elementi, acqua, aria, fuoco e terra si fondono in uno. Ritorno al primo giorno della creazione..."
"L'imperturbabile costanza" della vita di suo padre, era per l'imperatore romano Marco Aurelio un punto di riferimento affidabile per il corso della propria vita, premessa dell'autocritica catartica delle sue famose "Autocontemplazioni".
Le scienze naturali e le scienze umanistiche erano ancora inscindibili per i grandi eruditi alla soglia tra Medio Evo ed Età Moderna. Essi si ponevano la domanda sul ruolo dello uomo nell'infinitezza dell'universo. "Cos'è quindi l'uomo nell'universo?", si chiedeva il fisico, matematico e filosofo della religione Blaise Pascal, che nel 1654, dopo molti anni di vita mondana a Parigi e Rouen, fu scosso da una visione mistica di Dio che lo trasformò da libero pensatore in un difensore del Cristianesimo: "Tutte le cose sono scaturite dal nulla e toccano l'infinito. Chi seguirà questi straordinari passi?..."
Due secoli più tardi Alexander von Humboldt compì viaggi di ricerca nel vecchio e soprattutto nel nuovo mondo. Nella sua opera principale "Cosmos" egli fornì un modello completo di una "Descrizione fisica del mondo", presentando il mondo come un tutto unico mosso ed animato da forze interne sul quale gli uomini possono fare affidamento per comprendere fiduciosamente "lo spirito della natura".
I filosofi e gli scienziati del nostro XX secolo esprimono invece dubbi "sull'uomo saldo nella fede o nella irreligiosità" del mondo di oggi. Il "Dio che un tempo confermava l'ordine ben saldo dei valori... e che era considerato il 'coronamento' di questo ordine" non era più considerato esistente. Questo è quanto ha appurato il filosofo polacco Leszek Kolakowski dopo l'esperienza di due guerre apocalittiche e di un caos mondiale che schernisce ogni ordine fidato.
Giunti alla fine del millennio gli scienziati ed i pensatori evocano nuovamente un "ordine globale": "Solo chi vede il Tutto.. è in grado di circoscrivere per sè un suo campo parziale per mantenere in vita il Tutto." Si instaura la nuova immagine divina di un Cristo cosmico e la si postula quale religione del futuro.
E riguardo alla "Questione del futuro" Carl Friedrich von Weizsäcker ha affermato recentemente nel suo opus magnum: "Le immagini della vita terrena e della vita celeste offerteci dalle religioni tramandate, non erano solo ingenue... La strada delle percezioni meditative è aperta. Nuove esperienze, nuove immagini mentali ci attendono... La via che conduce alle esperienze passerà per pene profonde. E per felicità profonda."

Hans Dollinger

Les voies qui mènent à la fiabilité à travers un ordre global

Préface

Les forces omnipotentes de la nature, lever et coucher du soleil, semence et récolte, croissance et décroissance, marée basse et marée haute, symbolisent toujours pour les humains la fiabilité et ses synonymes, la sûreté, l'incorruptibilité, la probité, l'exactitude ou la stabilité, à l'exemple de l'ordre de la Création: "Quand le soleil couchant s'approche de l'horizon, les quatre éléments: eau, air, feu et terre se fondent en un seul. C'est le retour au premier jour de la Genèse..."
L'empereur romain Marc Aurèle voyait dans la "stabilité inébranlable" qui caractérisait la vie de son père un motif conducteur pour le chemin à suivre dans sa propre vie, condition de rigueur pour une autocritique clarifiante dans ses célèbres "Pensées". A la fin du Moyen-Age, au seuil des temps modernes, sciences naturelles et sciences humaines restèrent inséparables pour les grands savants qui posèrent des questions critiques sur le rôle de l'homme dans l'infini de l'Univers. "Qu'est donc l'homme en ce monde?" demanda le physicien, mathématicien et philosophe des religions Blaise Pascal, libre-penseur ayant longtemps mené une vie mondaine à Paris et à Rouen, qu'une expérience mystique vécue en 1654 transforma en défenseur de la chrétienté: "Toutes les choses proviennent du Néant et se perdent dans l'infini. Qui suivra ces pas étonnants?..." Deux siècles plus tard, Alexander von Humboldt partit à la découverte de l'ancien et surtout du nouveau monde; dans son principal ouvrage "Cosmos", il exposa un large "Concept d'une description physique du monde", présentant le monde comme un Tout mû et animé par des forces internes auquel l'homme pouvait se fier, encourageant ce dernier à "saisir" sans plus de façons "l'esprit de la nature".

Des doutes sur l'"homme ancré dans la foi ou l'incroyance" dans le monde d'aujourd'hui apparurent dans l'esprit des philosophes et scientifiques de notre 20ème siècle. On prétendit que ce "Dieu qui confirmait... autrefois l'ordre bien établi des valeurs et qui était considéré comme la voûte de cet ordre" n'existait pas. Le philosophe polonais Leszek Kolakowski en arriva à cette thèse suite à l'expérience de deux guerres apocalyptiques et d'un chaos mondial raillant tout ordre fiable.
Les savants et penseurs contemporains, au seuil du changement de millénaire prochain, soutiennent à nouveau l'existence d'un "ordre du Tout": "Seul celui qui voit le Tout... est en mesure de se limiter à son domaine particulier pour préserver le Tout". On érige la nouvelle statue d'un Christ cosmique, postulat de la religion de l'avenir. Et pour ce qui est de la "question du futur", Carl Friedrich von Weizsäcker remarquait récemment dans son oeuvre principale "Temps et savoir": "Les images d'ici-bas et de l'au-delà qu'on offertes les religions que nous connaissons n'étaient pas purement naïves... La voie des perceptions méditatives est ouverte. De nouvelles expériences, de nouvelles visions intellectuelles nous attendent... La voie qui mène à ces expériences passera par une profonde souffrance. Et par un profond bonheur."

Hans Dollinger

Предисловие

Всемогущие силы природы, восход и заход солнца, сев и уборка, рост и гибель, отлив и прилив — все это для нас, людей, по-прежнему символы надежности, синонимы которой — неуклонность, неподкупность, добросовестность или постоянство на примере мирового порядка вещей: «Когда заходящее солнце приближается к горизонту, все четыре стихии — вода, воздух, огонь и земля — сливаются в одно. Возвращение к первому дню творения...»
В «незыблемом постоянстве» образа жизни своего отца римский император Марк Аврелий видел надежную путеводную нить для собственного жизненного пути, предпосылку для критического отношения к себе в знаменитом «Наедине с собой».
Естественные и гуманитарные науки были на рубеже средневековья и нового времени для великих ученых неотделимы друг от друга. Они критически рассматривали роль человека в бесконечности Вселенной. «Что же есть человек в мире?» — спрашивал физик, математик и религиозный философ Блез Паскаль, который в 1654 г., пережив мистический опыт встречи с Богом, после долгой светской жизни в Париже и Руане обратился из вольнодумца в защитника христианства. «Все вещи вышли из ничего и простираются в бесконечность. Кто может проследить эти удивительные шаги?..»
Старый, но прежде всего Новый Свет, объездил и изучил два столетия спустя Александр Гумбольдт, давший в своем главном труде «Космос» всеобщий «Проект физического описания мира». Он изобразил мир как единое целое, которым движут и который оживляют внутренние силы; люди на это целое могут положиться и приобщиться к «духу природы».
Сомнения «в человеке, укрепленном в вере или неверии» слышатся у философов и ученых 20-го века. «Бога, который когда-то подтверждал установленный порядок вещей... и который считался венцом этого порядка», признали несуществующим. К этому выводу пришел польский философ Лешек Колаковский на основании опыта двух апокалиптических войн и всемирного хаоса, в котором немыслим никакой порядок.
На рубеже нового тысячелетия ученые и мыслители вновь взывают к «порядку целого»: «Только тот, кто видит целое, способен ограничиться в своей области, чтобы целое сохранить». Создается новый образ Бога — космического Христа — и постулируется как религия будущего.
А по поводу «вопроса о будущем» Карл Фридрих Вайцзеккер в своем главном труде «Время и знание» констатировал: «Картины посюстороннего и потустороннего мира, которые предлагали традиционные религии, не были простой наивностью... Путь к медитативному восприятию открыт. Новый опыт, новые мысленные картины ждут нас... Путь к опыту будет проходить через глубокое страдание. И через глубокое счастье.»

Ганс Доллингер

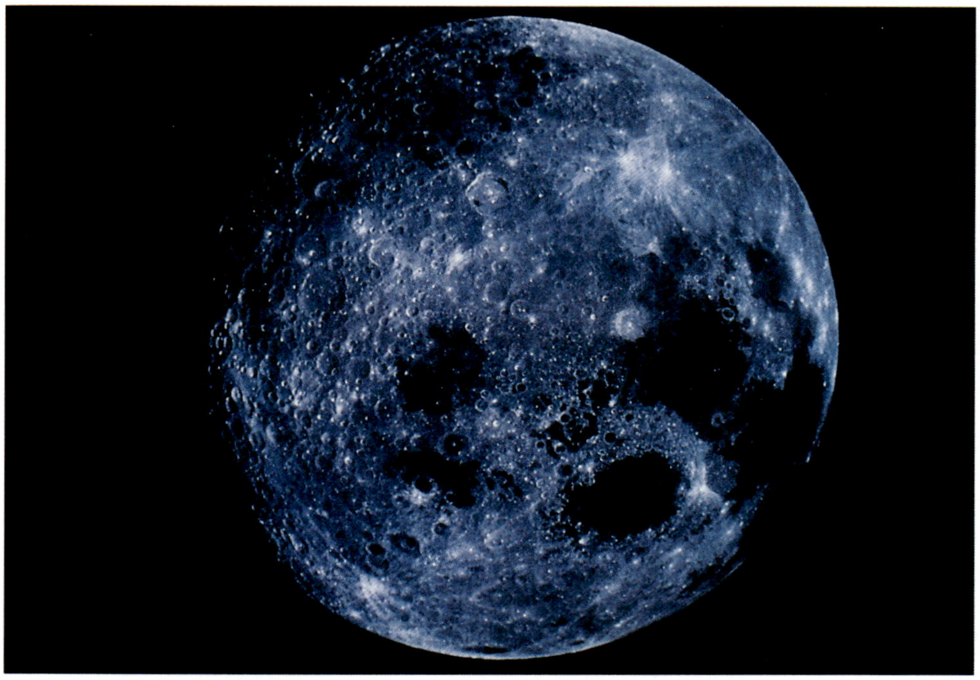

Meine Insel im Norden

Mit den im Osten und Westen vorgelagerten Sand- und Muschelbänken erreicht die Insel Juist eine Länge von siebzehn Kilometern, das bedeutet: siebzehn Kilometer weißer Sandstrand zum offenen Meer hin und siebzehn Kilometer Watt zum Festland hin. An ihrer schmalsten Stelle ist die Insel nicht breiter als dreihundert Meter. Ihre höchste Erhebung beträgt zwanzig Meter überm nahen Meeresspiegel.
In den Schaufenstern sieht man alte Stiche von „Juyst" im Querformat, auch die Ansichtskarten im Querformat, selbst dann noch mehr Himmel als Land. Der Maler Nolde gilt als der Erfinder der „Himmelschaften" über der Nordsee. „Mir ist so nach Nolde", hörte ich einen weiblichen Kurgast mit heiterem Seufzen in der Inselbuchhandlung sagen.
Wenn sich die untergehende Sonne dem Horizont nähert, verschmelzen die vier Elemente Wasser, Luft, Feuer und Erde zu einem. Rückkehr zum ersten Schöpfungstag. Augenblicke für Maler, Fotografen und Dichter. In der rechten Manteltasche eine Handvoll Muscheln, in der linken eine Handvoll Gedichtzeilen, beides sandig und salzig; abends räumt man die Taschen aus.
Ebbe und Flut. Einatmen und Ausatmen.
Das Meer kommt mit wenigen, tiefen Atemzügen am Tag aus; bei jedem Atemzug wechseln Sandbänke und Muschelfelder den Besitzer, gehören das eine Mal dem Meer, dann wieder dem Land. Dünen entstehen und Dünen verschwinden. Die Insel wächst sichtbar nach Osten hin, verliert im Westen merklich an Land. Sandstaub steht in hellen Wolken überm Strand, Sand fegt über die Dünen und die Deiche, auf denen der grüne „Helm" wächst, ein Strandhafer, von Menschenhand gesetzt, um den Schöpfungsprozeß zu beschleunigen. Noch wächst hier kein Stein, aber im Schutz der Dünen bildet sich Humus.
Kleine Wälder wurden angepflanzt, nach Ausmaß und Größe der Bäume, die der Wind niedrig hält, heute liebevoll „Erlenwäldchen" und „Kiefernwäldchen" benannt. Hier muß die Natur vor der Natur geschützt werden. Sanddorn, Ginster, duftendes Geißblatt, Brombeeren und Holunder bilden eine undurchdringliche Macchia der Nordseeinseln; in allen Gärten die halbwilden Zaunrosen, die den Sommer über blühen und im Herbst pralle rote, nicht nur den Vögeln wohlschmekkende Rosenäpfel tragen. Man pflückt bittere Schlehen und süße Brombeeren; was hier blüht und reift, ist kostbar.
Krähen und Drosseln übernehmen die Aufforstung, säen und düngen zugleich in einem vereinfachten Verfahren. Sandige, verschlungene Pfade im Windschutz der Dünen, in einer buckligen, schlupfwinkligen Welt, für die Ausmaße des Menschen zu winzig, eher für Karnickel geeignet, für Kinder und Vögel. Auf Juist halten die Zugvögel ihre Herbstmanöver ab; sie kommen von Norden, machen hier Rast, nähren sich von Mehlbeeren und den Früchten des Sanddorns und des Holunders. Erschöpfte Goldhähnchen, deren Müdigkeit größer ist als die Angst, hocken für ein paar Stunden im Schutz der Hauswände. Wildgänse ziehen durch, Brandentenzüge dicht überm Meeresspiegel, Stare formieren sich zu neuen Zeichen am Himmel.

Christine Brückner

Mijn eiland in het Noorden

Samen met de in het Oosten en Westen gelegen zand- en mosselbanken bereikt het eiland Juist een lengte van zeventien kilometer, dat betekent: zeventien kilometer wit zandstrand naar de open zee en zeventien kilometer wad naar het vaste land. Aan het smalste punt is het eiland niet breder dan driehonderd meter. De hoogste verheffing bedraagt twintig meter boven de nabij gelegen zeespiegel.
In de etalages liggen oude gravures van "Juyst" in dwarsformaat, ook de ansichtkaarten in dwarsformaat, zelfs dan nog meer lucht dan water. De schilder Nolde geldt als de uitvinder van de "luchtenschappen" boven de Noordzee. "Ik verlang zo naar Nolde", hoorde ik een vrouwelijke badgast opgeruimd in de boekhandel van het eiland zuchten. Als de ondergaande zon de horizont nadert, smelten de vier elementen water, lucht, vuur en aarde samen. Terugkeer naar de eerste scheppingsdag. Ogenblikken voor schilders, fotografen en dichters. In de rechter jaszak een handvol met schelpen, in de linker een handvol gedichtregels, beide zandig en zout: 's avonds haalt men de zakken leeg.
Eb en vloed. Inademen en uitademen.
De zee komt daags met weinig, diepe ademhalingen uit: bij elke ademtocht verwisselen de zandbanken en mosselvelden van eigenaar, zijn de ene keer van de zee, dan weer van het land. Duinen ontstaan en duinen verdwijnen. Het eiland groeit zichtbaar in richting oosten, verliest in het westen duidelijk land. Zandstof hangt in lichte wolken boven het strand, zand schuift over de duinen en dijken waar de groene "helm" groeit, een helmgras, door mensenhand geplaatst om het scheppingsproces te versnellen. Nog groeit hier geen steen, maar onder bescherming van de duinen onstaat humus.
Kleine bossen worden aangeplant, al naar

afmeting en grootte van de bomen, die de wind laag houdt, vandaag de dag liefdevol "Elzenbosje" en "Dennenbosje" genoemd. Hier moet de natuur tegen de natuur worden beschermd. Duindoorn, heidebrem, geurende kamperfoelie, braambessen en vlierbessen vormen een ondoordingbaar maquis van de Noordzee-eilanden; in alle tuinen de halfwilde hegrozen die in de zomer bloeien en in de herfst dikke rode rozebottels dragen die niet alleen de vogels goed smaken. Men plukt bittere sleeën en zoete braambessen; wat hier bloeit en rijpt is kostbaar.
Kraaien en lijsters nemen de bebossing over, zaaien en bemesten in een vereenvoudigd procces. Zandige, kronkelige paden, door de duinen beschut tegen de wind, in een hobbelige wereld vol met schuilhoeken, voor de omvang van de mens te klein, beter geschikt voor konijnen, voor kinderen en vogels. Op Juist houden de trekvogels hun herstmanoeuvre: ze komen uit het noorden, rusten hier uit, voeden zich met lijsterbessen en de vruchten van de duindoorn en vlierbes. Uitgeputte goudhaantjes, wier vermoeidheid groter is dan de angst, zitten een paar uur door de huismuren beschut. Wilde ganzen trekken voorbij, een zwerm bergeenden dicht boven de zeespiegel, spreeuwen vormen nieuwe tekens aan de lucht.

Christine Brückner

My island in the North

Including the offshore sandbanks and shell beds in the east and west, the island of Juist attains to a length of seventeen kilometers. That means: seventeen kilometers of white sandy beaches towards the open sea and seventeen kilometers of mud flats in the direction of the mainland. At its narrowest point the island is no wider than three hundred meters. Its highest elevation is twenty meters above the nearby sea-level.
In the shop windows one sees old oblong-size engravings of "Juyst", and also broadside picture postcards - both with more sky than land. The painter Nolde is looked upon as the originator of these "sky-scapes" over the North Sea. "I feel like a bit of Nolde," I heard a female visitor to the health-resort say, as she sighed happily in the island bookshop.
When the setting sun nears the horizon, the four elements, water, air, fire and earth, merge into one. A return to the first day of creation. Special moments for painters, photographers and poets. In one's right-hand coat pocket a handful of shells, in the left-hand pocket a handful of lines of poems, both sandy and salty; in the evening the pockets are emptied.
Ebb and flood. Breathing in and breathing out. The sea can make do with a few deep breaths during the day; with each breath, sandbanks and shell beds change owners, sometimes they belong to the sea, sometimes they are a part of the land again. Dunes come into being and dunes disappear. The island is clearly growing towards the east and losing territory in the west. Sand dust hovers in bright clouds above the beach, sand sweeps over the dunes and the dikes on which the green "helmet" grows, a form of beach grass, sowed by human hand to speed up the processes of creation. Nothing else grows here yet, but in the shelter of the dunes humus is formed. Small wooded areas have been planted. Reflecting their size and the type of tree planted, trees kept low in size by the wind, they are lovingly called "little alder woods" or "little pine woods". Here nature must be protected from nature. Sea buckthorn, broom, fragrant honeysuckle, brambles and elderberry form an impenetrable macchia in the North Sea islands; in all the gardens the half-wild rose bushes which bloom throughout the summer and in the autumn carry those big red rose apples which do not just taste good to the birds alone. People pick sour sloe plums and sweet blackberries; what blooms and matures here is precious.
Crows and thrushes take over the task of afforestation, sowing and fertilizing at the same time in a more simplified process. Twisting, sandy paths in the shelter of the dunes, in an undulating world full of nooks and crannies, more suited for rabbits, and children and birds. Migrant birds stage autumn manoeuvres over the island; they come from the north, rest here, and feed on the hawthorn berries and the fruits of the sea buckthorn and the elder. Exhausted marigold finches, whose tiredness is greater than their fear, pause for a few hours in the shelter of the walls of the houses. Wild geese pass through, lines of shelducks fly just over the surface of the water, starlings take up ranks to set new signs in the sky.

Christine Brückner

La mia isola al Nord

Con i banchi di sabbia e conchiglie antistanti i versanti orientale ed occidentale, l'isola di Juist raggiunge una lunghezza di diciassette chilometri, questo significa: diciassette chilometri di bianca spiaggia sabbiosa dalla parte del mare e, dalla parte della terra ferma, diciassette chilometri di 'watt', dove il fondo marino affiora dall'acqua. Nel suo punto più stretto, l'isola non è più larga di trecento metri. La sua altura maggiore misura venti metri sul livello del vicino mare.
Nelle vetrine si vedono vecchie incisioni di "Juyst" dal formato oblungo, anche le cartoline hanno un formato oblungo e anche così si vede più cielo che terra. Il pittore Nolde è considerato l'inventore dei "paesaggi di cielo" del Mar del Nord. Una volta nella libreria dell'isola sentii una bagnate che sospirando allegramente disse "Ho tanta voglia di Nolde". Quando al tramonto il sole si avvicina all'orizzonte, i quattro elementi, acqua, aria, fuoco e terra si fondono in uno. Ritorno al primo giorno della creazione. Momenti per pittori, fotografi e poeti. Nella tasca destra del cappotto una manciata di conchiglie, in quella sinistra una manciata di versi, entrambi sabbiosi e salati; di sera si vuotano le tasche.
Flusso e riflusso. Inspirare ed espirare.
Il mare riesce a superare la giornata con pochi respiri: ad ogni respiro i banchi di sabbia ed i campi di conchiglie cambiano padrone, una volta appartengono al mare, poi di nuovo alla terra. Dune che si formano, dune che scompaiono. L'isola cresce visibilmente verso est, ad ovest perde visibilmente terreno. Polvere di sabbia è nelle chiare nuvole sulla spiaggia, la sabbia spazza le dune e le dighe sui quali cresce il 'casco" verde, una specie di ammofila piantata dall'uomo per accelerare il processo di creazione. Qui ancora non crescono pietre, ma al riparo delle dune si forma humus. Sono stati piantati piccoli boschi che oggi, date le dimensioni e l'altezza degli alberi, tenuti bassi dal vento, vengono chiamati affettuosamente "Boschetto d'ontani" e "Pinetina". Qui la natura deve esere protetta dalla natura. Olivello spinoso, ginestre, caprifoglio profumato, more e sambuco formano la macchia impenetrabile delle isole del Mar del Nord; in tutti i giardini le rose semiselvatiche lungo le staccionate che fioriscono in estate e che in autunno portano pomi rossi e sodi che non piacciono solo agli uccelli. Si raccolgono le amare susine selvatiche e le dolci more; ciò che fiorisce e matura qui è prezioso.
Cornacchie e tordi provvedono al rimboschimento, seminano e concimano al contempo seguendo una procedura semplificata. I sentieri sabbiosi e tortuosi che le dune riparano dal vento, in un mondo ondulato e pieno di nascondigli, troppo minuscolo per le dimensioni umane, adatto piuttosto ai conigli, ai bambini ed agli uccelli. A Juist gli uccelli migratori fanno le loro manovre autunnali; vengono dal nord, fanno sosta qui, si nutrono di bacche di farinaccio, delle susine selvatiche e del sambuco. Fiorrancini sfiniti, la cui stanchezza supera la paura, stanno accovacciati per un paio d'ore al riparo dei muri delle case. Le oche selvatiche attraversano l'isola, processioni di tadorne basse sullo specchio del mare, storni si mettono in formazione a comporre nuovi segni contro il cielo.

Christine Brückner

Mon île du Nord

Avec ses bancs de sable et de coquillages la bordant à l'Est et à l'Ouest, l'île de Juist atteint une longueur de dix-sept kilomètres; cela signifie dix-sept kilomètres de plage de sable blanc regardant le large et dix-sept kilomètres de hauts-fonds regardant le continent. A son endroit le plus étroit, l'île n'a pas plus de trois cents mètres. Sa plus haute élévation est de vingt mètres au-dessus du niveau de la mer.
Dans les vitrines, on découvre d'anciennes gravures de "Juyst" de format oblong; les cartes postales ont ce même format oblong et là aussi on y voit plus de ciel que de terre. Le peintre Nolde est considéré comme l'inventeur des "paysages célestes" au-dessus de la mer du Nord. Dans la librairie de l'île, j'entendis une curiste dire un jour dans un soupir serein: "J'ai Nolde dans l'âme".
Quand le soleil couchant s'approche de l'horizon, les quatres éléments, eau, air, feu et terre se fondent en un seul. Retour au premier jour de la Genèse. Instants de prédilection pour les peintres, photographes et poètes. Dans la poche droite du manteau, une poignée de coquillages, dans la poche gauche une poignée de poésie, les deux pleines de sable et de sel; le soir, on vide ses poches.
Marée haute et marée basse. Inspirer, souffler. Quelques inspirations profondes chaque jour suffisent à la mer; à chaque respiration, les bancs de sable et les champs de coquillages changent de propriétaires, appartenant à la mer puis de nouveau à la terre ferme. Des dunes surgissent et des dunes disparaissent. L'île s'agrandit visiblement vers l'Est et perd sensiblement du pays à l'Ouest. De la poussière de sable forme des nuages clairs au-dessus de la plage, le sable balaye les dunes et les digues où poussent les "brins" verts plantés par la main de l'homme pour accélérer le processus de la Création.

Là, aucune pierre ne pousse encore, mais, protégé par les dunes, l'humus se forme. On a planté de petites forêts qu'on appelle aujourd'hui avec amour "boqueteaux d'aulnes" et "boqueteaux de pins", étant donné le nombre et la hauteur des arbres que le vent empêche de grandir. Ici, il faut protéger la nature contre la nature. Arbousiers, genêt, chèvrefeuille odorant, mûres et sureau composent le maquis impénétrable des îles de la mer du Nord; dans tous les jardins, les églantiers à demi sauvages fleurissent pendant tout l'été et portent en automne des baies rouges et pleines que les oiseaux ne sont pas les seuls à apprécier. On cueille des prunelles amères et des mûres sucrées; ce qui fleurit et mûrit ici est précieux.
Corneilles et grives se chargent du reboisement, sèment et répandent l'engrais en un seul acte simplifié. Des sentiers sablonneux et sinueux, protégés du vent par les dunes, dans un monde bosselé, aux recoins paisibles, trop minuscule pour les hommes, convenant plutôt aux lapins, aux enfants et aux oiseaux. Sur l'île de Juist, les oiseaux migrateurs font leurs manoeuvres d'automne; ils viennent du Nord, s'arrêtent, se nourrissent de baies et des fruits de l'arbousier et du sureau. Des roitelets huppés épuisés, dont la fatigue est plus grande que la peur, se reposent quelques heures, abrités par les murs des maisons. Des oies sauvages ne font que passer, des vols de canards rasent la surface de l'eau, des étourneaux se regroupent pour former de nouveaux signes dans le ciel.

Christine Brückner

Мой остров на севере

С песчаными и ракушечными отмелями на востоке и западе остров Юист достигает длины 17 км, что означает: семнадцать километров белого песчаного пляжа в сторону моря и семнадцать километров ваттов в сторону суши. В самом узком месте остров имеет ширину не более трехсот метров. Самая высокая точка лежит на двадцать метров над уровнем близкого моря.
В витринах выставлены старые гравюры с видами острова в горизонтальном формате и открытки, тоже в горизонтальном формате, но даже и на них неба больше, чем земли. Художник Нольде считается автором жанра «небесных ландшафтов» Северного моря. «У меня настроение под Нольде», — услышала я как-то вздох курортницы в книжном магазине.
Когда заходящее солнце приближается к горизонту, все четыре стихии — вода, воздух, огонь и земля — сливаются в одно. Возвращение к первому дню творения. Время для художников, фотографов и поэтов. В правом кармане пальто горсть раковин, в левом — горсть стихотворных строчек, то и другое в песке и соли; вечером вытряхиваешь карманы.
Отлив и прилив. Вдох и выдох.
Днем морю хватает нескольких глубоких вдохов и выдохов; при этом песчаные отмели и поля ракушек меняют хозяина, попеременно принадлежат то морю, то опять суше. Дюны появляются и дюны исчезают. Остров на глазах растет на восток и заметно убывает на западе. Песчаная пыль стоит светлыми облаками над пляжем, песок проносится над дюнами и над дамбами, на которых растет зеленый шлемник — трава, посеянная человеческой рукой, чтобы ускорить процесс творения. Здесь пока не растет камень, но в укромных местах за дюнами образуется перегной. Насадили карликовые леса, сообразно с высотой деревьев, которым ветер не дает расти вверх; теперь они любовно называются «ольховый лесок» и «сосновый лесок». Здесь природу надо защищать от природы. Облепиха, дрок, душистая жимолость, ежевика и бузина образуют на островах Северного моря непролазные заросли; во всех садах растет полудикий шиповник, он летом цветет, а осенью приносит ядреные красные плоды, которые по вкусу не только птицам. Собирают горький терн и сладкую ежевику; то, что здесь растет и плодоносит, имеет особую ценность.
Вороны и дрозды взяли лесонасаждение на себя, сеют и удобряют одновременно, по упрощенному методу. Песчаные перепутанные тропинки под защитой дюн, в горбатом укромном мирке, слишком маленьком для человека, подходящем разве что для кроликов, детей и птиц. На Юисте перелетные птицы собираются на осенние маневры; они летят с севера, отдыхают здесь, питаясь рябиной, плодами облепихи и шиповника. Выбившиеся из сил корольки, у которых усталость побеждает страх, на несколько часов садятся под защиту стен домов. Пролетают дикие гуси, над самой водой тянутся вереницы пеганок, стаи скворцов рисуют узоры на небе.

Кристина Брюкнер

Das Leben meines Vaters

Das Leben meines Vaters war für mich eine Schule der Milde und doch zugleich auch unerschütterlicher Beständigkeit in allem, wofür er sich einmal nach reiflicher Überlegung entschieden hatte. Er war unempfindlich gegen jede Eitelkeit auf anscheinende Ehrenbezeigungen, ein Freund der Tätigkeit und unverdrossen darin, hörte gern gemeinnützige Vorschläge anderer an, ließ sich durch nichts abhalten, jeden nach Verdienst zu behandeln, wußte recht wohl, wo man die Zügel anziehen und wo nachlassen mußte... Seine Freunde suchte er sich zu erhalten und wurde ihrer weder überdrüssig, noch war er unvernünftig für sie eingenommen. In jeder Lage zufrieden, war er stets heiter; auf die Zukunft nahm er von ferne schon Bedacht und machte ohne viel Aufhebens sich auf die geringsten Vorfälle gefaßt. Alles Zujauchzen und jede Schmeichelei wies er zurück. Auf die Staatsbedürfnisse war er jederzeit wachsam und haushälterisch beim Ausgeben öffentlicher Gelder und ließ den Tadel solcher Grundsätze willig über sich ergehen. Um die Gunst der Götter buhlte er ebensowenig auf abergläubische Weise als um die Gunst der Menschen durch Künste der Gefallsucht oder durch Begünstigung des Pöbels; vielmehr war er in allem nüchtern und fest, nirgends unanständig, noch neuerungssüchtig. Die Güter, welche das Leben angenehm machen und die ihm das Glück in Fülle darbot, benutzte er ebenso fern von Übermut als von Ausflüchten und genoß daher das Vorhandene ebenso ungesucht, als er das Fehlende vermißte... Insbesondere aber ist an ihm das zu rühmen, daß er Männern, welche in etwas eine vorzügliche Stärke besaßen, wie in der Beredsamkeit, der Gesetzeskunde, der Sittenlehre oder in anderen Fächern, ohne Neid den Vorrang einräumte und ihnen sogar dazu behilflich war, daß jeder nach dem Maße seiner besonderen Geschicklichkeit Anerkennung fand. Obgleich er alles gemäß den Einrichtungen der Vorfahren behandelte, so vermied er doch selbst den Schein der Anhänglichkeit an dieselben... Nichts Unfreundliches noch auch Ungeziemliches, Ungestümes, noch etwas war an ihm zu entdecken, wovon man hätte sagen können: „Es war im Übermaß", sondern alles wohl und gleichsam bei guter Muße überlegt, unerschütterlich geordnet, fest und mit sich selbst übereinstimmend. Und so konnte man denn auf ihn anwenden, was von Sokrates berichtet wird, daß er Dinge zu entbehren und zu genießen gewußt habe, bei deren Entbehrung sich viele schwach und bei deren Genuß sie sich unmäßig verhalten. Dort aber mutig zu ertragen, hier nüchtern zu bleiben, verrät einen Mann von vollendeter Geistesstärke...

Marc Aurel

Het leven van mijn vader

Het leven van mijn vader was voor mij een school van mildheid en toch tegelijk ook van onwrikbare bestendigheid in alles waarvoor hij eenmaal na rijp beraad een besluit genomen had. Hij was ongevoelig voor ijdelheid gebaseerd op vermoedelijk eerbetoon, een vriend van bezigheid en hierin onvermoeid, luisterde graag naar voorstellen ten behoeve van het algemeen belang, hij liet zich door niets weerhouden een ieder naar verdienste te behandelen, wist zeer wel waar men de teugels strakker aanhalen en waar men ze moest laten vieren... Hij trachtte zijn vrienden te behouden en kreeg noch genoeg van hen, noch was hij onredelijk met hen ingenomen. In alle omstandigheden tevreden, was hij steeds blijmoedig; aan de toekomst schonk hij reeds van verre aandacht en bereidde zich zonder veel ophef op de geringste gebeurtenissen voor. Toejuichingen en elke vorm van vleierij wees hij af. Over de overheidsbelangen waakte hij te allen tijde en was spaarzaam bij het uitgeven van openbare middelen en verdroeg berispingen van zulke principes geduldig. Naar de gunst van de goden dong hij net zo min op bijgelovige wijze als naar de gunst van de mensen door behaagzuchtige vaardigheden of door het gepeupel te bevoordelen; hij was veeleer altijd nuchter en vast; nergens onfatsoenlijk, noch belust op veranderingen. De goederen die het leven aangenaam maken en die het geluk hem in overvloed bood, benutte hij evenzo ver van hoogmoed als van uitvluchten en genoot daarom het beschikbare net zo ongezocht als hij het ontbrekende vermiste... Maar vooral is aan hem te prijzen dat hij mannen, die in iets bepaalds uitstekend waren, zoals in de welsprekendheid, de geneeskunde, de zedenleer of in andere vakken, zonder afgunst voorrang gaf en hen zelfs daarbij behulpzaam was, dat een ieder naar mate van zijn bijzondere

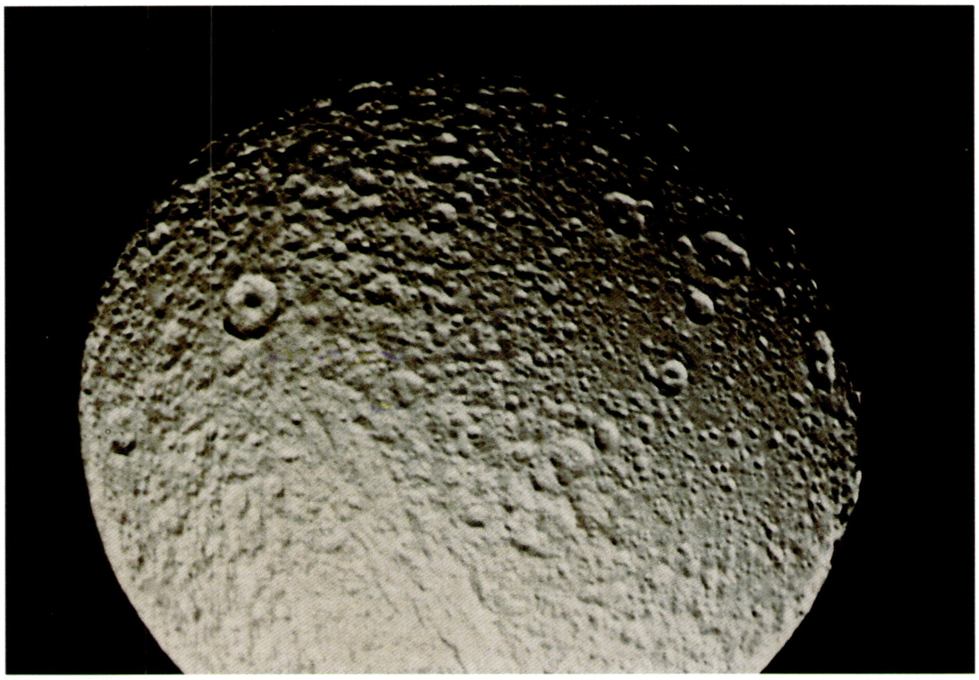

vaardigheid waardering ondervond. Hoewel hij alles volgens de regeling van de voorouders behandelde, vermeed hij toch zelf de schijn aan dezelfden verknocht te zijn... Niets onvriendelijks, noch onbetamelijks, ontstuimigs, noch iets kon aan hem worden ontdekt, waarvan men had kunnen beweren: "Hij is buitensporig", maar alles goed en in zekere zin met innerlijke rust bedacht, onwrikbaar geordend, vast en met zichzelf in overeenstemming. En zo kon men op hem toepassen wat over Sokrates wordt bericht, dat hij dingen kon ontberen en genieten zonder zich als velen bij de ontbering zwak en het genot onmatig te gedragen. Zich hier moedig te gedragen, hier nuchter te blijven, geeft blijk van een man van voltooide geestkracht...

Marc Aurelius

The life of my father

The life of my father was for me a school of mildness and yet at the same time one of unshakeable steadfastness in everything that he, after careful deliberation, decided to embark upon. He was impervious to every kind of vanity in relation to marks of tribute, a man who got things done and who was unflagging in this respect; he liked to listen to constructive proposals for the common good made by others; allowed nothing to stop him treating everyone according to his merits; knew very well where to tighten the reins and when to let them hang loose... He tried to keep his friends and did not tire of them, nor was he unreasonably prejudiced in their favour. He was satisfied in every situation, always in a good mood; he was concerned about possible future developments long in advance and, without ado, he prepared himself even for the slightest of incidents. He rejected all forms of acclaim and every type of flattery. He was at all times intent on serving the interests of the state and was economical when it came to spending public money, and he willingly put up with the reproaches that were levelled at him as a result. He did not supersticiously court the favour of the gods, nor did he woo the favour of the common people by trying to please or by favouring the plebs; on the contrary, he was in all things sober and firm, never indecent, never overtly innovative. Possessions which made life pleasant and which could afford him fortune in excess, he used in an unpresumptious manner and without subterfuge and enjoyed what was to be had in the same unaffected manner as he missed what was not on hand. A merit of his, worthy of special praise, however, is that he conceded without envy priority to men who possessed outstanding prowess in some particular area, be it eloquence, knowledge of the laws, or of ethics or in other subjects and even did what he could to help them find recognition according to the degree of their special skills. Although he did everything as laid down by his ancestors, he himself avoided the impression of being too devoted in this respect... There was nothing unfriendly, nothing unseemly, nothing impetuous about him; indeed, there was nothing at all about him of which one could say it was "in excess". Everything was well-considered, imperturbably ordered, firm and tallying in all points. And so one could say of him what was said about Socrates: that he knew how to forgo things and to enjoy things, things which many people were weak about forgoing and things which many people enjoyed in excess. Putting up with things courageously on the one hand and remaining levelheaded on the other betrays a man of perfect strength of spirit....

Marcus Aurelius

La vita di mio padre

La vita di mio padre è stata per me una scuola di mitezza ma, al tempo stesso, anche di imperturbabile costanza in tutto ciò per cui, dopo matura rilessione, si era deciso. Egli era insensibile a qualsiasi tipo di vanità suscitata da atti di omaggio apparenti, era amico dell'attività ed indefesso in essa, ascoltava volentieri proposte di altri volte al bene comune, niente poteva trattenerlo dal trattare ciascuno secondo il proprio merito, sapeva bene quando tirare le briglie e quando allentarle... Cercava di mantenere i suoi amici e non si stancava di loro né era prevenuto insensatamente a loro favore. Soddisfatto in ogni condizione, era sempre sereno: si disponeva per tempo ad affrontare il futuro e si preparava senza fare tanto chiasso ai più piccoli eventi. Rifiutava ogni applauso ed ogni lusinga. Vigilava sempre sulle necessità dello Stato ed era economo nello spendere i denari pubblici sopportando docilmente il biasimo per questi principi. Non cercava di propiziarsi il favore degli dèi con pratiche superstiziose né di rendersi favorevoli gli uomini con le arti dell'allettamento o agevolando la plebe, ma era piuttosto sobrio e saldo in ogni cosa, mai sconveniente nè smanioso di innovazioni. I beni che rendono piacevole la vita e che la fortuna gli aveva donato in abbondanza, li adoperava senza cadere nella sfrenatezza o ricorrere a pretesti e godeva spontaneamente di ciò che possedeva senza rammaricarsi per ciò che mancava... In particolare però va elogiato in lui il fatto che egli concedeva senza invidia il primato a uomini che eccellevano particolarmente in una disciplina, come l'eloquenza, la pratica delle leggi, la morale, e li sosteneva persino affinchè a ciascuno andassero i riconoscimenti adeguati alla sua particolare abilità. Sebbene trattasse tutto secondo quanto avevano disposto gli antenati, evitava per sè di apparire attaccato ad essi... In lui non si sarebbe potuto scoprire niente di scortese né di sconveniente o impetuoso né alcuna cosa di cui si fosse potuto dire: "Era esagerato", tutto veniva invece ben ponderato e, per così dire, in tutta calma, in modo irremovibilmente ordinato, saldo ed in armonia con la sua natura. E così potrebbe valere per lui ciò che si riferisce di Socrate, vale a dire che sapeva rinunciare e godere di cose nella cui rinuncia molti erano deboli e nel cui godimento si comportavano smisuratamente. Ma il conservare animo nell'un caso ed il rimanere sobrio nell'altro, svela un uomo di forza d'animo compiuta...

Marco Aurelio

La vie de mon père

La vie de mon père fut pour moi une école de clémence, mais en même temps aussi de constance inébranlable dans toutes les décisions qu'il avait prises après mûre réflexion. Il était insensible à toute vanité tirée de témoignages honorifiques apparents, un ami de l'activité à laquelle il s'adonnait infatigablement; il écoutait volontiers les propositions d'intérêt général que lui soumettaient les autres; rien ne pouvait l'empêcher de traiter chacun selon son mérite et il savait bien où tenir la bride haute et où relacher les rênes... Il essayait de garder ses amis et ne se lassait pas d'eux, sans pourtant se laisser prendre par déraison dans leurs filets. Satisfait quelle que soit la situation, il était toujours serein; prévoyant, il avait égard au futur et se préparait sans faire beaucoup de bruit au moindre événement. Il refusait toute démonstration d'allégresse et toute flatterie. Il veillait toujours aux besoins de l'Etat et se montrait économe quand il s'agissait de dépenser des fonds publics, acceptant le blâme que lui amenaient de tels principes. Il ne briguait pas la faveur des Dieux par la superstition tout comme il ne briguait pas celle des hommes par la coquetterie ou par des privilèges accordés à la plèbe; il était au contraire sobre et ferme, n'était jamais malséant ni avide d'innovations. Les biens qui rendent la vie agréable et que la chance lui offrait à profusion, il s'en servait sans orgueil et sans faux-fuyants, savourant ainsi ce qu'il possédait aussi naturellement qu'il regrettait ce qui lui manquait... Mais il faut en particulier vanter le fait qu'il cédait sans jalousie le pas aux hommes qui disposaient dans un certain domaine d'un talent exceptionnel, comme l'éloquence, la connaissance des lois, la morale et autres; il s'efforçait même que chacun trouvât l'approbation correspondant au degré de son talent particulier. Bien qu'il traitât chaque chose en vertu des institutions de ses

ancêtres, il évitait lui-même de paraître attaché à ces dernières... On ne découvrait chez lui ni inclémence, ni inconvenance, ni impétuosité ou autre trait dont on aurait pu dire: "C'était exagéré"; au contraire, chez lui, tout était bien réfléchi, dans la détente et l'équilibre, ordonné de manière inébranlable, fermement, en harmonie avec sa personne. C'est ainsi qu'on pouvait dire de lui ce qu'on rapporte au sujet de Socrate: qu'il avait su se passer de certaines choses et en apprécier d'autres, alors que la privation rendait bien des gens faibles et que la jouissance les poussait à un comportement immodéré. Supporter avec courage d'un côté, rester objectif de l'autre, traduisent chez un homme une force spirituelle accomplie...

Marc Aurèle

Жизнь моего отца

Жизнь моего отца была для меня школой доброты и в то же время незыблемого постоянства во всем, на что он по зрелому размышлению решался. Он был невосприимчив к любому тщеславному желанию внешних почестей, был другом деятельной жизни без уныния, охотно выслушивал общеполезные советы других, не допускал, чтобы ему что-то мешало относится к каждому по его заслугам, очень хорошо знал, где следует натянуть поводья, а где отпустить... Друзей старался сохранять, не утомляясь от них и не проявляя неразумных симпатий. Довольный любым положением, был неизменно весел; о будущем думал заблаговременно и, не привлекая внимания, готовился к любым неожиданностям. Восторженные приветствия и лесть он отвергал. К государственным нуждам был всегда внимателен, а при расходе общественных денег рачителен, добровольно вынося порицания за подобные принципы. Благосклонности богов домогался суетной верой столь же мало, сколь и расположения людей искусством нравиться или благоволением плебсу; напротив, был во всем трезв и тверд, избегая всякой непристойности, как и страсти к новшествам. Блага, делающие жизнь приятною, которые счастье в изобилии предлагало ему, он принимал равно как без высокомерия, так и без отговорок, и поэтому наслаждался имеющимся без вожделения, как и не сожалел об отсутствующем... Особенно же достойно в нем похвалы то, что людям, имеющим в чем-то неоспоримое превосходство, как то: в красноречии, знании законов, этики или прочих предметов, без зависти уступал первенство и даже помогал им, дабы каждый получил признание в меру своего сугубого умения. Хотя он все делал согласно установлениям предков, тем не менее сам избегал видимой приверженности этому... Ничего недружелюбного, ни неприличного, буйного не было в нем, о чем можно было бы сказать: «Это через меру», но все хорошо и неспешно продумано, незыблемо упорядочено, прочно и в согласии с самим собой. Итак, применимо к нему то, что сообщают о Сократе, который умел терпеть недостаток и наслаждаться вещами, отсутствие которых делало многих слабыми, а наслаждение — неумеренными. Умение же в одном случае мужественно терпеть, а в другом — сохранять трезвость, выдает мужа совершенного силой духа...

Марк Аврелий

Zwischen Unendlichkeit und Nichts

Denn was ist nun der Mensch in der Welt? Ein Nichts vor dem Unendlichen, ein All vor dem Nichts, eine Mitte zwischen Nichts und All. Unendlich entfernt davon, die Extreme zu begreifen, sind ihm das Ende der Dinge und ihr Ursprung unüberwindlich verborgen in einem undurchdringlichen Geheimnis; er ist gleichermaßen unfähig, das Nichts zu sehen, aus dem er gezogen ist, und das Unendliche, in das er verschlungen ist. Was bleibt ihm also zu tun, außer etwa einen Schein von der Mitte der Dinge wahrzunehmen, in einer ewigen Verzweiflung, weder ihren Ursprung noch ihr Ende zu erkennen? Alle Dinge sind hervorgegangen aus dem Nichts und reichen bis ins Unendliche. Wer wird diesen erstaunlichen Schritten folgen? Der Urheber dieser Wunder begreift sie. Kein anderer vermag es.

Blaise Pascal

Tussen oneindigheid en niets

Want wat is de mens nu eenmaal in de wereld? Een niets voor het oneindige, een alles voor het niets, een midden tussen niets en alles. Niet in het minst in staat het extreme te begrijpen, zijn voor hem het einde der dingen en hun oorsprong onoverkomelijk verborgen in een ondoordringbaar geheim: hij is evenzeer niet bij machte het niets te zien waaruit hij ontstaan is en het oneindige, waarmee hij verstrengeld is. Wat blijft hem dus anders over dan bij benadering een schijnsel van het midden der dingen waar te nemen, in eeuwige wanhoop noch diens oorsprong, noch diens einde waar te nemen? Alles is uit het niets ontstaan en reikt tot in de oneindigheid. Wie zal deze verwonderlijke stappen volgen? De schepper van deze wonderen begrijpt het.
Geen ander is er toe in staat.

Blaise Pascal

Between eternity and nothing

What is Man in this world? A nothing in the face of the eternal, a universe in the face of nothing, a middle between nothing and the universe. Boundlessly far away from understanding the extremes, the end of all things and their origin are unsurmountably concealed in an impenetrable secret; Man is just as uncapable of seeing the void out of which he originates as he is the endlessness in which he is enveloped. What other choice has he but to perceive an illusion of the middle of all things, in eternal despair that he will never recognize either beginning or end. Everything has emerged from nothing and stretches to infinity. Who can fathom these astounding steps? The creator of this miracle understands them. No-one else is in a position to do so.

Blaise Pascal

Fra l'Infinito e il Nulla

Cos'è poi l'uomo nel mondo? Un nulla davanti all'Infinito, un Tutto davanti al Nulla, una via di mezzo fra Nulla e Tutto. Infinitamente lontano dal comprendere gli estremi, un segreto invincibile e impenetrabile gli nasconde la fine e il principio delle cose. Non sa vedere né il Nulla dal quale è stato tratto, né l'Infinito che lo avvolge. Cos'altro puó fare se non percepire un bagliore del centro delle cose, eternamente disperando di poterne mai cogliere sia la fine che il principio? Tutte le cose sono scaturite dal Nulla e toccano l'Infinito. Chi seguirà questi straordinari passi? Colui che ha originato questa meraviglia li comprende. Nessun altro ne è capace.

Blaise Pascal

Entre l'infini et le néant

Car, qu'est donc l'homme en ce monde? Un rien devant l'infini, un tout devant le Néant, un milieu entre le Néant et le Tout. Infiniment loin de comprendre les extrêmes, la fin des choses et leur commencement lui restent cachés dans un mystère impénétrable; il est également incapable de voir le Néant d'où il est sorti et l'infini qui l'engloutit. Que lui reste-t-il donc à faire si ce n'est percevoir une lueur au milieu des choses, dans son éternel désespoir de ne reconnaître ni son origine ni sa fin?
Toutes les choses proviennent du Néant et se perdent dans l'infini. Qui suivra ces pas étonnants? L'auteur de ces miracles les comprend. Personne d'autre ne le peut.

Blaise Pascal

Между бесконечностью и ничем

Ибо что же есть человек в мире? Ничто перед бесконечным, вселенная перед ничем, середина между ничем и вселенной. Бесконечно далек от того, чтобы постигать крайности, ибо непреодолимо скрыты от него в непроницаемой тайне конец вещей и их начало; равным образом неспособен он увидеть ничто, из которого извлечен, и бесконечность, которая его поглощает. Так что же ему остаетя делать, кроме как воспринимать видимость середины вещей, в вечном отчаянии, не познавая ни их начала, ни конца? Все вещи вышли из ничего и простираются в бесконечность. Кто может проследить эти удивительные шаги? Творец этих чудес понимает их. Никто другой этого не может.

Блэз Паскаль

Den Geist der Natur ergreifen...

Die Natur ist für die denkende Betrachtung Einheit in der Vielheit, Verbindung des Mannigfaltigen in Form und Mischung, Inbegriff der Naturdinge und Naturkräfte, als ein lebendiges Ganzes. Das wichtigste Resultat des sinnigen physischen Forschens ist daher dieses: in der Mannigfaltigkeit die Einheit zu erkennen; von dem Individuellen alles zu umfassen, was die Entdeckungen der letzteren Zeitalter uns darbieten; die Einzelheiten prüfend zu sondern und doch nicht ihrer Masse zu unterliegen: der erhabenen Bestimmung des Menschen eingedenk, den Geist der Natur zu ergreifen, welcher unter der Decke der Erscheinungen verhüllt liegt. Auf diesem Wege reicht unser Bestreben über die enge Grenze der Sinnenwelt hinaus; und es kann uns gelingen, die Natur begreifend, den rohen Stoff empirischer Anschauung gleichsam durch Ideen zu beherrschen.
Wenn wir zuvörderst über die verschiedenen Stufen des Genusses nachdenken, welchen der Anblick der Natur gewährt, so finden wir, daß die erste unabhängig von der Einsicht in das Wirken der Kräfte, ja fast unabhängig von dem eigentümlichen Charakter der Gegend ist, die uns umgibt. Wo in der Ebene, einförmig, gesellige Pflanzen den Boden bedecken und auf grenzenloser Ferne das Auge ruht; wo des Meeres Wellen das Ufer sanft bespülen und durch Ulfen und grünenden Seetang ihren Weg bezeichnen: überall durchdringt uns das Gefühl der freien Natur, ein dumpfes Ahnen ihres „Bestehens nach inneren ewigen Gesetzen". In solchen Anregungen ruht eine geheimnisvolle Kraft; sie sind erheiternd und lindernd, stärken und erfrischen den ermüdeten Geist, besänftigen oft das Gemüt, wenn es schmerzlich in seinen Tiefen erschüttert oder vom wilden Drange der Leidenschaften bewegt ist. Was ihnen Ernstes und Feierliches beiwohnt, entspringt aus dem fast bewußtlosen Gefühle höherer Ordnung und innerer Gesetzmäßigkeit der Natur; aus dem Eindruck ewig wiederkehrender Gebilde, wo in dem Besondersten des Organismus das Allgemeine sich spiegelt; aus dem Kontrast zwischen dem sittlich Unendlichen und der eigenen Beschränktheit, der wir zu entfliehen streben. In jedem Erdstriche, überall wo die wechselnden Gestalten des Tier- und Pflanzenlebens sich darbieten, auf jeder Stufe intellektueller Bildung sind dem Menschen diese Wohltaten gewährt.

Alexander von Humboldt

De geest van de natuur grijpen

De natuur is voor de denkende beschouwing eenheid in de veelheid, verbinding van het veelsoortige qua vorm en menging, het toonbeeld van de natuurdingen en natuurkrachten, als levend geheel. Het belangrijkste van het zinvolle fysieke onderzoeken is daarom: in de veelsoortigheid de eenheid te herkennen; van het individuele alles te omvatten wat de ontdekkingen van de afgelopen eeuwen ons voordragen; de details kritisch te scheiden en het toch niet voor de massa onderdoen; gedachtig aan de verheven bestemming van de mens de geest van de natuur te grijpen die met de mantel der verschijnselen bedekt is. Langs deze weg reikt ons streven verder dan de nauwe grens van de zinnenwereld en we kunnen er in slagen de natuur begrijpend, de ruwe stof van de empirische beschouwing als het ware door ideeën te beheersen.
Als wij allereerst over de verschillende trappen van het genot nadenken die het schouwspel van de natuur verschaft, vinden wij dat de eerste onafhankelijk van het zicht op de totstandkoming van de machten, ja bijna onafhankelijk van het kenmerkende karakter van de omgeving is die ons omringt. Waar in de vlakte, eenvormig, in groepen levende planten de bodem bedekken en op grenzeloze wijde horizont de blik rust, waar de golven van de zee de oever zacht bespoelen en zeewier hun weg markeert: overal dringt het gevoel van de vrije natuur tot ons door, een flauw vermoeden van haar "bestaan volgens innerlijke, eeuwige wetten". In zulke prikkelingen rust een geheimzinnige kracht; zij zijn vermakelijk en kalmerend, sterken en verkwikken de vermoeide geest, doen het gemoed bedaren als het pijnlijk in zijn diepten aangetast of door de wilde drift van de hartstocht bewogen is. De ernst en plechtigheid die zij teweegbrengen, ontspringt uit het bijna bewustloze gevoel van hogere geordend-

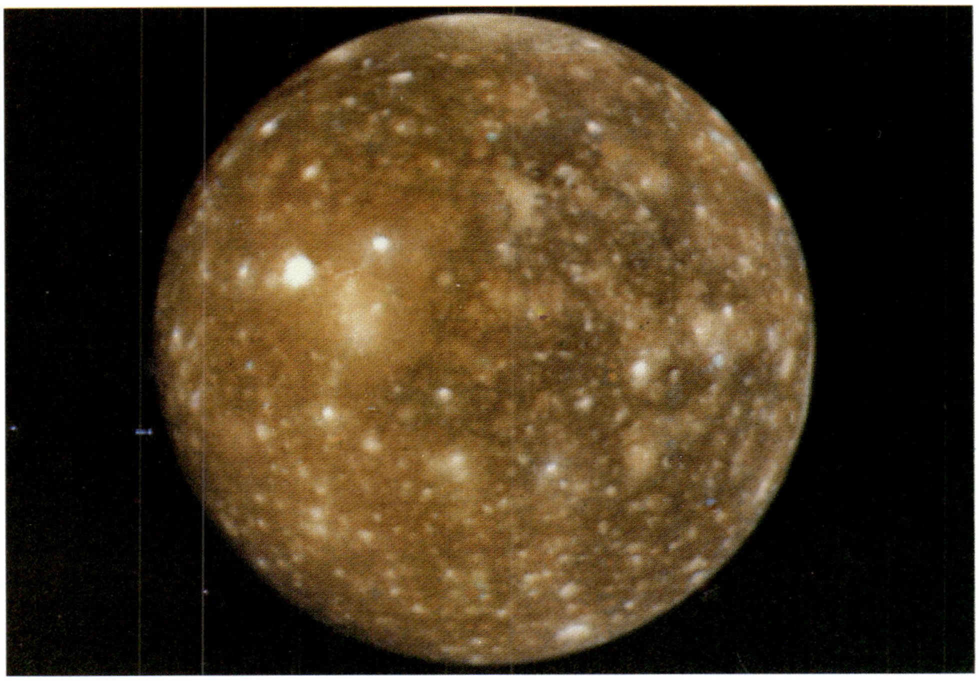

heid en een innerlijke wetmatigheid van de natuur; uit de indruk van zich eeuwig herhalende structuren, waar het algemene zich in het bijzondere van het organisme weerspiegelt; uit het contrast tussen het moreel oneindige en de eigen beperking die wij trachten te ontvluchten. In elke luchtstreek, overal waar de afwisselende gedaanten van dieren- en plantenwereld zich vertonen, op elk niveau van intellectuele vorming is de mens van deze weldaden verzekerd.

Alexander von Humboldt

To understand the spirit of nature...

Nature is, for those who contemplate it, unity in diversity, the intertwining of the manifold in form and variety, the embodiment of natural things and natural forces, a living Whole. The most important result of thoughtful physical research is thus the following: the ability to recognize the unity in all the multifariousness; to comprise all the individual elements which the discoveries of recent times offer us; to scrutinize the details and not to succumb to their sheer mass; and, bearing in mind the sublime vocation of Man, to understand the spirit of nature which lies concealed under the cover of appearances. In this fashion our striving goes beyond the narrow limits of the world of the senses; and we may succeed, as we understand more, to, as it were, hold sway over the raw material of empirical observation, through ideas.
If we first think of the different stages of enjoyment which the sight of nature grants us, then we find that the first is independent of the insight into the forces which are at work, indeed almost independent of the actual character of the area which surrounds us. Where in flat, open countryside, uniform, gregarious plants cover the ground and the eye rests over boundless distance; where the waves of the ocean gently wash the shore and mark their path by lines of green seaweed, here we sense everywhere the special aura of free, open nature, a vague feeling of its "existing according to eternal, inner laws". In such stimulating moments there rests a mysterious power; they are cheering and soothing, they strengthen and refresh the tired spirit, calm one's mood when one is shaken to one's depths by pain or moved by the wild urgings of passion. What is grave and solemn about such moments derives from the almost sub-conscious feeling of a higher order and an inner law of nature; from the expression of eternally recurring forms, where general principles are mirrored in the special features of the organism; out of the contrast between moral infinity and our own limitations which we are striving to leave behind. On every piece of earth, everywhere where the changing forms of plant and animal life are to be found, at every stage of intellectual development, these blessings are granted to Man to partake of.

Alexander von Humboldt

"Comprendere" lo spirito della Natura

Per il pensiero che la contempla, la Natura è unità nella molteplicità, è unione del multiforme in una forma e in una mescolanza, è quintessenza delle cose e delle forze naturali in quanto parti di una totalità vivente. Il risultato più importante di una ricerca fisica sensata è pertanto il seguente: comprendere l'unità nella multiformità; abbracciare tutto ciò che ci offrono le scoperte delle epoche più recenti sull'ente individuale; distinguere, esaminandoli, i dettagli senza per questo soggiacere alla loro quantità; memori della destinazione sublime dell'uomo "comprendere" lo spirito della natura che giace nascosto sotto il manto dei fenomeni. Percorrendo questo cammino il nostro studio supererà gli stretti confini dell'universo sensibile; comprendendo la natura potremo riuscire a dominare con la forza dell'intelletto la materia grezza dell'osservazione empirica. Se riflettiamo anzitutto sui diversi gradi del piacere che ci dà la vista della natura scopriamo che il primo di essi è indipendente dalla conoscenza delle sue forze, che è quasi indipendente dal carattere peculiare del luogo che ci circonda. Sia in pianura, dove in modo uniforme, piante amene ricoprono il suolo e dove l'oc-chio riposa in uno spazio senza confini; sia dove le onde del mare lambiscono dolcemente la riva segnando il proprio cammino: ovunque ci penetra il sentimento della libertà della natura e un'indistinta percezione del suo "esistere regolato da leggi eterne e intrinseche". In certe impressioni dimora una forza segreta: esse rallegrano e alleviano, rafforzano e rinfrescano lo spirito stanco, spesso addolciscono l'animo quando è profondamente turbato dalla sofferenza o scosso dall'impeto selvaggio delle passioni. Tutta la serietà e la solennità presenti in certe impressioni hanno origine dal sentimento quasi inconscio dell'ordine superiore e delle leggi interne della natura, dall'impressione dello eterno ritorno di forme in cui il generale si riflette nel più specifico degli organismi; dal contrasto fra l'Infinito morale e la propria finitudine, alla quale desideriamo sfuggire. In ogni lembo di terra, ovunque si incontrino le mutevoli forme della vita animale e vegetale a ogni livello culturale all'uomo è dato fruire di tutti questi benefici.

Alexander von Humboldt

Saisir l'esprit de la nature...

La nature qui forme un Tout animé est, pour l'observation pensante, l'unité dans la multitude, la réunion d'une multiplicité de formes et de mélanges, l'incarnation des choses et des forces naturelles. Le résultat essentiel de la recherche physique judicieuse est donc le suivant: reconnaître l'unité dans la multiplicité; cerner dans l'individuel tout ce que nous offrent les découvertes des derniers siècles; trier les détails après examen et ne pas succomber sous le poids de leur masse; tenant compte de l'auguste vocation de l'homme, saisir l'esprit de la nature qui repose, caché, sous la couverture des phénomènes naturels. Ainsi, nos efforts dépassent l'étroite limite du monde sensoriel et nous pouvons réussir, en comprenant la nature, à dominer en quelque sorte par nos idées la matière brute de la conception empirique. Si nous réfléchissons avant tout sur les différents stades du plaisir que nous procure la vue de la nature, nous trouvons que le premier est indépendant de l'examen de l'effet des forces naturelles, qu'il est même pratiquement indépendant du caractère particulier de la région qui nous entoure. Là où dans la plaine, uniformément, des plantes sociables recouvrent le sol et l'oeil repose dans l'infini du lointain; là où les vagues de la mer baignent paisiblement le rivage et marquent leur chemin de varech verdoyant: partout, nous sommes pénétrés du sentiment d'une nature libre, sourde intuition de son "existence conforme à des lois internes, éternelles". Une force mystérieuse repose au fond de telles impulsions; elles sont rassérénantes et soulageantes; elles tonifient et réconfortent l'esprit fatigué, apaisent souvent l'âme quand elle est douloureusement ébranlée dans ses profondeurs ou agitée par les pulsions déchaînées des passions. Ce qu'elles renferment de sérieux et de solennel naît du sentiment presque inconscient d'un ordre supérieur et de

lois inhérentes à la nature, de l'impression de l'éternel retour d'une structure où le général se reflète dans le particulier de l'organisme, du contraste entre l'infini moral et nos propres limites auxquelles nous essayons d'échapper. Dans chaque contrée, partout où les formes changeantes de la faune et de la flore s'offrent à nos yeux, à chaque échelon de la formation intellectuelle, ces bienfaits sont accordés à l'homme.

Alexander von Humboldt

Овладеть духом природы

Природа для мыслящего наблюдателя есть единство в разнообразии, соединение многообразия в форме и смешении, символ природных вещей и природных сил как живое единство. Поэтому самый важный результат для чувственного физического изучения следующий: увидеть в многообразии единство; охватить все индивидуальное, что предлагают открытия последних эпох; критически отделить детали и не потеряться в их массе; помня о возвышенном назначении человека, овладеть духом природы, скрытым под покровом явлений. На этом пути наше устремление выходит за тесные границы чувственного мира; нам может удастся, понимая природу, также с помощью идей овладеть и сырым материалом эмпирических взглядов.
Если мы в первую очередь задумаемся о тех различных ступенях удовольствия, которые приносит созерцание природы, то мы обнаружим, что первая не зависит от понимания действия сил, даже почти не зависит от своеобразия местности, которая нас окружает. Будь то на равнине, где земля однообразно покрыта многими растениями и взгляд устремляется в безграничную даль; будь то на море, где волны мягко омывают берег и обозначают свой путь зеленым фукусом: везде нас пронизывает чувство свободной природы, неясное предположение о ее «существовании по вечным внутренним законам». Такие побуждения заключают в себе таинственную силу; они бодрят и смягчают, укрепляют и освежают утомленный дух, нередко успокаивают душу, если она бывает потрясена в своей глубине или смущена необузданным натиском страстей. Что в вас есть серьезного и торжественного, происходит из почти неосознанного чувства высшего порядка и внутренней закономерности природы, из впечатления вечно повторяющихся структур, где в особенностях организма отражается общее; из контраста нравственно бесконечного и собственной ограниченности, которой мы стремимся бежать. В любом месте земли, везде, где предстают переменчивые картины животной и растительной жизни, на каждой ступени духовного образования сподобляется человек подобных благотворных действий.

Александр Гумбольдт

Die Welt, in der wir leben...

Die Welt, in der wir leben, ist aber keine Welt der in Gewißheit befriedigten, vergnügten, im Glauben oder Unglauben befestigten Menschen. Es ist vielmehr ein Zeitalter der Vertriebenen, der Flüchtlinge, der Exilanten, der Herumirrenden, der „ewigen Juden" auf der Suche nach verlorenem - geistigem oder physischem - Heimatland. In diesem Nomadenleben ist nichts mehr sicher, nichts gewährleistet, nichts endgültig festgesetzt, nichts - außer dem Wandern - fraglos gegeben. Ein Gott, der einst die wohletablierte Ordnung der Werte, der sozialen Verhältnisse, der Regeln des Denkens, des physischen Weltalls bestätigte und der als Gewölbe dieser Ordnung galt, ist nicht mehr da, weil keine solche Ordnung mehr sichtbar ist. Solange die Menschen der Dauerhaftigkeit dieser Ordnung vertrauen konnten, hatten auch die Gottlosen ihren Ort darin (ich habe immer nur die christlich-europäische Zivilisation im Auge). Mochten sie auch als Irrende, als Wahnsinnige oder als Sendboten der Hölle gelten, so war ihr Platz innerhalb der anerkannten Weltordnung doch bestimmt. Mochten sie auch verfolgt, bestraft, zum Tode verurteilt werden, so waren sie doch in einem guten Sinne glücklich, weil nicht nur ihrer Sache sicher, sondern auch geistig sorgenlos: es gibt keinen Gott, somit auch keinen Himmel und keine Hölle, und so ist es gut. Aber auch das ist vorbei. Mit dem Selbstvertrauen des Glaubens ist das Selbstvertrauen des Unglaubens zerbrochen. Im Gegensatz zu der gemütlichen, durch die wohlwollende, freundliche Natur geschützten Welt des aufklärerischen Atheismus wird die gottlose Welt von heute als ein bedrückendes, ewiges Chaos wahrgenommen... Was ist geschehen? Warum hat sich das Geschick des Atheismus mit dem des Christentums auf eine so merkwürdige Weise verbunden, daß die beiden Feinde in ihrem Unglück und in ihrer Unsicherheit einander begegnen? ... Inständig baten wir Gott, daß Er die Welt verlasse. Das hat Er gemacht, auf unser Verlangen. Ein gähnendes Loch ist geblieben. Wir beten immerfort zu diesem Loch, zum Nichts. Niemand antwortet. Wir sind wütend oder enttäuscht. Ist das ein Beweis der Nichtexistenz Gottes? ...

Leszek Kolakowski

De wereld waarin wij leven...

De wereld waarin wij leven is echter geen wereld van door zekerheid bevredigde, opgewekte, in het geloof of ongeloof gesterkte mensen. Het is veeler een tijdperk van de ontheemden, de vluchtelingen, de in ballingschap levenden, de rondzwervenden, de "eeuwige Joden" op zoek naar het verloren - geestelijke of fysieke - vaderland. In dit nomadenleven is niet meer zeker, niets gewaarborgd, niets onherroepelijk vastgelegd, niets - behalve het lopen - onbetwijfelbaar gegeven. Een god, die ooit de zo gevestigde orde van de waarden, de sociale verhoudingen, de regels van het denken, het fysieke heelal bevestigde en die als gewelf van deze orde gold, is er niet meer omdat een zulke orde niet meer zichtbaar is. Zolang de mensen op de duurzaamheid van deze orde konden vertrouwen, hadden hierin ook de goddelozen hun plaats (ik heb altijd alleen de christelijke Europese beschaving voor ogen). Al golden zij als dwalenden, als waanzinnigen of als afgezanten van de hel, hun plaats in de erkende wereldorde was bepaald. Ook al werden zij vervolgd, bestraft of ter dood veroordeeld, in zekere zin waren zij toch gelukkig omdat zij niet alleen zeker van hun zaak waren maar ook geestelijk zorgeloos: er bestaat geen god en dus ook geen hemel en geen hel en zo is het goed. Maar ook dat is voorbij. Met het zelfvertrouwen van het geloof is het zelfvertrouwen van het ongeloof verbroken. In tegenstelling tot de gemoedelijke, door de welgezinde, vriendelijke natuur beschermde wereld van het rationalistische atheïsme wordt de goddeloze wereld van vandaag als een bedrukkend, eeuwig chaos waargenomen... Wat is er gebeurd? Waarom gaat het lot van het atheïsme met dat van het christendom op zo'n merkwaardige wijze samen dat de beide vijanden zich in hun ongeluk en in hun onzekerheid ontmoeten? Wij smeekten

God de wereld te verlaten. Dat heeft Hij gedaan, op ons verzoek. Een gapende leegte is gebleven. Wij bidden onafgebroken tot deze leegte, tot niets. Niemand antwoordt. Wij zijn woedend of teleurgesteld.
Bewijst dat de niet-existetie van God?..

Leszek Kolakowski

The world in which we live...

The world in which we live is not a world of people who are satisfied in the certainty of their knowledge, who are merry, who are anchored in faith or unbelief. It is rather an age of expellees, of refugees, of people living in exile, of those who wander around bewildered, of "eternal Jews" searching for their lost - spiritual or physical - homeland. Nothing in this nomadic life is certain any more, nothing - except the wandering - is definite. A god who once confirmed the well-established order of the values, of the social conditions, of the rules of thought, and of the physical universe and who was looked upon as the canopy of this order, is no longer there, because no such order as now visible. As long as people could trust in the lasting nature of this order, the godless also had their place in it (I am referring here throughout only to the Christian-European civilization). Regardless of whether they were viewed as those who had lost their way, or as madmen or as the envoys of hell, their place within the recognized world order was fixed. No matter whether they were persecuted, or punished or condemned to death, they were in one good sense happy, not just because they were sure of what they were doing but also because they were spiritually free of worries: there is no god and so there is no heaven and no hell and it is good so. But this, too, is a thing of the past. Now that the self-confidence of faith has been shattered, so also has the self-confidence of unbelief. In contrast to the cosy world of enlightened atheism protected by a well-meaning, friendly nature, the godless world of today is perceived as an oppressive eternal chaos... What has happened? Why has the fate of atheism become so remarkably linked with that of Christianity that both these foes come together in their unhappiness and their insecurity?... We beseeched god to leave the world. And he did so - at our request. A gaping hole remains. We pray continually to this hole, to nothing. No-one answers. We are raging or disappointed. Is this a proof of the non-existence of God?...

Leszek Kolakowski

Il mondo in cui viviamo...

Il mondo nel quale viviamo non è il mondo di uomini allegri, soddisfatti delle proprie certezze, saldi nella fede o nell'irreligiosità. Si tratta pittosto dell'epoca degli esiliati, dei fuggiaschi, degli esuli, degli smarriti, degli "ebrei erranti" alla ricerca della patria perduta - spirituale o psichica. In questa vita da nomadi non c'è più niente di sicuro, niente di garantito, niente di definitivamente stabilito, niente di dato per certo salvo il vagabondare.
Il Dio che un tempo confermava l'ordine ben saldo dei valori, dei rapporti sociali, delle regole del pensiero, del cosmo fisico e che rappesentava il coronamento di questo ordine non esiste più, dal momento che un tale ordine non è più visibile. Fintantoché gli uomini potevano riporre fiducia nella stabilità di quest'ordine anche gli atei vi trovavano il proprio posto (penso sempre e soltanto alla civiltà europea e cristiana). Anche se venivano considerati vagabondi, folli o emissari dell'inferno, il loro posto all'interno dell'ordine riconosciuto era una cosa certa. Anche se venivano perseguitati, puniti, condannati a morte, essi erano ciononostante felici in senso positivo, non solo perché erano sicuri del fatto loro, ma anche perché erano spiritualmente sereni: non esiste nessun Dio, quindi non esistono né cielo né inferno ed è bene che sia così. Ma anche questo è finito. Con la sicurezza della fede si è infranta anche la sicurezza della miscredenza. In opposizione al mondo comodo dell'ateismo illuministico, protetto da una benevola natura amica, il mondo di oggi viene percepito come un eterno caos che opprime... Che cosa è successo - perché il destino dell'ateismo si è legato a quello del Cristianesimo in un modo tale che i due nemici si incontrano nella comune disgrazia e nella comune incertezza? ...
Abbiamo pregato Dio con fervore perché abbandonasse il mondo. E Dio lo ha fatto su nostra richiesta. E' rimasto un vuoto assoluto.Noi continuiamo a pregare questo vuoto, questo nulla. Nessuno risponde. Allora siamo furiosi o delusi. E'questa una prova della non esistenza di Dio?

Leszek Kolakowski

Le monde dans lequel nous vivons...

Le monde dans lequel nous vivons n'est pas un monde où l'assurance donne aux humains satisfaction et gaîté, où ils sont ancrés dans la foi ou l'incroyance. Notre époque est plus celle des bannis, des réfugiés, des exilés, des sans repos, des "Juifs Errants" à la recherche de la patrie perdue, spirituelle ou physique. Dans cette vie de nomade, rien n'est plus sûr, rien n'est garanti, rien n'est fixé définitivement, rien (à par les promenades) n'est accordé sans poser de questions. Un Dieu qui confirmait autrefois l'ordre bien établi des valeurs, des conditions sociales, des règles de la pensée, de l'espace physique et qui était considéré comme la voûte de cet ordre, n'est plus présent parce qu'un tel ordre n'est plus visible. Tant que les hommes pouvaient avoir confiance dans la durabilité de cet ordre, les impies y avaient également leur place (je parle uniquement de la civilisation chrétienne européenne). Qu'on les ait qualifiés de vagabonds, de fous ou de messagers de l'enfer, ils disposaient cependant d'une place déterminée à l'intérieur de l'ordre du monde généralement admis. Qu'on les ait persécutés, punis, condamnés à mort, ils étaient malgré tout heureux au sens propre du mot, car ils étaient sûrs d'avoir raison et de plus, aucun souci ne tourmentait leur esprit; Dieu n'existe pas, donc le Ciel et l'Enfer n'existent pas non plus et c'est bien ainsi. Mais cette ère est également révolue. La confiance en soi que donnait l'incroyance a été brisée par la confiance en soi que donne la foi. Au contraire de l'agréable monde protégé par la nature bienveillante et aimable de l'athéisme rationaliste, le monde aujourd 'hui privé de Dieu est perçu comme un chaos oppressant, éternel... Que s'est-il passé? Pourquoi le destin de l'athéisme a-t-il si curieusement rejoint celui de la chrétienté, de sorte que les deux ennemis se rencontrent dans leur malheur et leur incertitude?... Nous avons ardemment

prié Dieu de quitter le monde. C'est ce qu'll a fait, répondant à notre voeu. Un trou béant demeure. Nous prions constamment en direction de ce trou, du Néant. Personne ne nous répond. Nous sommes furieux ou déçus. S'agit-il bien d'une preuve de la non-existence de Dieu?...

Leszek Kolakowski

Мир, в котором мы живем...

Мир, в котором мы живем, это не мир удовлетворенных, довольных в своей уверенности, укрепленных в вере или неверии людей. Это, скорее, эпоха изгнанников, беженцев, эмигрантов, бесцельно блуждающих, «вечных жидов» в поисках потерянной — духовной или физической — родины. В этой кочевой жизни нет ничего надежного, ничего обеспеченного, ничего окончательно установленного, ничего несомненно данного, кроме странствования.
Бога, который когда-то подтверждал установленный порядок ценностей, социальных отношений, правил мышления, физического космоса и который считался венцом этого порядка, больше нет, потому что никакого такого порядка больше не видно. Пока люди могли доверять долговечности этого порядка, безбожники тоже имели свое место в нем (я имею в виду только христианско-европейскую цивилизацию). Хотя бы они и считались заблуждающимися, безумными или посланцами ада, их место в признанном мировом порядке было определено. Хотя бы их и преследовали, наказывали, приговаривали к смерти, тем не менее они в добром смысле были счастливы, потому что были не только уверены в своей правоте, но и духовно беззаботны: никакого Бога нет, то есть нет никакого Неба и никакого ада, и это хорошо. Но и это миновало. Вместе с уверенностью веры погибла и уверенность неверия. В противоположность уютному, охраняемому мирволящей человеку природой миру просвещенного атеизма нынешний безбожный мир воспринимается как гнетущий, вечный хаос... Что случилось? Почему судьба атеизма так странно соединилась с судьбой христианства, что эти враги встретили друг друга в своем несчастьи и своей неуверенности?.. Настоятельно молили мы Бога, чтобы Он покинул мир. И Он это сделал по нашему требованию. Осталась зияющая дыра. Мы продолжаем молиться этой дыре, этому ничему. Никто не отвечает. Мы приходим в ярость или испытываем разочарование. Это ли доказательство небытия Божия?..

Лешек Колаковский

Einheit und Erneuerung

Unser Handeln, auch in den Fragen des Tages, muß stets ausgerichtet bleiben auf das gemeinsame Ziel. Nur dann lassen sich unsere Einzelfallentscheidungen und die Fülle der einzelnen Maßnahmen zu einer größeren Ordnung zusammenfügen. Nur dann ist die Übereinstimmung der Teile mit dem politischen Ganzen gewährleistet, die das Ziel jeder Politik sein muß, die in Ordnungen denkt. Die Ordnung des Ganzen kann nur dann ihre den Grundkonsens der Bürger stiftende, das Gemeinwesen befriedende Wirkung entfalten, wenn es ihr gelingt, sich immer wieder gegen die Widersprüchlichkeit und Unordnung zu behaupten, die entstehen, wenn es der Fülle der Teilinteressen gelingt, sich zu Lasten des Ganzen durchzusetzen. Nur wer das Ganze sieht und an die Ziele und den Sinn des Ganzen glaubt, das Ganze und seine Ordnung also für richtig hält, ist in der Lage, sich in seinem Teilbereich zu begrenzen, um das Ganze zu erhalten, dessen Wohlergehen uns allen zum Nutzen gereicht.

Kurt H. Biedenkopf

Eenheid en vernieuwing

Ons handelen, ook in de dagelijkse situaties, moet altijd op het gemeenschappelijke doel georiënteerd blijven. Alleen zo kunnen onze beslissingen in op zichzelf staande gevallen en de overvloed afzonderlijke maatregelen tot een grotere orde worden verenigd.
Alleen dan is de overeenkomst van de delen met het politieke geheel gewaarborgd, wat het doel van elke politiek moet zijn die in orde denkt. De orde van het geheel kan alleen dan een door de burger in te stellen basisconsensus, een voor de gemeenschap bevredigende werking ontplooien, als deze er in slaagt, zich steeds weer tegen de tegenstrijdigheid en wanorde door te zetten die ontstaan als de massa deelbelangen er in slaagt zich ten koste van het geheel door te zetten. Alleen wie het geheel ziet en aan de doelen en de zin van het geheel gelooft, het geheel en zijn orde dus goedkeurt, is in staat, zich tot zijn deelgebied te beperken om het geheel te behouden waarvan het welzijn in ons aller voordeel is.

Kurt H. Biedenkopf

Unity and renewal

Our doings, also in relation to daily questions, must always be directed towards the common goal. Only then can our individual decisions and the wealth of separate measures merge to form a larger order. Only then is the concurrence of the parts with the political whole guaranteed; and this must be the objective of every form of politics which thinks in terms of order. The order of the whole can only then unfold its effect, one that contributes to the basic consent of the citizens and satisfies the community, if it repeatedly succeeds in holding its ground in the face of the contradictions and disorder that arise when all the sub-interests succeed in getting their way at the expense of the whole. Only those who see this whole and who believe in the goals and the purpose of the whole and who therefore consider its order to be correct are in a position to limit its subordinate interests and to preserve the whole, the welfare of which we all profit from.

Kurt H. Biedenkopf

Unità e rinnovamento

Il nostro agire, anche in occasioni quotidiane, deve sempre restare rivolto verso il fine comune. Solo allora le nostre decisioni circa casi particolari e le tante misure singole potranno comporsi a formare un ordine superiore. Solo allora sarà garantita la corrispondenza delle parti con il Tutto politico che deve essere l'obiettivo di ogni politica che pensa secondo categorie ordinate. L'ordine del Tutto può sviluppare il suo effetto che crea il consenso di base dei cittadini e che pacifica la collettività solo se riesce ad affermarsi contro l'incoerenza ed il disordine che si creano quando la molteplicità di interessi particolari riesce ad avere la meglio sul Tutto. Solo chi vede il Tutto e crede agli obiettivi ed al senso del Tutto, che quindi ritiene giusto il Tutto ed il suo ordine, sarà in grado di circoscrivere per sè un suo campo parziale per mantenere in vita il Tutto, il cui benessere va a vantaggio di noi tutti.

Kurt H. Biedenkopf

Unité et renouveau

Notre action, même dans le quotidien, doit toujours rester centrée sur notre but commun. C'est le seul moyen de réunir dans un ordre supérieur nos décisions individuelles et l'abondance des différentes mesures prises. C'est le seul moyen d'assurer la concordance des différents éléments avec le Tout politique, concordance qui doit être le but de chaque politique dont la pensée est régie par un ensemble d'ordres. L'ordre du Tout ne peut développer son effet établissant un accord fondamental et apaisant la chose publique que s'il réussit à s'affirmer encore et toujours face à la contradiction et le désordre qui naîssent lorsque la masse des intérêts particuliers parvient à s'imposer au détriment du Tout. Seul celui qui voit le Tout et qui croit aux buts et au sens de ce Tout ainsi qu'au bien-fondé du Tout et de son ordre, est en mesure de se limiter dans le domaine individuel afin de préserver ce Tout dont la prospérité nous est profitable à tous.

Kurt H. Biedenkopf

Единство и обновление

Наши действия, даже в повседневных делах, всегда должны быть направлены к общей цели. Только тогда наши отдельные решения и вся масса отдельных мероприятий могут быть приведены в больший порядок. Только тогда будет обеспечено согласование всех частей с политическим целым, что должно быть целью любой политики, если она задумывается о порядке. Порядок общего может проявить свое действие, выражающееся в единодушии всех граждан по основным вопросам и успокаивающее общество, лишь в случае, если ему удастся противостоять противоречивости и беспорядку, которые возникают, когда массе частных интересов удается, взять верх над интересами целого. Только тот, кто видит целое и верит в цели и смысл целого, то есть считает целое и его порядок правильными, тот в состянии ограничить себя в своей области, чтобы сохранить целое, благо которого идет на пользу всем нам.

Курт Биденкопф

Der kosmische Christus

Der Glaube an den kosmischen Christus und die entsprechende Theologie sind eine mögliche Antwort auf die neue Epoche holistischen, das heißt ganzheitlichen Denkens und Verhaltens, die angebrochen ist. Die weltweite soziale und ökologische Krise ist nur zu verstehen und kann, wenn überhaupt noch, nur bestanden werden, wenn wir von der Hypothese ausgehen: Das gesamte Universum, nicht nur die Erde oder die Menschheit, ist eine einzige, zusammenhängende Wirklichkeit, die sich ähnlich einem Organismus, wie wir ihn aus dem Bereich des Lebendigen kennen, in gerichteter Entwicklung befindet, die jedoch äußerst empfindlich ist gegenüber Gleichgewichtsstörungen, wie der Mensch sie zur Zeit in immer größerem Ausmaß verursacht...

„Gott" nennen seit Menschengedenken die Religionen den letzten „Grund des Universums". Dabei hat Gott je nach Kulturen und Traditionen die unterschiedlichsten Namen angenommen: „Das Eine, Brahman, die Leere, das Tao, die ewige Vereinigung von Shiva und Shakti, die Heilige Dreieinigkeit" (Rupert Sheldrake)...

Soll das Gottesbild einer Kultur für die jeweiligen Menschen der praktischen Orientierung dienen und Überlebenswissen bereitstellen, so muß es den Bedürfnissen des Wissens und Glaubens der Epoche entsprechen. Das läßt sich auch an den Gottesbildern im Christentum ablesen, die sich gemäß dem Wort Jesu „Wer mich gesehen, hat den Vater gesehen" (Joh 14,9), in den jeweiligen Christusbildern konkretisieren.

Auf vereinfachte Formeln gebracht - die historische Wirklichkeit ist komplexer, farbiger, auch widersprüchlicher -, wurde Christus in der Antike teils als der Gute Hirte, teils als der Weltenschöpfer und Allherrscher gesehen; im Mittelalter ist Christus der Heerführer und siegreiche König, aber auch der erlösende Leidensmann und mystische Freund der Seele; die Renaissance sieht in Christus den antik-schönen Weisheitslehrer der Menschheit, die Barockzeit die himmlisch-irdische Heldengestalt in dramatischer Auseinandersetzung mit dem Bösen; in der Aufklärung erscheint Christus als der Anwalt des gesunden Menschenverstandes, in der Romantik und im Biedermeier als der milde, dem Wandel der Zeiten enthobene Nazarener; im 19. und 20. Jahrhundert greift man, je nach Bedürfnis, zweifelnd und suchend auf historische Christusbilder zurück oder gibt Christus expressionistisch die Züge persönlicher oder kollektiver Glaubensnöte.

Auf die Frage, welches Gottesbild, welches Christusbild nun den Bedürfnissen der Zukunft, die sich bereits in unserer Gegenwart andeuten, entspricht, hat der Naturwissenschaftler und Theologe Pierre Teilhard de Chardin bereits 1916 geantwortet: das Gottesbild des kosmischen Christus. Dabei war er davon überzeugt, nur einen bis dahin - weil es dafür keine existentielle Notwendigkeit gegeben hatte - vernachlässigten Strang der biblischen Ordnung aufzugreifen. „Wie aktuell es immer erscheinen mag - dieses Evangelium vom kosmischen Christus, darin vielleicht das Heil unserer Zeit liegt, ist wahrhaft und wirklich das unseren Vätern vom Himmel gesandte Wort, der neue Schatz, der vorsorglich neben den alten Werten niedergelegt ist..."

In den Worten Teilhards klingt an, wieso das Gottesbild des kosmischen Christus geeignet ist, die Bedürfnisse des gegenwärtigen und zukünftigen holistischen Weltbildes zu erfüllen. Zunächst wird die Gespaltenheit des Christen, vermutlich auch vieler religiöser Menschen in anderen Religionen, zwischen Gottes- und Weltliebe, zwischen religiösem und weltlichem Engagement aufgehoben: Weltdienst ist Gottesdienst. Das hat angesichts der ökologischen Krise dringend notwendige praktische Konsequenzen: Der Mensch ist auch aus religiösen Gründen gehalten, sich der Rettung der Natur zu widmen - nicht nur um des Menschen, sondern um ihrer selbst willen -, denn sie ist der in Entwicklung begriffene kosmische Leib Christi...

Trotz der Widerstände innerhalb der christlichen Kirchen und Theologien gegen einen Glauben an den kosmischen Christus scheint dieser Glaube die Religion der Zukunft zu werden. Die neueren religiösen Bewegungen bedienen sich vieler Vorstellungen und Formulierungen, die denen über den kosmischen Christus nahekommen. Der einzelne spürt immer mehr den göttlichen Impulsen nach, die er in seinem Innern verspürt, und erfährt ihre Resonanz im Gespräch mit anderen und im Umgang mit der Natur. Auf der Suche nach den letzten Prinzipien der Wirklichkeit nähern sich Naturwissenschaftler immer mehr der Gaia-Hypothese, der Auffassung des Universums als eines lebendigen Zusammenhangs, dessen Funktionen von einem Prinzip garantiert werden, dem frühere Generationen den Namen »Weltseele« gegeben haben...

Angesichts solcher Erfahrungen und Überlegungen kann sich der Christ wie jeder religiöse Mensch der Frage nicht entziehen, ob und in welcher Weise der Gott, an den man glaubt, damit zu tun hat. Nur ein Gottesbild, das so großartig ist, daß es die angedeuteten Phänomene in sich begreift und dadurch uns begreiflich macht - in den Grenzen menschlichen Begreifens -, hat Aussicht, auch in Zukunft Gläubige zu finden. Das Gottesbild des kosmischen Christus scheint mir von der Art zu sein.

Günther Schiwy

De kosmische Christus

Het geloof aan de kosmische Christus en de overeenkomstige theologie zijn een eventueel antwoord op het nieuwe tijdperk van holistisch, d.w.z. van globaal denken en gedrag dat is aangebroken. De wereldwijde sociale en ecologische crisis kan alleen worden gevat en kan, indien überhaupt nog, alleen worden doorstaan, als wij van de hypothese uitgaan dat het hele universum, niet alleen de aarde of de mensheid, een grote, samenhangende werkelijkheid is die zich net als een organisme, zoals wij het van het levende kennen, in een gerichte ontwikkeling bevindt die echter uitermate gevoelig is voor evenwichtsstoornissen, zoals de mens die momenteel in steeds grotere omvang veroorzaakt...

"God" noemen sinds mensenheugenis de geloofsovertuigingen de laatste "grond van het universum". Daarbij heeft God al naar culturen en tradities de meest uiteenlopende namen aangenomen: "Het Ene, Brahma, de Leegte, het tao, de eeuwige Vereniging van Shiva en Shakti, de Heilige drieëenheid" (Rupert Sheldrake)...

Moet het godsbeeld van een cultuur voor de betreffende mensen tot de praktische oriëntering dienen en overlevenskennis gereed houden, dan moet het aan de behoeften van de kennis en geloof van het tijdperk beantwoorden. Dat kan ook uit de godsbeelden in het christendom worden opgemaakt die zich volgens het woord van Jezus: "Wie mij gezien heeft, heeft de Vader gezien" (Joh.14:9), in de heersende christusbeelden concretiseren. Kort en raak geformuleerd - de historische werkelijkheid is complexer, kleuriger, ook tegenstrijdiger -, werd Christus in de klassieke oudheid als Goede Herder, deels als wereldschepper en universele heerser betracht: in de middeleeuwen is Christus de legeraanvoerder en zegevierende koning, maar ook de verlossende martelaar en mystische vriend van de ziel; de renaissance ziet in Christus de mooie meester-wijsgeer van de mensheid uit de klassieke oudheid, de baroktijd de hemels-aardse held in dramatische strijd met de Boze; in de Verlichting verschijnt Christus als de verdediger van het gezonde mensenverstand, in de romantiek en de biedermeiertijd als de milde, de van de wisseling der tijden bevrijde Nazarener; in de 19de en 20ste eeuw grijpt men, al naar behoefte, twijfelend en zoekend op historische christusbeelden terug of verleent Christus expressionistisch de trekken van persoonlijke of collectieve geloofscrises.

Op de vraag, welk godsbeeld, welk christusbeeld nu aan de behoeften van de toekomst, die zich reeds in onze tijd aankondigen, beantwoordt, heeft de natuurwetenschapper en theoloog Pierre Teilhard de Chardin reeds in 1916 geantwoord: het godsbeeld van de kosmische Christus. Daarbij was hij er van overtuigd, slechts een tot dan toe - omdat er daarvoor geen existentiële noodzaak voor had bestaan, verwaarloosde streng van de bijbelse orde op te nemen. Hoe actueel het ook lijkt, is waarlijk en werkelijk het door onze vaders vanuit de hemel gezonden woord, de nieuwe schat die uit voorzorg naast de oude waarden neergelegd is.

In de woorden van Teilhard klinkt door hoezo het godsbeeld van de kosmische Christus geschikt is de behoeften van het huidige en toekomstige holistische wereldbeeld te bevredigen. Allereerst wordt de gespletenheid van de christen, vermoedelijk ook tal van religieuze mensen in andere religiën, tussen gods- en werelds engagement opgeheven: werelddienst is godsdienst. Dat heeft ook met het oog op de ecologische crisis dringend noodzakelijke, praktische consequenties: de mens is ook om religieuze redenen gedwongen zich aan de redding van de natuur te wijden - niet alleen ter wille van de mens maar ter wille van de natuur zelf - want de natuur is het kosmische lichaam van Christus in wording ...

Ondanks het verzet binnen de christelijke kerken en godsgeleerdheden tegen een geloof in de kosmische Christus, schijnt dit geloof de religie van de toekomst te worden. De nieuwe religieuze bewegingen gebruiken veel voorstellingen en formules die de voorstellingen en formules over de kosmische Christus benaderen. De enkeling volgt steeds meer de goddelijke impulsen die hij diep in zijn hart bespeurt en ervaart de resonantie in het gesprek met anderen en in de omgang met de natuur. Op zoek naar de laatste principes van de werkelijkheid komen de natuurwetenschappers steeds dichter bij de gaia-hypothese, de interpretatie van het universum als een levend verband waarvan de functies door een principe gewaarborgd worden dat vroegere generaties "wereldziel" noemden.

Met het oog op zulke ervaringen en overwegingen kan de christen zich evenals elk religieus mens niet aan de vraag onttrekken of en op welke wijze de God, in die men gelooft, daarmee te maken heeft. Alleen een godsbeeld dat zo groots is dat het de aangeduide fenomenen omvat en ons daardoor duidelijk maakt - binnen de grenzen van het menselijke bevattingsvermogen - heeft kans ook in de toekomst gelovigen te vinden. Het godsbeeld van de kosmische Christus lijkt mij van dien aard te zijn.

Günther Schiwy

The Cosmic Christ

The belief in the Cosmic Christ and the corresponding theology are a possible answer to the developments of a new epoch of holistic thinking and behaving which has been ushered in. The reasons for the worldwide social and ecological crisis are easily understandable and this crisis can only be overcome if we accept the hypothesis that the entire universe, not just the Earth or Mankind, is one single, interlocking reality which, similar to an organism as we know it from the sphere of all living things, is involved in a set development, one which is however extremely sensitive to disturbances in its balance, disturbances which Man is at present causing on an ever increasing scale...
"God" is what since time immemorial the religions have called the final "reason of the universe", whereby god has taken on the most differing names depending on the culture or tradition involved: "The One, Brahman, the Void, Tao, the eternal union of Shiva and Shakti, the Holy Trinity" (Rupert Sheldrake)...
If the picture of god drawn up by a culture is to serve as a practical orientation in life and to provide knowledge of survival, then it must suit the needs of the knowledge and the faith of the epoch. This can be seen from the pictures of god drawn up in Christianity which take on concrete shape in the respective pictures of Christ, according to the word of Jesus: "Who has seen me has seen the Father" (John 14,9).
Reduced to a simple formula - the historical reality is more complex, more colourful, almost more contradictory. In the ancient world, Christ was viewed partly as the Good Shepherd and partly as the Creator of the Universe and as the great ruler; in the Middle Ages, Christ is the victorious king and leader of his army but also the redeeming man of suffering and the mystical friend of the soul; the Renaissance sees in Christ the classically beautiful teacher of wisdom; the Baroque period the heavenly-earthly hero figure involved in dramatic conflict with Evil; in the Age of Enlightenment, Christ appears as the advocate of common sense; in Romanticism and the Biedermeier era as the mild Nazarene, elevated above the changing times; while in the 19th and 20th centuries, as needs require, people turn doubtingly or searchingly to historical pictures of Christ or impart to him in Expressionistic fashion the traits of a personal or collective crisis of faith. Asked which picture of god, which picture of Christ would now suit the needs of the future, needs which are already becoming apparent in our day and age, the scientist and theologist Pierre Teilhard de Chardin answered in the year 1916 already: the picture of god as the Cosmic Christ. In saying so, he was convinced - because at that time there was no particular existential need - that he had done nothing other than to take up an up to then neglected line of the Biblical order. "However topical it may appear - this gospel of the Cosmic Christ, in which perhaps the salvation of our age lies, is in truth the word sent from Heaven to our fathers, the new treasure, which was recorded as a precaution alongside the old values..."
From Teilhard's words we can see why the picture of the Cosmic Christ is so suited to fulfilling the needs of the present and the future holistic picture of the world. First it does away with the dilemma that the Christian and presumably many other religious people in other religions suffer under when it comes to deciding between religious and worldly commitment: serving the world means serving god. In the light of the existing ecological crisis this has urgently necessary practical consequences: Man is also obliged for religious reasons to dedicate himself to the task of saving nature - not just for his own sake but for the sake of nature itself - for it is the developing cosmic body of Christ...
In spite of resistance within the Christian churches and theologies to belief in the Cosmic Christ, this faith seems to be becoming the religion of the future. The new religious movements are using many ideas and formulations which come close to those used in connection with the Cosmic Christ. Individuals are increasingly trying to find out more about the divine impulses which they feel inside and are finding an echo in their conversations with others and in their dealings with nature. In their search for the final principles of reality, natural scientists are coming ever closer to the Gaia hypothesis, the idea of the universe as a living organism the function of which is guaranteed by a principle to which previous generations gave the name "world soul"...
In the light of such experiences and considerations, the Christian, like every other religious person, cannot avoid the question whether and in what way the God in which he believes has something to do with this. Only a picture of god which is so magnificent that it contains the intimated phenomena and makes them comprehensible - within the limits of human understanding - has a chance of finding believers in the future as well. The picture of god as the Cosmic Christ seems to me to be of this kind.

Günther Schiwy

Il Cristo cosmico

La fede nel Cristo cosmico e la rispettiva teologia sono una possibile risposta alla nuova epoca olistica, vale a dire di pensiero e comportamento globale, che è iniziata. La crisi sociale ed ecologica mondiale è comprensibile e può essere superata, ammesso che sia ancora possibile, solo se partiamo dall'ipotesi: l'intero universo, non solo la terra o l'umanità, è un'unica realtà continua che, simile ad un organismo come lo conosciamo dal campo del vivente, si trova in un'evoluzione mirata che tuttavia è estremamente sensibile ai disturbi di equilibrio come quelli causati attualmente dall'uomo in misura sempre maggiore...
Da sempre le religioni chiamano "Dio" l'ultima "causa dell'universo". Ma a seconda delle culture e delle tradizioni dio ha assunto i nomi più diversi: "L'Uno, Brahman, il Vuoto, il Tao, l'eterna unione di Shiva e Shakti, la Santissima Trinità" (Rupert Sheldrake)...
Se l'immagine che una cultura ha di Dio deve servire ai rispettivi uomini ad orientarsi praticamente e fornire nozioni per la sopravvivenza, allora deve corrispondere alle esigenze del sapere e della fede dell'epoca. Questo può essere dedotto anche dalle immagini di Dio del Cristianesimo che, secondo la parola di Gesù "Chi mi ha visto, ha visto il Padre" (Joh 14.9) si concretizzano nelle rispettive immagini di Cristo.
Riducendo il tutto ad una formula semplificata - la realtà storica è più complessa, più variopinta e anche più contraddittoria -Cristo fu visto nell'antichità in parte come il Buon Pastore, in parte come il Creatore del mondo e Signore onnipotente; nel Medioevo Cristo è condottiero e re trionfante, ma anche il liberatore dolente e l'amico mistico dell'anima; il Rinascimento vede in Cristo un filosofo dell'umanità di bellezza antica, l'Epoca barocca la figura eroica terrena e celeste, impegnata in una lotta drammatica con il male; durante l'illuminismo Cristo appare come l'avvocato del buon senso, nell'epoca romantica e durante l'epoca del Biedermeier è visto come il Nazareno mite, non toccato dal mutare dei tempi; nel XIX e nel XX secolo, si ricorre, a seconda delle necessità, fra dubbi e ricerche, ad immagini storiche di Cristo oppure si danno espressionisticamente a Cristo i connotati delle crisi religiose personali o collettive.
Alla domanda, quale immagine di Dio, quale immagine di Cristo corrisponda alle necessità del futuro che già si stanno delineando nel nostro presente, lo scienziato e teologo Pierre Teilhard de Chardin ha risposto già nel 1916: l'immagine del Cristo cosmico. Egli era convinto di raccogliere un unico filone dell'ordine biblico fino ad allora trascurato per la mancanza di una domanda di tipo esistenziale. "Per quanto attuale possa apparire, questo Vangelo del Cristo cosmico, nel quale risiede forse la salvezza del nostro tempo, è veramente e realmente il Verbo inviato ai nostri padri dal Cielo, il nuovo tesoro premurosamente riposto accanto agli antichi valori..."
Le parole di Teihard preannunciano perchè l'immagine divina del Cristo cosmico sia adatta a soddisfare le esigenze della concezione del mondo olistica attuale e futura. Dapprima viene eliminato il conflitto del Cristiano, probabilmente anche di molti seguaci di altre religioni, fra l'amore per Dio e l'amore per il mondo, fra impegno religioso e impegno mondano: il servizio nel mondo è servizio a Dio. Di fronte alla crisi ecologica questo ha delle conseguenze pratiche immediate: l'uomo è tenuto anche per motivi religiosi a dedicarsi al salvataggio della natura per amore non solo dell'umanità, ma anche di se stesso, dal momento che la natura è il corpo cosmico di Cristo concepito nel suo sviluppo...
Nonostante le resistenze interne alla Chiesa cattolica e le dottrine teologiche contrarie a una fede nel Cristo cosmico, quest'ultima sembra essere la religione del futuro. I movimenti religiosi più recenti si servono di molte immagini e modi di dire vicini a quelli che definiscono il Cristo cosmico. Il singolo segue sempre di più la percezione degli impulsi divini nella propria interiorità e ne sente la risonanza nel dialogo con gli altri e nel rapporto con la natura. Alla ricerca degli ultimi principi della realtà gli scienziati si avvicinano sempre di più "all'ipotesi di Gaia", ovvero alla concezione dell'universo come una totalità vivente, le cui funzioni sono garantite da un principio al quale le generazioni precedenti hanno dato il nome di "Anima Mundi"... Davanti a tali esperienze e a tali riflessioni il cristiano, così come ogni uomo religioso, non può sottrarsi alla domanda, se e in quale modo il Dio nel quale si crede possa avere una parte in tutto questo. Solamente una immagine di Dio che sia tale da comprendere in sé i fenomeni suddetti e da renderli in questo modo comprensibili anche a noi - nei limiti dell'umana capacità di capire -, ha speranza di trovare credenti anche nel futuro. L'immagine divina del Cristo cosmico mi sembra essere di questo tipo.

Günther Schiwy

Le Christ cosmique

La foi en un Christ cosmique et la théologie correspondante sont une réponse possible à l'aube de cette nouvelle époque de pensée et de comportement holistiques, autrement dit basés sur l'intégralité. La crise sociale et écologique mondiale ne peut être comprise et, autant que faire se peut, surmontée que si nous partons de l'hypothèse suivante: l'ensemble de l'Univers, et non pas uniquement la Terre ou l'Humanité, est la seule réalité cohérente qui, comme l'un des organismes que nous connaissons dans le monde vivant, se trouve en pleine évolution mais se montre extrêmement sensible aux troubles d'équilibre tels que l'homme les provoque actuellement dans une large mesure...
Depuis les temps immémoriaux, les religions appellent "Dieu" la dernière "raison de l'Univers". Selon les cultures et les traditions, Dieu a des noms différents: "l'Un, Brahma, le Vide, le Tao, l'union éternelle de Shiva et Shakti, la Sainte trinité" (Rupert Sheldrake)...
Quand l'image qu'une civilisation se fait de Dieu doit servir d'orientation pratique à ses membres et leur fournir un savoir leur permettant de survivre, il faut qu'elle corresponde aux besoins de savoir et de foi de l'époque. On s'en aperçoit en regardant les images chrétiennes de Dieu qui concrétisent dans les différentes représentations du Christ les paroles de Jésus: "Celui qui m'a vu a vu le Père" (Jean 14,9).
Ramené à des formules simplifiées (la réalité historique est plus complexe, plus colorée mais aussi plus contradictoire), les croyants de l'Antiquité voyaient le Christ soit en "Bon Pasteur", soit en Créateur du monde et en Tout-Puissant; au Moyen-Age, le Christ est le chef des armées et le roi victorieux, mais aussi l'homme qui souffre pour la Rédemption et l'ami mystique de l'âme; la Renaissance voit dans le Christ un maître d'une beauté antique qui enseigne la sagesse à l'humanité; le baroque fait de Lui un héros à la fois céleste et terrestre menant un combat dramatique contre le Mal; au siècle des lumières, le Christ apparaît comme le défenseur du bon sens; à l'époque du romantisme et au Biedermeier, c'est le clément Nazaréen libéré de l'évolution des temps; aux 19e et 20e siècles, on en revient, selon les besoins, qu'il s'agisse du doute ou de la recherche de Dieu, aux images historiques du Christ ou bien on donne au Christ, à la manière expressioniste, les traits des doutes personnels ou collectifs.
A la question de savoir quelle image de Dieu, quelle image du Christ répondrait aux besoins du futur qui se profilent déjà présentement, le scientifique et théologue Pierre Teilhard de Chardin à répondu dès 1915: l'image divine du Christ cosmique. Il était convaincu se consacrer à une seule partie de l'ordre biblique négligée jusqu'à présent (parce qu'il n'y avait pas eu de nécessité existentielle). "Aussi actuel qu'il puisse paraître, cet Evangile du Christ cosmique, qui contient peut-être le salut de notre époque, est vraiment et véritablement la parole envoyée du Ciel à nos pères, le nouveau trésor posé par précaution auprès des anciennes valeurs..."
On devine dans les mots de Teilhard pourquoi l'image du Christ cosmique est propre à répondre aux besoins de la vision holistique, présente et future, de notre monde. Premièrement, la scission entre les Chrétiens sera abolie, probablement aussi celle entre les nombreux croyants d'autres religions, entre l'amour divin et terrestre, entre l'engagement religieux et laïc: le service laïc est un service religieux. Etant donné la crise écologique, des conséquences pratiques instamment nécessaires en découlent: l'homme est tenu, entre autres pour des raisons religieuses, de se vouer à la sauvegarde de la nature, non seulement pour l'humanité mais pour la nature elle-même, car elle est le corps cosmique du Christ en train de se développer...
Malgré la résistance au sein des églises et théologies chrétiennes face à une foi en un Christ cosmique, cette croyance semble être la religion du futur. Les récents mouvements religieux se servent de représentations et de formules qui se rapprochent de celles concernant le Christ cosmique. Chaque être humain s'intéresse de plus en plus aux impulsions divines qu'il ressent au plus profond de son âme et fait l'expérience de sa résonance quand il communique avec son prochain ou avec la nature. A la recherche des derniers principes de la réalité, les scientifiques s'approchent de plus en plus de l'hypothèse de Gaia qui conçoit l'Univers comme une connexion vivante dont les fonctions sont garanties par un principe auquel les générations précédentes ont donné le nom d'"âme du monde"...
Face à ces expériences et à ces réflexions, le Chrétien, comme tout croyant, ne peut pas se soustraire à la question de savoir si et dans quelle mesure le Dieu auquel on croit prend part à tout cela. Seule une image de Dieu si grandiose qu'elle renferme les phénomènes évoqués et qu'elle les nous explique (dans les limites de l'intelligence humaine), a des chances de trouver aussi des adeptes dans le futur. L'image divine du Christ cosmique me semble être de cette nature.

Günther Schiwy

Космический Христос

Вера в космического Христа и ее богословие являются возможным ответом на наступившую эпоху холического, т. е. целостного мышления и поведения. Всемирный общественный и экологический кризис может быть понят и — если это вообще возможно — преодолен, если мы будем исходить из следующего предположения: вся Вселенная, а не только Земля или человечество, является единственной и единой действительностью, и подобно организму, известному нам из области живой материи, находится в направленном развитии, которое тем не менее весьма восприимчиво к нарушению равновесия, вызываемому сейчас человеком во все возрастающей мере...

С незапамятных времен все религии называют «богом» последнее «основание Вселенной». При этом бог в разных культурах и традициях принимал самые разные имена: «Единственный, Брахман, Пустота, Дао, Вечное Слияние Шивы и Шакти, Св. Троица» (Руперт Шелдрейк)...

Если представление о Боге в какой-либо культуре должно служить людям для практического ориентирования и нести знание, необходимое, чтобы выдерживать испытания, оно должно отвечать потребностям веры и знания определенной эпохи. Это видно и по представлениям о Боге в христианстве, которые, по слову Иисуса: «Видевший Меня видел Отца» (Ин. 14, 9), конкретизируются в разных представлениях о Христе.

Сводя все к упрощенным формулам — историческая действительность имеет более комплексный характер, разнообразнее и противоречивее — в Христе в античное время видели и Доброго Пастыря, и Творца мира, и Вседержителя; в средневековье Христос — Архистратиг и победоносный Царь, а также искупляющей Страдалец и мистический Друг души; Ренессанс видит в Христе прекрасного античного Учителя мудрости, барокко — небесного и земного Героя в драматической борьбе со злом; в эпоху Просвещения Христос появляется как представитель человеческого разума, в романтизме и бидермейере — как кроткий, не знающий перемен Назарянин; в 19 и 20 вв., в зависимости от потребности, в сомнениях и поисках обращались к историческим образам Христа или придавали Христу экспрессионистские черты, отражавшие личный или коллективный кризис веры. На вопрос о том, какое представление о Боге, какое представление о Христе будет отвечать потребностям будущего, которое уже проявляет себя в настоящем, естествоиспытатель и философ Пьер Тейяр де Шарден ответил уже в 1916 г.: образ космического Христа. При этом он был убежден, что взял только одно ответвление библейского порядка, которое было в забвении из-за отсутствия жизненно важной потребности в нем.

«Каким бы современным оно ни казалось — Евангелие о космическом Христе, в котором, возможно, заложено спасение нашего времени, есть истинно данное нашим отцам с Неба слово, новое сокровище, предусмотрительно положенное рядом с со старыми ценностями...»

В словах Тейяра слышится ответ на вопрос, почему представление о космическом Христе отвечает потребностям холического мировоззрения настоящего и будущего. Сперва снимается раздвоенность христиан — и, по-видимому, многих верующих других религий — между любовью к Богу и любовью к миру, между религиозной и светской деятельностью: служение миру есть служение Богу. Ввиду экологического кризиса это может иметь крайне необходимые практические следствия: человек должен и по религиозным соображениям посвящать себя спасению природы, причем не ради человека, а ради ее самой, ибо она есть развивающееся космическое тело Христа. Несмотря на сопротивление вере в космического Христа внутри христианских Церквей и богословских учений, эта вера, как кажется, станет религией будущего. Новейшие религиозные движения пользуются многими представлениями и формулировками, приближающимися к понятию о космическом Христе. Человек все чаще исследует божественные побуждения, которые он ощущает внутри себя, и находит отзыв в беседах с другими и в опыте обращения с природой. В поисках окончательных принципов действительности ученые все ближе подходят к идее Геи, представлению о Вселенной как о живой связи, функции которой гарантированы одним принципом, которому прежние поколения дали наименование «мировая душа»...

Ввиду такого опыта и таких соображений, христианин, как и любой верующий человек, не может уйти от вопроса, имеет ли Бог, в которого он верует, отношение к этому, и, если да, то какое. Только такое величественное представление о Боге, которое включает в себя указанные феномены и делает их тем самым понятными, — в рамках человеческого понимания — имеет шанс найти сторонников в будущем. Представление о космическом Христе, как мне кажется, именно такого рода.

Гюнтер Шиви

Diesseits und Jenseits

Wir entgehen der Frage nach der Zukunft nicht. Was werden wir wahrnehmen? Was sollen wir tun?
Die beiden Fragen sind trennbar, ihre Antworten nicht. Neue Wahrnehmung ändert die Handlungstendenzen, neues Handeln bringt neue Wahrnehmung hervor...
Was werden wir wahrnehmen? „Wir", das soll jetzt nicht nur heißen die Wissenschaftler und Philosophen, sondern wir Menschen in unserer Kultur.
Prognosen sind ungewiß. Wir werden alsbald sehen, daß die Frage im Grunde lautet: Was sollten wir wahrnehmen? Nicht nur aus Gründen der Moral und der Nützlichkeit, sondern in der simplen Wahrheitssuche. Wie stellen wir uns die Wirklichkeit vor, die wir wahrnehmen sollten?
Es ist sehr wohl möglich, daß die heutige Weltkultur sich zugrunde richten wird. Die Frage, die wir jetzt stellen, ist aber: Wenn sie überlebt, welcher inneren Logik müßten dann ihre Wahrnehmungen folgen?
Wohl in allen überlieferten Kulturen gibt es einen Unterschied, den ich versuchsweise durch die abendländischen Begriffe Diesseits und Jenseits bezeichnen will. Diesseits heißt, was unserer Wahrnehmung zugänglich ist, Jenseits heißt, was wir glauben oder vermuten, ohne es doch, im allgemeinen, wahrzunehmen. Natürlich ist der Wahrnehmungsumfang verschiedener einzelner Menschen verschieden. Jeder nimmt nur einen Teil des Diesseits wahr; einigen ist etwas vom Jenseits zugänglich.
Diesseits und Jenseits sind insofern soziale Begriffe der jeweiligen Kultur, in denen die Kultur sich ihre eigene Möglichkeit auslegt. Im Weltbild der Wissenschaft zeigt sich eine neue, vielleicht bescheidenere Schranke: diesseits und jenseits der sinnlichen Wahrnehmbarkeit. Der antike Atomismus kannte ein rein „materielles" Jenseits der Sinne: die Atome. Die Chemie des 19. Jahrhunderts hat dieses Bild definitiv wiederaufgenommen. Und dieses Jenseits ist rational erkennbar: die Atome genügen den Naturgesetzen. Unser Denken trägt weiter als unsere Sinnlichkeit.
Dem „Jenseits im Kleinen" steht schon im antiken Atomismus ein „Jenseits im Großen" gegenüber. Unsere Welt ist für die antike Astronomie durch die Fixsternsphäre begrenzt. Sie ist eine endliche Kugel, ein Diesseits des Himmelsgewölbes. Aber jenseits davon sind, wenigstens im Weltbild der Atomisten, in der unendlichen Leere unendlich viele ähnliche Welten. Und wenn es denn Götter geben soll, führen sie in den Zwischenräumen zwischen den Welten ein privates Leben. Die Astronomie der Neuzeit hat auch dieses Jenseits rational aufgenommen. Die Sterne selbst sind die fernen Welten, geordnet in Galaxien, den Naturgesetzen unterworfen.
Aber die Physik des 20. Jahrhunderts bewahrheitet diese Bilder nur, indem sie ihre Naivität wieder auflöst. Denken wir die Atome und den Raum der Galaxien richtig, wenn wir sie uns gemäß den Strukturen unseres engen sinnlich wahrnehmbaren Diesseits vorstellen? Wenn aber nicht, so wird diese Erkenntnis - wie es stets die Bilder des Jenseits vermocht haben - auch unser Bild des Diesseits im Prinzip verändern. Vielleicht bezeichnen Strukturen, mit denen wir das sinnlich Wahrnehmbare beschreiben, nur die Oberfläche einer tieferen Wirklichkeit. Schon die reale Sinnlichkeit enthält noch andere Qualitäten als nur diese mathematischen Strukturen; vielleicht verbirgt oder enthüllt sie auch teilweise andere Gestalten. Die Grenze zwischen Diesseits und Jenseits wird wieder durchlässig...
Die Erfüllung des von der heutigen Wissenschaft gebotenen Umrisses ist aber nicht die volle Aufgabe des theoretischen Denkens. Die Bilder vom Diesseits und Jenseits, welche die überlieferten Religionen boten, waren nicht bloße Naivität, auch soweit sie offenbar mythische Bilder waren. Welche Gestalt wurde in der Schaffung dieser Bilder eigentlich wahrgenommen? Der Weg der meditativen Wahrnehmung ist offen. Neue Erfahrungen, neue gedankliche Bilder warten auf uns.
Der Weg aber zu diesen Erfahrungen wird ohne die Bereitschaft zu verantwortlichem Handeln nicht beschritten werden. Der Fortschritt ist ambivalent. Der Weg zu den Erfahrungen wird durch tiefes Leid gehen.
Und durch tiefes Glück.

Carl Friedrich von Weizsäcker

Het aardse leven en het hiernamaals

Wij ontkomen niet aan de vraag naar de toekomst. Wat zullen wij waarnemen? Wat zullen wij doen? De beide vragen zijn te scheiden, hun antwoorden niet. Nieuwe waarneming wijzigt de handelstendensen, nieuw handelen brengt nieuwe waarneming voort.
Wat zullen wij waarnemen? "Wij", daarmee worden niet alleen wetenschappers en filosofen bedoeld, maar wij mensen in onze cultuur.
Prognosen zijn onzeker. Wij erkennen al gauw dat de vraag eigenlijk luidt: wat zouden wij moeten waarnemen? Niet alleen om morele of nuttige redenen, veeleer terwille van de simpele waarheidsbevinding. Hoe stellen wij ons de werkelijkheid voor die wij zouden moeten waarnemen?
Het is goed mogelijk dat de huidige wereldcultuur zichzelf te gronde richt. De vraag die wij nu stellen is echter: als zij overleeft volgens welke innerlijke logica moesten haar waarnemingen dan aflopen?
Vermoedelijk kennen alle overgeleverde culturen een verschil dat ik bij wijze van proef met de westerse begrippen het aardse leven en het hiernamaals zal aanduiden. Het aardse leven is dat, wat voor onze waarneming toegankelijk is, het hiernamaals is dat, wat wij geloven of vermoeden, echter zonder het over het algemeen waar te nemen. Natuurlijk is het waarnemingsveld van verschillende afzonderlijke mensen verschillend. Iedereen neemt slechts een deel van het aardse leven waar; voor enkele is ook een gedeelte van het hiernamaals toegankelijk. Het aardse leven en het hiernamaals zijn in zoverre sociale begrippen van de betreffende cultuur, waarbinnen die cultuur haar eigen mogelijkheid bepaalt.
In het wereldbeeld van de wetenschap verschijnt een nieuwe, misschien bescheidener grens: het aardse leven en het hiernamaals van de zinnelijke waarneembaarheid. Het klassieke atomisme kende een puur "materieel" aards leven van het bewustzijn: de atomen. De chemie van de negentiende eeuw heeft dit beeld definitief hervat. En dit hiernamaals is rationeel duidelijk zichtbaar: de atomen voldoen aan de natuurwetten. Ons denken gaat verder dan onze zinnelijkheid.
Tegenover "het aardse leven in 't klein" stond in de klassieke astronomie al een "hiernamaals in 't groot". Onze wereld is voor de klassieke astronomie door de vaste ster sfeer beperkt. Zij is een eindige bol, een aards leven van het hemelsgewelf. Maar aan gene zijde daarvan zijn er, tenminste volgens het wereldbeeld van de atomisten, in de oneindelijke leegte oneindelijk veel soortgelijke werelden.
En als er dan goden bestaan, voeren zij in de tussenruimten tussen de werelden een privéleven. De astronomie van de nieuwe tijd heeft ook deze andere wereld rationeel opgenomen. De sterren zelf zijn de verre werelden, volgens galactische stelsels ingedeeld, gebonden aan natuurwetten.
Maar de fysica van de twintigste eeuw bewaarheidt deze beelden alleen door haar naïviteit weer op te heffen. Vormen wij ons van de atomen en de ruimte van de galactische stelsels een juist denkbeeld als wij ons deze volgens de structuren van ons nauwe, zinnelijk waarneembare aardse leven voorstellen? En indien niet zal dit kennen - zoals het de beelden van het hiernamaals altijd gelukt is - ook onze visie op het aardse leven in principe wijzigen. Misschien typeren structuren, waarmee wij het zinnelijk waarneembare beschrijven, alleen het oppervlak van een diepere werkelijkheid. Alleen al de reële zinnelijkheid heeft nog andere kwaliteiten dan alleen deze mathematische structuren; misschien verbergt of onthult zij ook gedeeltelijk andere gedaanten. De grens tussen het aardse leven en het hiernamaals wordt weer doorlatend...

De vervulling van de door de huidige wetenschap geboden omlijning is echter niet de volledige opgave van het theoretische denken. De beelden van het aardse leven en het hiernamaals, die de traditionele godsdiensten boden, waren niet pure naïviteit, ook zover deze blijkbaar mythische beelden waren. Welke gedaante werd bij de schepping van deze beelden eigenlijk waargenomen? De weg naar de meditatieve waarneming is geëffend. Nieuwe ervaringen, nieuwe abstracte beelden wachten op ons.
Maar de weg naar deze ervaringen kan zonder de bereidheid tot verantwoordelijk handelen niet worden bewandeld. De vooruitgang is ambivalent. De weg naar deze ervaringen leidt door diep leed.
En door diep geluk.

Carl Friedrich von Weizsäcker

"This world" and "the world beyond"

We cannot avoid the question of the future. What will we perceive? What are we to do? The two questions are inseparable; the answers to them not. New perception alters behavioural tendencies, new forms of behaviour produce new forms of perception...
What will we perceive? "We", this does not mean the scientists and philosophers, but we people in our culture.
Forecasts are uncertain. We will soon see that the question basically runs: What should we perceive? Not for reasons of morality and usefulness. How do we imagine the reality to be which we should perceive?
It is very possible that today's world culture will destroy itself. The question which we now put is, however: If it survives then according to which inner logic would its perceptions then have to function?
In all traditional cultures there is a difference which I will attempt to describe by the western concepts "this world" and "the world beyond". "This world" means that which is accessible to our perception. "The world beyond" means what we believe or suppose, without in general being able to perceive it. Of course the degree of perception of different individuals varies. Each of us is only aware of a part of this world; some have access to some elements of the world beyond. This world and the world beyond are in this respect social terms used in the respective culture involved, in which this culture lays out its own possibility. In the world-view of science, a new, perhaps more modest barrier reveals itself: the two sides of sensual perception. The ancient philosophical theory of Atomism knew only a purely "material" world beyond the senses: the atoms. 19th-century chemistry took up this picture again. And this "other world beyond" is rationally recognizable: the atoms comply with the natural laws. Our thinking takes us further than our sensual powers.
In ancient Atomism, this other world beyond is confronted by a larger version of the same. In ancient astronomy, our world is bounded by the sphere of the fixed stars. It is a limited sphere, a life this side of the heavenly canopy. But beyond it, at least in the view of the Atomists, there are infinitely many similar worlds in the infinite void. And if there are gods, then they lead a private life in the areas between the worlds. Modern astronomy has also taken up this idea of a world beyond. The stars themselves are the far worlds, ordered in galaxies, subject to the natural laws.
Twentieth-century physics, however, confirms these pictures only by depriving them of their naivety. Do we conceive of the atoms and the space of the galaxies rightly when we imagine them according to the structures of our narrow sensually perceptible world? If this is not the case, then this recognition - as was always the case where pictures of the world beyond were concerned - will also in principle alter our picture of this world. Perhaps the structures with which we describe what is perceptible by means of the senses only describe the surface of a deeper reality. Even real sensuality contains other qualities than these purely mathematical structures; perhaps it conceals or shrouds in part other forms. The boundary between this world and the world beyond this becomes a flowing, pervious one again...
The fulfilment of the model drawn up by modern science is, however, not the sole task of the theoretical "thinker". The pictures of this world and the world beyond, which were put forward by the traditional religions were not just mere naivety, even when they were obviously mythical pictures. What figure or form was actually perceived when these pictures were created? The path of meditative perception is open. New experiences, new ideational pictures are awaiting us.
However, the way to these experiences will not be embarked on without displaying readiness for responsible action. Progress is ambivalent. The way to these experiences will involve great suffering but it will also bring great happiness.

Carl Friedrich von Weizsäcker

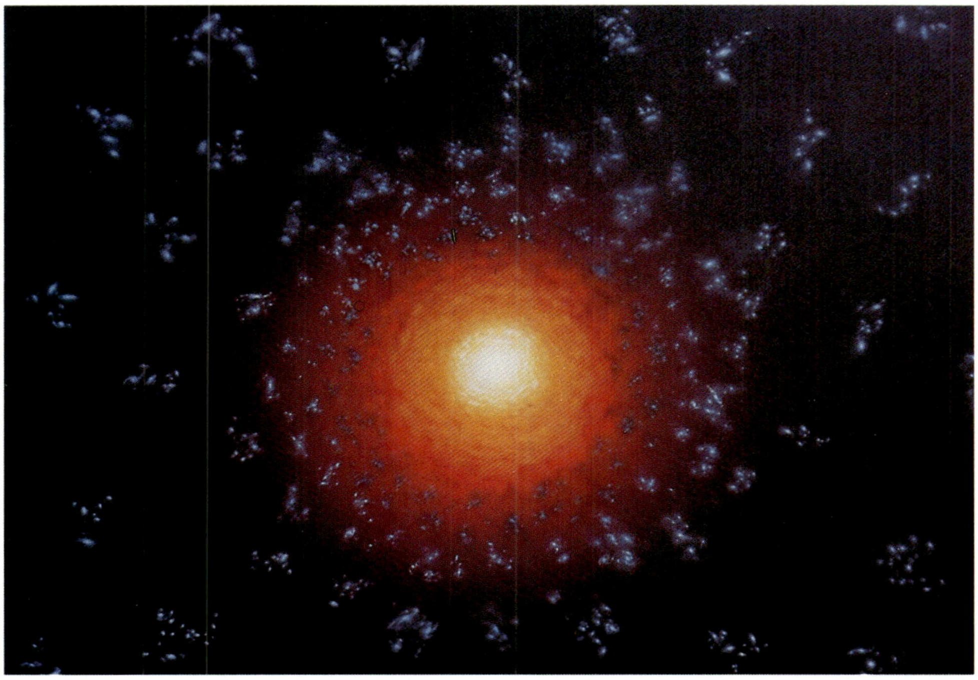

Vita terrena e aldilà

Non possiamo sottrarci alla questione del futuro. Cosa percepiremo? Cosa dobbiamo fare? Le due domande sono separabili, le risposte no. Una nuova percezione muta le tendenze dell'azione, un nuovo agire produce nuove percezioni...
Cosa percepiremo? "Noi", questo non significa solo gli scienziati ed i filosofi, ma noi uomini nella nostra cultura.
Le prognosi sono incerte. Presto vedremo che la domanda in fondo è: cosa dovremmo percepire? Non per motivi di morale o di utilità, ma per la semplice ricerca della verità. Come ci immaginiamo la realtà che dovremmo percepire?
È ben possibile che l'odierna cultura mondiale mandi se stessa in rovina. La domanda che ora poniamo è però: se sopravviverà, quale logica interna dovrebbero seguire le sue percezioni?
In quasi tutte le religioni tramandate si fa una distinzione che voglio definire servendomi dei termini occidentali di vita terrena e di aldilà. Vita terrena indica quanto è accessibile alla nostra percezione, aldilà significa ciò che crediamo o supponiamo tuttavia senza generalmente percepirlo. Naturalmente l'entità delle percezioni di diversi singoli individui è diversa. Ciascuno percepisce solo una parte della vita terrena; per alcuni è accessibile qualcosa dell'aldilà. La vita terrena e l'aldilà sono quindi concetti sociali della rispettiva cultura nei quali la cultura spiega le sue proprie possibilità.
Nella concezione del mondo della scienza si presenta una nuova barriera forse più modesta: vita terrena e aldilà della percettibilità dei sensi. L'atomismo dell'antichità conosceva un aldilà dei sensi puramente "materiale": gli atomi. La chimica del XIX secolo ha definitivamente ripreso questa concezione. E questo aldilà è razionalmente riconoscibile: gli atomi soddisfano le leggi naturali.
Il nostro pensiero ha una portata più lunga dei nostri sensi.
"All'aldilà in piccolo" si contappone già sin dall'atomismo dell'antichità l'"aldilà in grande". Per l'astronomia antica il nostro mondo è limitato dalla sfera delle stelle fisse. Questa è una sfera finita, una vita terrena della volta celeste. Ma al di là di essa vi sono, almeno nella concezione del mondo degli atomisti, nel vuoto infinito, innumerevoli mondi simili. E se mai debbano esserci degli dèi, essi conducono la loro vita privata negli spazi liberi tra i mondi. L'astronomia dell'Era Moderna ha assunto razionalmente anche questo aldilà. Le stelle stesse sono i mondi lontani, ordinati in galassie, soggetti alle leggi naturali.
Ma la fisica del XX secolo conferma queste concezioni annullandone l'ingenuità. Pensiamo gli atomi e lo spazio delle galassie nella maniera giusta quando ce li immaginiamo secondo le strutture della nostra ristretta vita terrena, percepibile con i sensi? Se però non è così, questa nozione dovrebbe cambiare anche la nostra immagine della vita terrena così come hanno potuto sempre farlo anche le nostre concezioni dell'aldilà. Forse, le strutture con le quali descriviamo ciò che è sensibilmente percettibile definiscono solo la superficie di una realtà più profonoda. Già la sensibilità reale contiene altre qualità che solo queste strutture matematiche; forse essa nasconde o svela in parte anche altri aspetti. Il limite tra vita terrena e aldilà diventa nuovamente penetrabile...
L'adempimento del profilo tracciato dalla scienza di oggi non è però l'intero compito del pensiero teorico. Le concezioni di vita terrena e aldilà offerte dalla religioni tramandate non erano solo ingenuità anche quando si trattava evidentemente di immagini mitiche. Quale aspetto è stato percepito per creare queste concezioni? La strada delle percezioni meditative è aperta. Nuove esperienze, nuove immagini mentali ci attendono.
Ma la strada che conduce a queste esperienze non potrà essere percorsa senza la disponibilità ad agire responsabilmente. Il progresso è ambivalente. La via che conduce alle esperienze passerà per pene profonde. E per felicità profonda.

Carl Friedrich von Weizsäcker

En- deçà et au-delà

Nous n'échappons pas à la question du futur. Que percevrons-nous? Que devons-nous faire?
Ces deux questions peuvent être posées séparément, mais leurs réponses sont inséparables. Une perception nouvelle modifie les tendences de l'action; un nouveau mode d'action entraîne une nouvelle perception...
Que percevrons-nous? "Nous", ce ne sont pas seulement les scientifiques et les philosophes, mais tous les êtres humains appartenant à notre civilisation. Les prévisions sont incertaines. Nous verrons tout de suite qu'en fait, il nous faut demander: Que devons-nous percevoir? Non seulement pour des raisons de morale et d'utilité, mais aussi dans la simple recherche de la vérité. Comment nous représentons-nous la réalité que nous devrions percevoir?
Il est très possible que la civilisation mondiale actuelle se détruise un jour. Mais la question que nous posons maintenant est la suivante: Si elle survit, à quelle logique devront obéir ses perceptions? Dans toutes les civilisations traditionnelles, il existe certes une différence que j'aimerais essayer de qualifier par les notions occidentales d'en-deçà et d'au-delà. En-deçà désigne ce qui est accessible à notre perception; au-delà signifie ce que nous croyons ou supposons, sans toutefois généralement le percevoir. Bien entendu, l'ampleur de la perception diffère en fonction de l'homme. Chaque être humain ne perçoit qu'une partie de l'en-deçà; certains accèdent à une partie de l'au-delà. Dans cette mesure, l'en-deçà et l'au-delà sont des notions sociales de la civilisation dans lesquelles la civilisation s'explique sa propre possibilité. Dans la vision scientifique du monde, une nouvelle limite peut-être plus modeste se dessine: l'en-deçà et l'au-delà de la perception sensible. L'atomisme antique connaissait un en-deçà purement "matériel" des sens: les atomes. La chimie du XIXe siècle a repris définitivement cette image. Et cet en-deçà est reconnaissable de manière rationnelle: les atomes suffisent aux lois de la nature. Notre pensée va plus loin que notre faculté sensitive. A "l'en-deçà petit format", l'atomisme antique oppose déjà un "en-deçà grand format". Pour l'astronomie antique, notre monde est borné par la sphère des étoiles fixes; c'est une boule limitée, un endeçà de la voûte céleste. Mais au-delà, c'est du moins la vision du monde des atomistes, il existe dans le vide infini un nombre illimité de mondes semblables. Et si les Dieux existent, ils mènent une vie privée dans les espaces intermédiaires entre les mondes. L'astronomie des temps nouveaux a aussi adopté rationellement cet en-deçà. Les étoiles elles-mêmes sont les mondes lointains, organisés en galaxies, soumis aux lois de la nature.
Mais la physique du XXe siècle ne confirme ces images qu'en supprimant leur naïvité. Notre représentation des atomes et de l'espace des galaxies est-elle correcte quand nous les imaginons conformes aux structures de notre étroit en-deçà perceptible par nos sens? Si ce n'est pas le cas, cette prise de conscience modifiera en principe notre vision de l'en-deçà, comme l'ont toujours fait les visions de l'au-delà. Les structures avec lesquelles nous décrivons ce qui est perceptible par la faculté sensitive ne caractérisent peut-être que la surface d'une réalité plus profonde. La sensibilité effective contient déjà d'autres qualités que ces structures mathématiques; peut-être masque-t-elle ou enveloppe-t-elle en partie d'autres formes. La frontière entre l'en-deçà et l'au-delà redevient perméable... La réalisation des contours proposés par la science actuelle ne représente cependant pas toute la mission de la pensée théorique. Les images de l'en-deçà et de l'au-delà que nous offraient les anciennes religions n'étaient pas purement naïves, même s'il s'agissait apparemment d'images mythiques. Quel fut l'élément ou l'être effectivement perçu à la création de ces images? La voie de la perception méditative est ouverte. De nouvelles expériences, de nouvelles visions intellectuelles nous attendent. S'engager sur le chemin de ces expériences sera néanmoins impossible si nous ne sommes pas disposés à peser les conséquences de nos actes. Le progrès est ambivalent. Le chemin qui mène à ces expériences passera par une profonde souffrance. Et par un profond bonheur.

Carl Friedrich von Weizsäcker

По эту и по ту сторону

Мы не уйдем от вопроса о будущем. Что мы будем воспринимать? Что мы должны делать?
Оба вопроса можно разделить, но не ответы на них. Новое восприятие изменяет характер поступков, новые действия порождают новое восприятие...
Что мы будем воспринимать? «Мы» означает не только ученые и философы, а мы, люди нашей культуры.
Прогнозы неопределенные. Мы очень скоро увидим, что этот вопрос по сути звучит так: что мы должны воспринимать? Не только по соображениям нравственности и полезности, но в простых поисках истины. Как мы представляем себе действительность, которую мы должны воспринимать?
Вполне возможно, что нынешняя мировая культура погубит себя. Однако вопрос, который мы сегодня задаем, звучит так: если она выживет, то какой внутренней логике должно следовать восприятие?
Пожалуй, во всех известных культурах есть различие, которое я попробую обозначить западными понятиями посюстороннего и потустороннего. Посюстороннее — это то, что доступно нашему восприятию, потусторонее обозначает то, во что мы верим и что мы предполагаем, в целом не воспринимая. Объем восприятия у разных людей, конечно же, различный. Каждый воспринимает только часть посюстороннего мира, некоторым доступно что-то из потустороннего.
В этой связи посюстороннее и потустороннее суть социальные понятия той или иной культуры, в которых культура сама трактует свои возможности.
В научном мировоззрении проявляется новое, пожалуй, более скромное ограничение: по сю и по другую сторону чувственного восприятия. Античный атомизм знал только чисто «материальное» потустороннее явление для чувств — атомы. Химия 19-го века подхватила эту картину. И это потустороннее рационально различимо: атомы подчиняются законам природы. Наше мышление простирается дальше чувственных ощущений.
Уже в античном атомизме «потустороннему малому» противостоит «потустороннее в великом». Для античной астрономии наш мир ограничен сферой неподвижных звезд. Она представляет собой ограниченный шар, по эту сторону небосвода. Однако по ту сторону его, хотя бы согласно мировоззрению атомистов, в бесконечное множество пустоте находится бесконечно много подобных миров. И, если боги есть, то они в пространствах между мирами ведут свою частную жизнь. Астрономия нового времени рационально восприняла и этот потусторонний мир. Даже звезды представляют собой далекие миры, упорядоченные в галактики, подчиненные законам природы.
Однако физика 20-го века подтверждает эти представления постольку, поскольку убирает их наивность. Правильно ли мы мыслим об атомах и галактическом пространстве, если представляем их себе в соответствии с нашим узким, чувственно воспринимаемым посюсторонним миром? Если же нет, то этот вывод, как всегда случалось с представлениями о потустороннем, в принципе изменит наше представление о посюстороннем. Может быть, структуры, которыми мы описываем чувственно воспринимаемое, представляют собой лишь поверхность более глубокой действительности. Уже реальная чувственность содержит и другие качества, чем только эти математические структуры; может быть, она скрывает или передает отчасти иные образы. Граница между посюсторонним и потусторонним миром опять становится проницаемой.
Однако наполение смыслом предлагаемой современной наукой схемы не является всей задачей теоретической мысли. Картины посюстороннего и потустороннего мира, которые предлагали традиционные религии, не были простой наивностью, даже если они были явно мифическими. Какой образ, собственно, воспринимался при создании этих картин? Путь медитативного восприятия   открыт. Новый опыт, новые мысленные картины ждут нас. Однако путь к этому опыту будет невозможен без готовности к ответственным действиям. Прогресс имеет двойственный характер. Путь к опыту будет проходить через глубокое страдание.
И через глубокое счастье.

Карл Фридрих Вайцзеккер

Captions

Cover	Milky Way Galaxy	Don Dixon
I/2	Galaxy rising	Don Dixon
I/10-11	Milky Way Galaxy	Don Dixon
I/12-13	Maelstrom	Don Dixon
I/14	Trifid nebula in Sagittarius	Kitt Peak Observatory
I/15	Cluster of spherical stars	Kitt Peak Observatory
I/16	Flare star	Don Dixon
I/17	Black hole	Don Dixon
I/18-19	Star clouds, nebulas and clusters of the Milky Way	Kitt Peak Observatory
I/20	Voyager / Saturn	NASA
I/21	Voyager / Saturn	NASA
I/22	Voyager / Saturn rings	NASA
I/23	Voyager / Saturn rings	NASA
I/24-25	Jupiter from Io Kazuaki Iwasaki	©Space Art International
I/26 top	Voyager / Jupiter, turbulent surroundings	NASA
I/26 bot.	Voyager / Jupiter Ganymed moon	NASA
I/27 top	Voyager / Jupiter from Equator	NASA
I/27 bot.	Magellan at Venus	NASA
I/28	False colour image of solar disc	NASA
I/29	Massive solar flare	NASA
I/30	Space vision	© Ron Miller
I/31 top	Halley's Comet	Finley Films
I/31 bot.	Halley's Comet	Finley Films
I/32-33	Voyager's flight to Jupiter	Finley Films
I/34	Mural at Houston Space Center	Anselm Spring
I/35	Fun at Houston Space Center	Anselm Spring
II/1-2	LEP-Tunnel / CERN, Genf	Anselm Spring
II/3-4	CERN-Hall, Genf	Anselm Spring
II/5-6	Accelerator at CERN	Anselm Spring
II/7	Tunnel view / CERN	Anselm Spring
II/8	Cable networks and details/ CERN	Anselm Spring
II/9-11	ALEPH-detector / CERN	CERN-Photo
II/12	Particle tracks/Computer-graphic	Drevermann/ CERN
II/13	Abstract 3-D-image of particle tracks in ALEPH-detector	CERN-Photo
II/14	Particle tracks	CERN-Photo
III/1-2	Fractal / Julia	A. Pasieka
III/3-4	Baumstamm	Anselm Spring
III/5	Fractal Mountain / Mandelbrot set	A. Pasieka
III/6	Erosion / Bryce Canyon, Utah / USA	Anselm Spring
III/7	Fractals Julia and Mandelbrot	A. Pasieka
III/8	A Datura blossom opens	Anselm Spring
III/9	Salt/silver crystallizations	Anselm Spring
III/10	Fractals Julia and Mandelbrot	A. Pasieka
III/11	Fractal 3-D-computer-graphic / Mandelbrot	A. Pasieka
III/12	River ramifications	Anselm Spring
III/13	Silver crystallizations	Anselm Spring
III/14	Apfelmännchen / Mandelbrot	A. Pasieka
III/15	Lichens	Anselm Spring
III/16	Fractals Apfelmännchen and Seepferdchen	A. Pasieka
III/17	Fractal 3-D-computer-graphic / Mandelbrot	A. Pasieka
III/18	Erosion in the south of Utah / USA	Anselm Spring
IV/1-2	Creation of Adam / Michelangelo	Archiv für Kunst und Geschichte, Berlin
IV/3-4	Elohim creates Adam / William Blake	Archiv für Kunst und Geschichte, Berlin
IV/5-6	The days of Creation / Julius Schnorr v. Carolsfeld	Archiv für Kunst und Geschichte, Berlin
IV/7	Title of the "Astronomicum Caesareum" / P. Apianus	C. Hansmann
IV/8	From the "Atlas Portatilis"	C. Hansmann
IV/9	"Systema Ideale Pyrophylaciorum"	C. Hansmann
IV/10	Baroque world schema	C. Hansmann
IV/11	Astronomer observing the stars	C. Hansmann
IV/12	Ptolemy with astrolabe	C. Hansmann
IV/13-14	Celestial map/ "Atlas Portatilis"	C. Hansmann
IV/15-16	Stage set from the "Zauberflöte" / F. Schinkel	C. Hansmann
IV/17	Tantrist painting / India	C. Hansmann
IV/18	Position of the signs of the zodiac and the planets on the hand	C. Hansmann
IV/19-20	Tantrist painting, India	C. Hansmann
IV/21-22	Temple-relief, Egypt	Anselm Spring
IV/23-24	"Papyrus Greenfield", British Museum London	C. Hansmann
V/1-2	Sculptor and model, London	Anselm Spring
V/3-4	Cardboard icons, Las Vegas	Anselm Spring
V/5-6	Buddha, Kuala Lumpur	Anselm Spring
V/7-8	Photo montages	Anselm Spring
V/9-10	Mural painting, Venice Beach	Anselm Spring
V/11	Stone relief, Dresden	Anselm Spring
V/12	Stone relief, Italy	Anselm Spring
V/13-14	Jean Harlow, waxworks, Hollywood	Anselm Spring
V/15	Bust of a sex-star, California	Anselm Spring
V/16	Billboard	Anselm Spring
V/17-18	Christ on Cross, Recife	Anselm Spring
V/19	Goethe Monument, Frankfurt	Anselm Spring
V/20	Detail in the Schiller Museum, Weimar	Anselm Spring
V/21	Superman, waxworks, Hollywood	Anselm Spring
V/22	Stalin, Churchill, Roosevelt, waxworks, San Francisco	Anselm Spring
V/23-24	Liberation Hall, Kelheim	Anselm Spring
VI/1-2	Grand Canyon, Arizona	Anselm Spring
VI/3-4	Indian mural paintings, Baha California, Mexico	Anselm Spring
VI/5-6	Mount Sinai	Anselm Spring
VI/7-8	Pyramids of Gizeh	Anselm Spring
VI/9-10	Jerusalem	Anselm Spring
VI/11-12	Church of the Nativity	Anselm Spring
VI/13	Church in Biasca, Ticino	Anselm Spring
VI/14	Church frescoes, Ticino	Anselm Spring
VI/15	Standing Rock, Utah	Anselm Spring
VI/16	Glastonbury, South of England	Anselm Spring
VI/17	Precolumbian Nazca culture triangles, Peru	Georg Gerster /Agt. Hamann
VI/18	Stonehenge, South of England	Georg Gerster /Agt. Hamann
VI/19-20	Skyline, Dallas, Texas	Anselm Spring
VII/1-2	Indian tepee, prolonged exposure	Anselm Spring
VII/3-8	3 generations of a Navajo-Family, Monument Valley	Anselm Spring
VII/9-10	Navajo reservation, Arizona	Anselm Spring
VII/11-12	Full moon, Henry Mountains, South Utah	Anselm Spring
VII/13	Indian pony in a storm	Anselm Spring
VII/14	Indian dance	Anselm Spring
VII/15-16	Fire in front of a loam sweat-house	Anselm Spring
VII/17-18	Shaman drum	Anselm Spring
VII/19-20	Indian petroglyphs, South Utah	Anselm Spring
VII/21	Navajo sand painting	Anselm Spring
VII/22	Modern thunderbird, Highway 66, Arizona	Anselm Spring
VII/23	Cult object, Boulder, Utah	Anselm Spring
VII/24	Blossom as a closed cosmos	Anselm Spring
VII/25-26	Rattlesnake as cosmic spiral symbol, Utah	Anselm Spring
VII/27 top	Mandan Chief, North Dakota	Anselm Spring
VII/27 bot.	Eagle's feather	Anselm Spring
VII/28 top	The raven	Anselm Spring
VII/28 bot.	The eagle - holy animal of the indians	Anselm Spring
VII/29	Parched earth, South Utah	Anselm Spring
VII/30	Calfcreek-Falls, South Utah	Anselm Spring
VII/31	Grand Canyon	Anselm Spring
VII/32	Eagle and rainbow	Anselm Spring
VII/33-34	Storm atmosphere, Utah	Anselm Spring
VIII/1-28	Las Vegas	Anselm Spring
IX/1-12	Satiric collages	© Steven Steinman, Atlanta, USA
X/1-19	All illustrations by Chesley Bonestell	©Space Art International
X/20-30	Cosmic interactions	Anselm Spring
XI/2	Supernova	Don Dixon
XI/3	Voyager I Liftoff	NASA
XI/4	Challenger Orbiting	NASA
XI/5	Moon	ESO
XI/6	Man on moon	NASA
XI/7	Space capsule / moon	NASA
XI/8	Space capsule	NASA
XI/9-10	Voyager II, Saturn Encounter	NASA
XI/11	Space capsule	NASA
XI/12	Space man	NASA
XI/13	Jupiter and its four moons	NASA
XI/14	Jupiter moon "Ganymede"	NASA
XI/15	Saturn	NASA
XI/16	Saturn moon "Enceladus"	NASA
XI/17	Jupiter moon "Io"	NASA
XI/18	Jupiter moon "Callisto"	NASA
XI/19	Saturn from Titanus	Don Dixon
XI/20	Saturn eclipsing sun	Don Dixon
XI/21	Eclipse by the Earth	Don Dixon
XI/22	Mars, vertical view	Don Dixon
XI/23	Cluster of stars	Kitt Peak Observatory
XI/24	Cluster of stars, Nebulas	Kitt Peak Observatory
XI/25	Mars landscape	Don Dixon
XI/26	Pluto and Charon	Don Dixon
XI/27	Laguna nebula (M8)	Kitt Peak Observatory
XI/28	Halley's Comet	Space Photo Library
XI/29	Untitled	©Space Art International
XI/30	Earth-crossing asteroid	Don Dixon
XI/31	Mars	Don Dixon
XI/32	Mars orbit insertion	Don Dixon
XI/33	Giotto and Halley's Comet	Don Dixon
XI/34	Comet "Nucleus"	Don Dixon
XI/35	Sunrise on Mercury	Don Dixon
XI/36	Yalles Kasei-Mars	Don Dixon
XI/37	Black hole	Don Dixon
XI/38	A cosmological mandala	Don Dixon
XI/39	Ice lead on Europa	Don Dixon
XI/40	Division of Saturn's rings	Don Dixon

* "Art on other planets - a visual book", published by Nexus Press, Atlanta, USA. The brochure-edition designed and published by Steven Steinman and containing 17 of his color-collages is limited to 300 signed and numbered copies.

Verehrte Leserin, verehrter Leser.

Dieses Werk aus der Werteschutz-Edition Ethica Humana fördert und präsentiert Grundwerte und Ideale unserer Gesellschaft. Dazu zählt natürlich auch der Schutz unserer Umwelt.
Deshalb wurde dieses Buch auf chlorfrei gebleichtem Papier unter Ausschluß schwermetallhaltiger Farbpigmente gedruckt. Die verwendeten Druckfarben sind gesundheitlich unbedenklich.
Sämtliche Verpackungsmaterialien entsprechen den gesetzlichen Vorschriften und sind problemlos zu entsorgen. Die von uns beauftragten Produktionsbetriebe arbeiten vorbildlich bei der Trennung und Entsorgung des anfallenden Restmaterials.

IFK Privat-Institut für Sport, Cultur und Natur Verlags-Gesellschaft für Werteschutz-Editionen AG, München

IFK - Institut für Kultur- und Sportförderung AG, Kilchberg-Zürich

GFK Gesellschaft für Kultur- und Sporteditionen Verlagsges. m.b.H., Wien

Geachte lezeres, geachte lezer,

Dit werk uit de waardenbehoud-editie Ethica Humana steunt en presenteert elementaire waarden en idealen van onze maatschappij. Daartoe wordt uiteraard ook het behoud van het milieu gerekend.
Daarom is dit boek op chloorvrij gebleekt papier gedrukt, zonder het gebruik van kleurpigmenten die zwaar metaal bevatten. De drukinkten vormen geen gevaar voor de gezondheid. Alle verpakkings-materialen voldoen aan de wettelijke voorschriften en zijn zonder problemen te verwijderen.
De door ons met de produktie belaste bedrijven werken voorbeeldig bij het scheiden en verwijderen van de reststoffen die erbij ontstaan.

S.C.N. B.V. SPORT, CULTUUR EN NATUR, Amstelveen

Dear reader,

This work from the Ethica Humana series concerned with the preservation of human values promotes and presents basic values and ideals of our society. This of course includes the protection of our environment.
For this reason this book was printed on cellulose free of chlorine bleach and without having resort to colour pigments containing heavy metals. The printer's colours therefore pose no form of health risk.
All the packing materials meet with legal requirements and can be disposed of easily.
The producing firms commissioned by us work in exemplary fashion in respect of the separation and disposal of the residual materials.

IFK Privat-Institut für Sport, Cultur und Natur Verlags-Gesellschaft für Werteschutz-Editionen AG, München

Gentili Lettrici, gentili Lettori,

quest' Opera, un' edizione ETHICA HUMANA - PROGETTO TUTELA DEI VALORI, promuove e presenta i valori fondamentali e gli ideali della nostra società compresa, ovviamente, la salvaguardia del nostro ambiente.
Per tale motivo questo Volume è stato realizzato con carta prodotta senza l'utilizzo di cloro e con inchiostri a pigmenti di colori privi di metalli pesanti e pertanto innocui per la salute.
Tutti i materiali d' imballaggio rispondono alle norme stabilite dalla legge e possono essere smaltiti senza problemi.
Le aziende di produzione da noi incaricate agiscono in maniera esemplare nella separazione e nello smaltimento dei materiali residui.

SPORT E CULTURA SpA, Segrate-Milano

IFK - Istituto di Promozione per la Cultura e lo Sport SA, Kilchberg-Zurigo

Cher lecteur, chère lectrice,

Cet ouvrage de l'édition Ethica Humana pour la protection des valeurs encourage et présente les valeurs fondamentales et les idéaux de notre civilisation. La protection de notre environnement en fait bien entendu partie.
C'est pourquoi ce livre a été imprimé sur du papier blanchi sans chlore, exempt de pigments colorés aux métaux lourds. Les encres d'impression ne sont pas nuisibles à la santé.
Tous les matériaux d'emballage répondent aux règlements légaux et sont facilement recyclables.
Les entreprises que nous chargeons de la production travaillent de manière exemplaire sur le plan de la séparation et du recyclage des matériaux restants.

IFK - Institut pour la Promotion de la Culture et du Sport SA, Kilchberg-Zürich

Уважаемые читатели!

Эта книга из серии „Охрана ценностей“ „Этика гумана“ призвана поддерживать и популяризовать основные ценности нашего общества. К ним относится, безусловно, и окружающая среда.
Поэтому эта книга отпечатана на бумаге, отбеленной без хлора, красками, не содержащими тяжелых металлов и совершенно безвредными для здоровья.
Все упаковочные материалы отвечают требованиям закона и могут быть без труда переработаны в утильсырье.
Предприятия, которым мы поручаем выполнение наших заказов, используют передовую технологию утилизации вторичного сырья.

ИФК - частный Институт спорта, культуры и природы Издательское общество серии „Охрана ценностей“ АГ